THE ORIGINS OF LIFE

VOLUME I

THE PRIMOGENITAL MATRIX OF LIFE AND ITS CONTEXT

ANALECTA HUSSERLIANA

THE YEARBOOK OF PHENOMENOLOGICAL RESEARCH

VOLUME LXVI

Editor-in-Chief:

ANNA-TERESA TYMIENIECKA

The World Institute for Advanced Phenomenological Research and Learning
Belmont, Massachusetts

For sequel volumes see the end of this volume.

THE ORIGINS OF LIFE

THE PRIMOGENITAL MATRIX OF LIFE AND ITS CONTEXT

Edited by

ANNA-TERESA TYMIENIECKA

The World Phenomenology Institute

Published under the auspices of

The World Institute for Advanced Phenomenological Research and Learning

A-T. Tymieniecka, President

KLUWER ACADEMIC PUBLISHERS

DORDRECHT / BOSTON / LONDON

Library of Congress Cataloging-in-Publication Data

The origins of life / edited by Anna-Teresa Tymieniecka.
p. cm. -- (Analecta Husserliana ; v. 66)
Includes the work of the 2nd International Congress, held at the Politechnical University of Gdansk, Poland.
Includes bibliographical references and index.
Contents: Book I. The primogenital matrix of life and its context.
ISBN 0-7923-6246-2 (hb : alk. paper)
1. Life--Congresses. 2. Phenomenology--Congresses. I. Tymieniecka, Anna-Teresa. II. Series.

B3279.H94 A129 vol. 66
[[BD431]]
142'.7--dc21

00-028393

ISBN 0-7923-6246-2
set ISBN 0-7923-6446-5

Published by Kluwer Academic Publishers,
P.O. Box 17, 3300 AA Dordrecht, The Netherlands.

Sold and distributed in North, Central and Latin America
by Kluwer Academic Publishers,
101 Philip Drive, Norwell, MA 02061, U.S.A.

In all other countries, sold and distributed
by Kluwer Academic Publishers,
P.O. Box 322, 3300 AH Dordrecht, The Netherlands.

Printed on acid-free paper.

Printed in the Netherlands.

TABLE OF CONTENTS

ACKNOWLEDGEMENTS ix

THE THEME / The Origins of Life: *The Primogenital Region of Sense* xi

INAUGURAL STUDY

ANNA-TERESA TYMIENIECKA / Origins of Life and the New Critique of Reason: *The Primogenital Generative Matrix* 3

OVERTURE
THE TREE OF LIFE IN AESTHETIC INSPIRATION

PATRICIA TRUTTY-COOHILL / Leonardo's *Sala delle Asse* and Sullivan's Organic Architecture 19

NANCY GOLDFARB / Creative Timber: Poets and Trees 35

SUZANNA B. SIMOR / The Tree of the Credo: *Symbolism of the Tree in Medieval Images of the Christian Creed* 45

SECTION I
THE DIALOGUE BETWEEN LIFE SCIENCES AND PHILOSOPHY

VIJAY SHANKAR RAI / The Origin of Life: Individuation and Evolutionism 57

BOGDAN OGRODNIK / On the Metaphysical Foundations of Life 73

A. J. ANTONITES / Creative Emergence and Complexity Theory 83

SPAS SPASSOV / Contemporary Life Sciences and the Scientific Worldview 97

PIOTR LENARTOWICZ, S. J. and JOLANTA KOSZTEYN / On Some Problems Concerning Observation of Biological Systems 107

RINALDS ZEMBAHS / Life-Space and Life-World: *Merleau-Ponty on Situations* 121

SECTION II
PRIMAL ORIGIN, INDIVIDUATION, INTERPLAY

BENGT TJELLANDER / The Construction of the Concept "The Omnividual" 131
GRZEGORZ GRAFF and KAROL DZIEDZIUL / The Mathematical Horizon of the Future 157
HENRYK SZABAŁA / The Individualism of Twentieth-Century Phenomenology and Existentialism 165
WŁODZIMIERZ PAWLISZYN / Is Phenomenology as a Science Possible? *Reading Heidegger's Viewpoint* 177
JAROMIR BREJDAK / Self-Interpretation of Time as a Rule of Individuation in Scheler's, Dilthey's and Heidegger's Concepts of Man 187

SECTION III
THE TRANSITIONS OF SENSE:
BODY, ORGANISM, CONSCIOUS LIFE

CARLA CANULLO / The Body and the Self-Identification of Conscious Life: *The Science of Man between Physiology and Psychology in Maine de Biran* 203
W. KIM ROGERS / The Reciprocity of Human Organism and Circumstance: *An Ecological Approach to Understanding the Actions and Experiences of a Human Organism in Its Environment* 225
EVA SYŘIŠŤOVÁ / Die Sprache des Traumes und der Traum der Sprache: *Beitrag zur Phänomenologie der Träume in den kritischen Lebenssituationen* 241
BRUNO CALLIERI / The Connection between Phenomenological Culture and the Clinical Practice of Psychiatry 253
E. V. ALTEKAR / The Dyadics of Complementarity: *Towards a New Vision of Reality* 261
DANIELA VERDUCCI / Giving Form to Life: *Processes of Functionalization and of Work in Max Scheler* 287
S. V. KOMAROV / The Consciousness-Corporeality Problem 297

JARI KAUPPINEN / Death as a Limit of Phenomenology: *The Notion of Death from Husserl to Derrida* 323
JERZY GRZESZCZUK / A Possible Reason for the 'Fatal Vision' of the Famous American Surgeon Jeffrey MacDonald 349
ALEXANDER KUZMIN / Reflexion and the Universal Structures of Consciousness 357
APPENDIX / Program of the Gdansk Congress 367

INDEX OF NAMES 377

At the Baltic Sea, Gdansk.

ACKNOWLEDGMENTS

It is with great pleasure that I present this volume, which includes the work of our second International Congress of Phenomenology/Philosophy and the Sciences of Life that was held in my native land of Poland (at the Polytechnical University of Gdansk). We thank those at the Polytechnical University for their collaboration, support and generous hospitality. In particular, Professor Dr. Alexander Kodziejczyk, President of the University, and the Dean, Professor Dr. Piotr Dominiak who presided over the Organization Committee, which was composed of members of the Department of Administration and Economics. Professors Adam Pawlak, Dr. Franciszek Blawat, and Boleslaw Garbacik among them, deserve our warmest thanks for their interest in and support of this important venture. We particularly appreciate the support of Dr. Mieczyslaw Migon, Secretary General of the Organization Committee, and Professor Dr. Adam Pawlak, for their tireless attention to logistical problems and their contribution to the expense of preparing this publication. To them we owe the marvelous success of the congress, the echo of which will continue to resound in the world.

First of all, however, our thanks and appreciation go to our old and new philosophical collaborators who came from all around the world to contribute their innovative ideas to the common pool of this pioneering work.

Isabelle Houthakker and Robert S. Wise, as usual, merit our recognition and gratitude for their expert editing and proofreading of the papers. My assistants, Jeffrey T. Hurlburt and Louis T. Houthakker, were of significant help in preparing this volume for publication.

Last but not least, the generosity of the NATA, a Polish company that produces soft drinks, and that gracefully supplied us with refreshments throughout the conference, cannot be forgotten!

A-T.T.

A group of participants in our 2nd International Congress of Phenomenology/Philosophy of Life and the Science of Life (the 46th International Congress of WPI) 1997, in Gdansk, Poland. In the front row: with Mieczyslaw Migon at the center, Marlies Kronegger at left, A-T. Tymieniecka at right, Daniela Verducci . . . second row: Christine Christine Berthold, Francesco Totaro, Walter Lammi, Robert Sweeney, . . . In the back: Edyta Supinska Polit, Alberto Carillo Cana, Adam Pawlak, among others.

THE ORIGINS OF LIFE

The Primogenital Region of Sense

We propose in this volume to approach life through an investigation of its origins. To understand life is to reach the network of rationalities that orient and sustain the play of its forces within the constructive design that they project. This means that we have to disentangle the knots of the innumerable meanings that constitute the networks of its manifestations and those of its functional operations, stations from whence significance is transmitted or novel sources of meaningfulness emerge. Then, along the lines of their indications, we have to seek the very meaningfulness of the "sense of life" through which they are infused. In short, the project of understanding life is convertible to a full critique of reason. Yet, even the most preliminary, naive perusal reveals not one but several original senses of life *per se*. How can we approach such a gigantic project of investigating life in its innermost workings when it spreads in a vast dynamic, ever-moving expanse, with all its proceedings, connections, effects, etc. mixing for the eye of the perceiver a vast and colorful canvas on which there is no obvious point of convergence to suggest a point at which the investigation could commence?

In our present period of thought, human beings having become aware of the historicity of their own beingness together with the mobile changeability of their life-world, we witness on all sides the turn to the "origins" as the point of departure from which to pursue understanding. While some philosophers disclaim the very validity of the question of origin, Leibniz claims that to understand anything we have to seek its origins, or at least the way of origination.

Accordingly, in our present phase of inquiry into life, we will adopt the Leibnizean dictum, which I have already revived in my book, *Leibniz' Cosmological Synthesis*.[1] This dictum is the very point of departure for our investigation of life. That is, we propose that, to understand life in its ungraspable variety, it must be approached from the point of view of its origin and pursued along the evolutive route of its unfolding.

This means that our ambition is not to grasp "what life is"; rather, it will be sufficient for the present purpose if we excavate some main rationalities that preside over its primogenital origination and bring to light further regions of sense as they carry the evolutive course of life onward.

Focusing now on the content of the present volume, we must clarify at the outset that the questioning of the origins of life in its organic/vital phase is also a focal point of interest of modern biological science. Philosophical inquiry into life goes hand-in-hand with scientific investigation; we find significant connections between them. According to the program of phenomenology/philosophy of life – integral and scientific – we pursue our investigation in a dialogue with the sciences of life.

As a point of fact, let us state that in both biological sciences and in the phenomenology/philosophy of life (as it explores the primogenital bio-vital phase of the origination of life) an axis of investigation comes to the fore: self-individuation of life as accounting for the formation of the living beingness, on the one hand, and the evolution of types of life, on the other.

However, in our pursuit of the origins of life we give to life its uniquely own characteristic: the enactment. It is along the lines of life's enactment that we find the rationalities constituting it in their concrete dynamic course. There is also a further characteristic of the life-enactment being proposed here: the various spheres of sense in the main trait of life which is "sharing". No individual emerges spontaneously alone; there is always a cluster of individuals, entities that share the common lot at their emergence as well as at their unfolding.

The various senses that the sharing-in-life acquires in the evolutive progress of forms of life touches the very nerve of life's rationalities.

Present-day scientific attempts to understand life are to honor two postulates. First, one must avoid unwarranted reductionism of one sphere of life (e.g. biological/vital) to another (intentional, cultural). Second, one must avoid the problem of infinite regress. This dictum raises for us the following question: how, in probing ever further into the biological origins of life, may we find a definitive point of departure?

In our approach, we attend to the first of the two postulates by paying attention to the four major spheres of rationalities of the sharing-in life which, in turn, give birth to spheres of life that are autonomous and inventive from within their routes.

The second postulate is satisfied by our conception of the generative model, or of the pattern of origin that I elaborated in discussions about scientific research. Philosophy/ontopoiesis of life comes to the fore here, offering an axis for such a model. This axis is an *onto-poietic design* that constitutes the backbone of and direction for the originary, constructive progress of life. We may circumscribe the pattern of the primogenital, that is, bio-vital, origination of the sense of life with reference not only to its constructive role, but also to its role as conductor of the transitions of the sense of life in its evolutive advance. Thus, life's various phases/spheres of sense – of modes to share-in-life – originate from the primogenital attunements and common modalities of participation in the circumambient condition until the occurrence of intelligible, intentional communication among conscious human beings in the societal sharing-in-life. In this volume we focus on the primogenital origination of life, which stands out from non-life and yet does not subsume but merely offers the groundwork for the forthcoming spheres of sense to which we just make here a "transitory" place, reserving their full treatment for the second volume.

Pursuing the unfolding of novel circuits of functions that lead to novel emergences of senses, or novel origins of senses of life, we pursue an essential route of the logos of life in its unfolding. This is a most significant task for the critique of reason, which I have elaborated elsewhere.[2]

In fact, the originations of life as it passes through several phases substantiate the new critique of reason through a significant clarification of the structure of reality: vegetal, animal, and human.

Hence the composition of our first volume dealing with the subject of the primogenital origination of life in its organic/vital significance. Towards the end of the volume, we will move toward intermediary zones of emerging significance, which will lead us to the second volume. That volume will be devoted to the issues concerning the three further originations of life: the animal sharing-in-life; the emergence of full intentional consciousness constituting the human reality of life, which is a surging, autonomous sphere of the societal sharing-in-life; and the origination of the creative work of the human being, exemplary in its originality and uniqueness, which carries the weight of the entire human significance of life in its incessant progress and transformation.

We give this quick outline of our two-volume collection here, in the first volume, to pinpoint the striking novelty of our approach to the origins of life. This approach via the critique of reason reveals the well-founded character of our view of the pluri-significant nature of life through its four autonomously originated spheres of significance.

Thus, our approach is not vulnerable to criticism about the reduction of life to one biological sphere. There are not one but several phases of the origins of life that, although founded in each other, stand out in their autonomy of meaning and their unique roles, as our inquiry shows. In fact, the reducing of one to the others reflects merely an unwarranted, naive prejudice.

Interest in life is quintessential to the human mind and heart. It runs through all the spheres of life, beginning with pragmatic concerns and reaching up to the contemplation of beauty.

It is, then, appropriate that we begin our inquiry with the aesthetic/artistic symbol that comprises it: the tree of life.

NOTES

[1] Anna-Teresa Tymieniecka, *Leibniz' Cosmological Synthesis* (Holland: Royal van Gorcum, 1963).

[2] Anna-Teresa Tymieniecka, *Logos and Life: Creative Experience and the Critique of Reason*, Analecta Husserliana, Vol. XXIV (Dordrecht: Kluwer Academic Publishers, 1988).

The inaugural session of the Congress: Dr. Kołodziejczyk, president of the Polytechnical University, Anna-Teresa Tymieniecka, Dean Piotr Dominiak speaking, Franciszek Blawat.

INAUGURAL STUDY

A Polish group from a pre-congress reunion at the Polish Academy of Science, Warsaw, including Maria Bielawka, A-T. Tymieniecka, Krystyna Szopowa, Agnieszka Nogal, Czesław Dmochowski Wiorka, Louis T. Houthakker, Pawel Polit.

ANNA-TERESA TYMIENIECKA

ORIGINS OF LIFE AND THE NEW CRITIQUE OF REASON

The Primogenital Generative Matrix

INTRODUCTION

Understanding Life through Its Origins

The extraordinary technological discoveries of the present time, together with the intellectual progress of humanity, have sharpened the human being's awareness of his/her own powers. However, left without direction as to the proper use of those powers, human beings have advanced millenial views about nature and human destiny while science and technology drill deeper and in more refined directions, disconnectedly and blindly driving onwards without orientation.

From all perspectives – scientific, social, cultural, personal – the present-day vitally left concerns and queries of humanity point to issues of life. It is life itself that becomes the central issue; life, which for the first time in history cannot be taken for granted, becomes the crux of all questioning.

But life itself still eludes all definitions; with its innumerable enigmas it is still a poorly chartered sea. I submit that the focal question toward which all hard sciences and other fields of scholarship tend is the question of the origin of life. Leibniz insisted that to understand a thing and to give an appropriate definition of it we have to indicate its origin, or at least its possible mode of origination. This postulate attains a unique significance in the context of the historicism of the present day and the associated historical approach to the world, society, and human person. Historical awareness of the human being, which reached its peak in this century, has, along with supporting the emergence of Darwin's theory of evolution, brought both a novel emphasis and significance to the question of origins and a significant re-orientation to the sciences of life and culture. Also, contemporary phenomenology (constitutive and genetic) and hermeneutics have offered a historical approach to the humanities as well as to the hard sciences (the life sciences). Both

A-T. Tymieniecka (ed.), Analecta Husserliana LXVI, 3–16.

approaches are grounded in the conviction that understanding an object requires the pursuit of its genesis.

Thus, the question of the origins of life essentially pertains to the quest for the understanding of life itself as the underlying factor common to the sciences, culture, and personal destiny. Although hard sciences like geology, biology, biochemistry and cosmology are particularly oriented to this quest, it is as relevant to human endeavors in general as it is to the human sciences. All human endeavors share this common aim, as well as the common trait that reflects their very *reason for being*: *the life enactment*. They perform this enactment in various specific modalities, but participate in each other's formation and unfolding in the historical process of culture, as well as in the individual processes of the life of plants, animals and humans.

Scientists themselves, as well as philosophers, have expressed this need to understand life by returning to its origins. Scientists like Francisco Varela call for a philosophical-scientific elucidation of life that honors the following two postulates: first, one must avoid a crude reductionism of life to matter; and second, one must avoid an infinite causal regress.[1] My claim is that philosophy may only fulfill such postulates by raising itself above the compactness of varying scientific requisites while remaining mindful of their work and its results.

However, contemporary philosophy, with its relativistic, discriminatory, fragmentary orientations, refuses principles, reasons, causes – in short, integration. Heidegger, Derrida and his followers "deconstructed", that is, deprived of sense, the very question of origin.

It would appear, then, that an interdisciplinary, philosophical research project into the origins of life would be an audacious enterprise to propose today. Indeed, in doing so one enters into a debate with the biological sciences, on one hand, and contemporary philosophy, on the other.

I do not pursue this theme haphazardly. It has emerged naturally from my philosophy – phenomenology of life – which has unfolded along with the Institute's programs throughout the last three decades. This question represents the next phase of our investigation.

Two tasks will structure my work in this paper. First, I propose a conception of the origins of life that satisfies the above-stated postulates. This is an ambitious endeavor. Second, in setting up the model of the origins of life, I must keep in mind two demands: first, the model must account for the origin of forms (or types) in their hierarchical (scalar) order; and second, it must account for the ordering of life itself.

Associated with the question of the origins of life are three further questions, which are of a universal nature. First, is there progress in evolution? As we know, scientists like Jay Gould deny such progress; besides, he reduces life to matter.[2] Second, does evolution continue, or is it finished? Scientists like R. Wilson believe that it is already finished and consider life a futile game.[3] Finally, what kinds of insights about life's prospects can we gain from exploring the origins of life?

These question are so widely discussed that they do not need to be specified any further. The clarification of the origins of life should shed light on them.

What Do We Mean by Origin?

This question cannot be treated in abstraction, even provisionally, without posing the notions of 'object', 'entity', 'event', or 'anything' that is 'generating' or 'originating'. There is an essential constitutive and material bind between the way of origination and the "something" that is supposedly originating. As is well known, a question already contains some major indications of how to seek an answer to it.

However, even if we, to begin with, bring out some universal features that would characterize each and every type of origination as it is differentiated in modalities in accordance with the types of its objectives, as in the case of a self-prompted crystallization of a novel entity or a "something", these "universal" features not only fall short of directing us toward the differentiation of the modalities of origination with respect to the types of the "something" that is originating, but also, if assumed as a point of departure of a query to the further features of origination proper, are altogether misleading. Assuming these two general features of origination as such – the origination itself and the "something" originating – allows us to treat the question of originations (plural) of life within its legitimate context: the universal ordering of life envisaged in the entire network of its spheres (biological/vital, psychic/conscious, conscious/mental, and creative/spiritual).

I have proposed elsewhere that there is a quintessential diversification of the circuits of life between the two first spheres, which establish a modus of the sharing-in-life, and the two latter spheres, which establish the creatively devised societal human existence.

If we wish to do justice to the question of the origins of life we must, first, as just mentioned, situate it within its proper universal context.

Second, we have to envisage origins in relation to the ordering functions that dominate or reign within each of the main spheres that only together constitute life. At this point, we are concerned only with indicating the intermediary phases.

Thus, we divide our discussion into two parts. In the first part we will discuss the understanding of origins within the biological/vital sphere and phase of life's unfolding. The discussion of the nature of origins within the sphere of the creative human condition – the societal/cultural sphere – will be the object of the second part of this study, which I reserve for the second volume devoted to this theme.

My aim in this paper is certainly not to solve the riddles of the biological sciences or of human life or to offer definitive answers to the questions delineated above. I only hope to propose, by presenting a discussion between science and philosophy, some basic ontological/metaphysical principles toward their adequate treatment. No reductionism or infinite regress will befall my account, but a key to the diversification and conjunction of the order of life will emerge.

1. ORIGINATION IN THE BIOLOGICAL/VITAL PHASE OF LIFE'S UNFOLDING

1.1. *Evidence and Insights Offered by the Sciences of Life: Post-Neo Darwinian Theory of Evolution, Microbiology, and Morphogenesis*

The first question that we must address is the primogenital pattern/model of the origin of life, simultaneously a concrete, singular surging of a beingness, event, or "something" in its crystallized form of individuality and in its "ontic" form as *linea entis*, that is, as beingness, event, or "something" as such.

At this "elementary" level we have to pursue our aim in a discussion with the biological sciences, which have brought a wealth of concrete information to this question.

Let us begin with the theory of evolution. As we know, Darwin saw as a crucial factor the evolution of types in the mechanism of selection. He believed that if we pursue in detail the nature of natural selection we will arrive at the "plan of creation". But we are still far from this plan. On the contrary, the Neo-Darwinians claimed that major determinants of the organismic states are forces external to natural selection. Thus, natural selection had to give way to external forces and adaptation in the very formation of the organism. However, this perspective on evolution by force is under serious attack from all sides of contemporary biology.

In his "Origins of Order in Evolution", Stuart Kaufman[4] emphasizes new evidence in the field of microbiology that indicates that:

1) Complex systems of the genesis of life forms exhibit high, spontaneous order. Such an order may limit the process of selection and even guide it. Thus, selection may not be the sole source of order in the organism. More specifically, a) the kind of spontaneous order in complex systems accounts for much of the order in organisms; b) this implies that selection is acting on systems that have their own inherent properties. The ultimate result, then, is a compromise between selection and the spontaneous properties of the system upon which it is acting.
2) Selection simultaneously acts on the environment or adaptive "landscape". Thus, it has to achieve entities that have internal properties to adapt. It is the first step toward sharing-in-life. Such properties, capable of adaptation, have to possess capacities of fitness and transformation.

To conclude this part of the paper, whether it be in ontogeny or in genetic systems, order emerges spontaneously. This order accounts for the origin and persistence of order in organisms. Selection works with systems that possess their own strongly self-organized properties, yet it manages to fashion them through their assuming "useful landscapes".

1.2. *Morphogenesis-Development-Hierarchy*

Moving now to another branch of biology, namely, to morphogenesis, or the evolution of generic forms, we find there, with Brian C. Goodwin, a significant corroboration of the self-promoting dynamics of life.[5] Here we shall see some modalities of the self-organization of life and its developmental mechanisms, in other words, "the rational dynamics of biological organization", which shapes determining aspects of development. Our inquiry has three organizing questions. First, how are forms initiated in systems with particular types of space-time organization?

Second, what are the basic building blocks out of which spatial forms are constituted? Lastly, what is the nature of the energetic relation between these building blocks?

The first principle of the morphogenetic theory is that symmetry comes through the breaking or bifurcation that results from a balance of forces acting within the system. The forces involved are those of chemical reactions. Genes, but not exclusively genes, generate patterns.[6] Genes only determine whether bifurcation occurs or not.

Second, we must consider the elements out of which a morphogenetic sequence is constructed. Turing's theory is a generally accepted, active reaction-diffusion theory[7] that may be of help to us here. The multiple and different types of field are characterized by different equations and describe spatio-temporal patterns with distinctive features of wave form and rate of pattern initiation and transformation. In the case of all fields, however, the solution of their equations is found in a *harmonic function*. The morphogenetic process proceeds in sequences: organisms move through generic morphological states during their development, which means that, in spite of constraints, forms that are generated arise naturally by a principle of least action. Ontogenesis, that is, the generation in development of living entities, is restricted to morphological patterns (e.g. cleavage sentences). Evolutionary variation arises from their limited possibilities. Natural selection cannot move organisms as genetic networks from the generic states of the morphogenetic sequence. Seen in the concrete, a morphogenetic sequence is, in my view, open on two sides: one on side, toward the conditions of the landscape, and on the other, toward the seminal genetic material of its field as it is energized and dynamized.[8]

The morphological sequence, with its constraints, stays, thus, in the center but without providing interior guidance.[9]

Third, and finally, let us consider the hierarchical nature of morphogenesis and its classification or taxonomy.

a. The latest investigations have shown that the basic aspect of morphogenesis is oriented in a progressive perpendicular line that proceeds from the general to the particular. We follow Brian Goodwin, who writes: "morphological complexity advances gradually as progressively more localized detail emerges within an early established global order".[10]

   This global order may be seen as a coordinate system, or as a *feature of the dynamics of morphogenesis* that gradually gives rise to increasingly finer patterns (e.g., embryonic development from a primary axis, in which spatial bifurcation of spatially periodic patterns develops finer and finer details at the intracellular level). Pattern-generating processes, we are told, result in combinatorial richness of terminal detail and specific gene products in different species with stable trajectories that lead to one or another species or form. We have, thus, a hierarchical progress of the living forms.

b. The properties of the building elements are related, on the one hand, to the *concrete particulars* of their composition (e.g. gene products) and, on the other, to principles of organization (e.g. principles of the morphogenetic field organization). Morphogenesis begins with the laying down of the elements of spatial order. Goodwin asks: "Is evolution merely a historical, contingent unfolding from the potential of the qualitative dynamics of the living state that would reveal, in radical challenge to Darwin, the rational plan of organic creativity?".[11]

Whatever the answer to this query, another question of even more paramount importance to us remains: What brings about the passage from one living form to another? In other words, what are the factors that bring about the novum? Which links do the novel emerging forms maintain with the former as well as with the latter? Our philosophical/ontopoietic intuitions will help us deal with these questions.

At this point, having situated the singularization and development of living entities – their ontogenesis – between the morphological principles, on the one hand, and the "landscape" (environment) of their unfolding and throwing their tentacles, on the other, let us emphasize that most contemporary biology focuses on the efforts of the developing individual to adapt to circumstances as much as it does on their transforming influence on that individual. Life does not throw itself onto a ready-made territory; its territory is existentially dependent on the emergence of life. They emerge together preparing the groundwork for sharing-in-life.

1.3. *Philosophical Premises of the Origin of Forms and Order (A Discussion with Bergson)*

From my succinct presentation of the insights and views of present-day biological sciences, which deal not directly with the origin of life but, rather, cautiously prepare vistas and lay down cornerstones toward it, we see that the main question animating our inquiry is that of the nature and origin of order. I have devoted a great amount of philosophical attention to this issue; in fact, I have proposed a new critique of reason. Now, with reference to my conception of the mind, we may enter into the field of contemporary philosophy in order to challenge some of its main tendencies. I will simultaneously attempt to offer some clarification of the scientific views discussed above. Hopefully, this will lead us to devise an adequate approach to the universal questions set forth at the outset of our investigation. Let us introduce the central argument through a discussion with Bergson, some of whose intuitions support mine (although I arrived at them independently and from quite a different angle, and although I ultimately diverge from the Bergsonian conception).[12]

My main point of agreement with Bergson lies in the basic understanding of rationality as the enactment of life. In other words, rationality (reason) is not the privileged fruit of cognition, understood as symbolic representations provided by the human mind. Bergson favors action *per se* over conceptual cognition, distinguishing the instinctual, intellectual and intuitive directives of life's enactment. He sees the evaluative progress of forms in the advancing of voluntary action over mere "fitness" in the coalescence of productive processes of life, as well as over instinctual choices. That is, he sees the evolution of forms as driven toward greater freedom for action. However, he neither analyzes the nature and differentiation in action – in life's enactment – itself nor defines the creative aspect that he attributes to life-enactment as such. In contrast to Bergson, I came to my intuitions about life from a detailed inquiry into creative action, specifically, human creative action, and I reserve the creative aspect of life's enactment for the human being alone. Furthermore, along these lines of differentiation, I distinguish more clearly than Bergson does the different modalities of life's enactment in different evolutive phases of life's forms/types, spontaneous fitness, evaluative inclinations to coalesce, instinctual choices, semi-voluntary opting for a choice, voluntary decisions, and, lastly, choice of creative evaluation and will.

All of these modalities are carried by "appetite", sensing, propensities, and so on. However, although they proceed at various levels of animal consciousness and establish the order of life and differentiate the steps of its advance, they do not reach the presentational, symbolic, conceptual level of sharing-in-life in conceptual communication until their advent in the evolution of types of the Human Creative Condition.

Cognition is the essential fruit of the human condition that may be the most fascinating, but it is merely one out of the many types of rationality. It remains, after all, in the process of innumerable diversifications.

In fact, we have seen in the biological sciences innumerable fields differentiated by the modalities of the life-enacting rationalities. The most significant for life's enactment can be seen in socio-biology.

This conception of rationality also presents a challenge to the cognitive sciences, which refer to symbolic representation of pre-given reality. Unfortunately, there cannot be a pre-given reality of the world, for symbols do not "re-present" but "present" life's forms as they are enacted through the entire network of life. These forms culminate in its filtering and shaping through the human creative orchestration of the mind.

In essence, my critique of reason centers on the view that the Logos of Life unfolding in the enactment of the progressive phases of life, its rationalities, punctuates distinctions among their forms and projects links between and amongst them.

This unfolding of the Logos of Life in life's enactment evolves around its main factor: *the ontopoietic-entelechial-design of life's self-individualization.* It is with reference to this model of rational expansion of life that the developmental construction of life, and the world that the Logos of Life carries on, proceeds. This model goes through all the evolutionary phases and types while differentiating in innumerable modalities while at work.

With these premises in mind, we may now pass on to our central endeavor, namely, the attempt to understand in what consists an origin and in what consists the origin of life.

1.4. *Understanding Life Through a Model of its Origination*

As mentioned above, we witness in science a quest after an ontopoietic, or, rather, onto-metaphysical, view that synthesizes its dispersed findings. The same quest underlies the societal, cultural and personal preoccupations of present-day humanity. Yet such a view is still extant.

Sciences of life begin to seek a philosophical-ontological, synthetic grasp for their findings. Such a grasp is extant, and yet seems to be receding from sight. The further that biology, genetics, and other sciences proceed with drilling into the intricacies of the unfolding of life and its pre-genealogical phases, the more they get lost in the minutia. They thus lose sight of unifying principles, modes, links, reasons, in short, of the overall constitution of reality.

Philosophy needs to offer a remedy. Even a quick perusal of the insightful character of biological findings reveals the possibility that philosophy may derive, by conjectural inference, scientific data that could serve as the cornerstone of a philosophical vision of life.

In the following, I propose that my models of the origination of life, differentiated according to the main spheres of rationality, may provide a point of departure for such a view.

The first model pertains to life's radical beginning. Summarizing the discussion so far, let us agree that the origination of life at its core translates into a spontaneous surging of a self-organizing complex of elements that carry with themselves a set of constructive virtualities, seminal germs, and a dynamic-energetic potential.

First, upon release into an appropriate environment, these elements coalesce into operative schemas that attune themselves to the conditions of the environment as well as transform it according to its needs. Second, their "falling together" releases dormant energies, prompting the germinal (seminal) propensities to unfold. They advance in operative sequences, constructing a distinctive, autonomous, and self-directing entity: a living being.

That is, however, the basic work of nature-life. It could not proceed constructively without a further factor that is intrinsic to the process of a complex enactment, that is, the axis of an overall constructive orientation. Giving due credit to morphogenesis, we must admit, assume, and acknowledge the presence of an intrinsic force that promotes the hierarchical differentiation of universal forms and types given us by the conjectural inference drawn from concrete givenness.

There remains, then, this inward axis, this intrinsic constructive vector, to account for. I have long argued for a life-intrinsic (intrinsic to the line of the life-enactment) carried by the basic elements and yet distinguished from them in its ontic role, entelechial design, or onto-poietic design. This accounts for the self-individualizing orientation of the basic dynamic as well as the self-prompting nature of genetic material that we discussed in the first part of this study.[13]

I would propose that it is this design that orients the morphogenetic sequences without identifying with its "material" forces.

In carrying out its role, the onto-poietic design is uniquely concrete because it works through the basic initial, elementary material that embodies it. It carries on its quasi-universal "message", crystalizing it into an order. This "message" is the scheme of directives for the self-individualizing ontic progress that proceeds in ontic sequences, because it constitutes the living entity in *linea entis* as much as it does the concrete living individual. Its directives are carried in ontic sequences ranging from the simplest to the most complex. This process occurs after the establishment within the system of life of an autonomous, distinctive beingness that is endowed with living virtualities toward the further propagation of the type in life.

With this last analysis I am submitting that the onto-poietic design embodied first in the basic, and then within elements of the advancing construction of a complex individual, accounts for the continuous yet discrete line of morphogenetic progress. Its spiral direction relates old and new in moving constructively onwards.

The most important feature of this process, which merits special emphasis, is that in *its own diversification* along a hierarchical line of progressing forms, it accounts for the rational diversification of evolutive types in their evolving features, types of rationality that I discussed above. That means that, with the conception of the onto-poietic design, we avoid the reduction of one type of rationality to another. That is, this account is grounded in neither unwarranted materialism nor spiritualism, which would threaten to engulf all but a gradation of types of beings from the cell to the highest creative works of the human spirit. Rather, each of these types has its own sui generis features that testify to its autonomous development from its own existential basis. This is the case not only for those common features, but also the divergent features, out of which life is created.

It is the self-prompting onto-poietic, entelechial fact that, through its spiral direction, brings forth new forms (new morphogenetic sequences), that is, new types of life. This is true along the scale of life's development: organic, vegetative individualization; the rational modes of instinct and sensing valuation; instinctive selection; conscious deliberation with a limited choice; and, finally, the presentational, inventive rationalities of the human creative orchestration of faculties that promote life.

Hence, our investigation of life, as the second part of this study as well as the composition of this volume shows, extends through, and embraces, the entire network of reason-life and its specific, differentiated bases.

But in this fashion we have already, in this first phase of our investigation, answered the first postulate pointed out at the start: a non-reductionistic conception, enlarging and deepening it considerably.

Answering the second question of our initial outline of the arguments, we might say, on the basis of our hitherto sketched outline of life's proceedings, that evolution is not finished with the human type, as some scientists (e.g. the sociobiologist Edward Wilson) claim. However, since living forms are not restricted to spatial, physical shapes but extend beyond them through the creative/inventive work of the human being, I claim that evolution is indeed finished with the human type. After all, we have witnessed the phenomena of great transformations of our planet earth that have occurred and continue to occur in the specifically human sphere and its extensions, including ecological and cultural transformation of the styles of life and its modalities, and evaluative transformations of criteria and thought.

Third, my proposal of a model of origination of nature/life satisfies the third postulate. It also anticipates the second part of this study, to appear in the second volume of *Origins*. There are, in this regard, three intimately interwoven levels that imply each other:

a) the field of spontaneous – energized – seminal elements dotted with constructive virtualities and carrying with itself means for a self-organization of structural/functional patterns;
b) these "universal" patterns in their generic schemes themselves;
c) the entelechial self-individualizing design running through them that serves as a point of reference for their developing individuality and embodies their work as their concrete dynamic core, which, in a spatial move onwards, points out their further course. This schema, constituting a dynamic web, manifests a field of a self-sufficient origination. It satisfies thereby the postulate of self-sufficiency by which we can avoid the problem of causal regress.

However, at this level, from the preceding first and primogenital outburst of life a novel line of questioning emerges. The originative model, its self-encircling circuit, is by no means a closed system. It points to the ontopoietic field of life, which is the circumambiant sphere of Nature/World/Life in view of their reciprocal interaction and influences. We may consider these spheres complements of the originative field proper.

Yet there cannot be overlooked an intuitive opening of the originative circuits in a further and opposite direction. Following the intuitions of life to their end, we have to move to a further, this time no longer ontic but meta-ontic, sphere. While on its efficient side, the self-prompting seminal elements of emerging life imply the ontic, entelechial self-individualizing design, so the constructive side implies the virtualities surging in a purposeful activation. The latter point out to virtual constructive modes "possible" forms of living entities to unfold (morphogenetic sequences also fall into this category). One could say that this intuitive opening revives Leibniz' metaphysical conception with its "reservoir" of possibilities. Yet this is not the case. The perspective opening toward a structural referential system does not indicate any existential mode of forms or their status.

Life's origination, consisting in its own self-organizing as well as referential system, does not need "the logic of supplement" and "causal regress". It remains, however, enigmatic, conjectural and inexplicit as to its "final reasons" (as Leibniz would call it), its "structural possibles" and "initial spontaneity" (as I would call it). The sphere of this properly metapoietic investigation has to be left for a further inquiry.

To conclude this analysis, let us return to the initial question: "may our inquiry contribute to the clarification of the question after the prospects of life?"

In light of my new critique of reason, I will agree with Bergson that the progress of evolution brings ever greater freedom of choice in the enactment of life.[14] The future of life lies in the exercise of this freedom with all that it comprises.

NOTES

[1] Francisco Varela and J-P. Dupuis, *Understanding Origins: an Introduction*, in *Understanding Origins* (Kluwer Academic Publishers, 1992).

[2] S. J. Gould.

[3] E. O. Wilson, *Consilience: The Unity of Knowledge* (New York: Knopf, 1998).

[4] Stuart A. Kaufmann, *Origins of Order in Evolution: Self-organisation and Selection*, pp. 153–181 in Fr. Varela and J-P. Dupuis, *Understanding Origins* (Kluwer Academic Publishers, 1992).

[5] *Ibid.*

[6] Brian Goodwin, *The Evolution of Generic Forms*, *ibid.*

[7] A. M. Turing, *The Chemical Basis of Morphogenesis*, 1952.

[8] This recognition in biology of formation systems organising by the spontaneous dynamisms of their elements meets squarely my first intuition of life, which I have expressed as "initial spontaneity on account of the origins of transcendental consciousness; cf. my study "Initial spontaneity" *Analecta Husserliana*, vol. II, 19.

[9] *Ibid.*, p. 220.
[10] *Ibid.*, pp. 222–225.
[11] *Ibid.*
[12] This recognition in biology of organization on the part of formation systems by the spontaneous dynamisms of their elements meets squarely my very own initial intuition about life, which I have called "initial spontaneity", first on account of the origin of transcendental consciousness.
[13] Henri Bergson, *L'Evolution Créatrice*, 1909, and *Matière et Memoire*, Paris, 1910, Felix Alcan, Ed.
[14] Daniel R. Brooks, *Incorporating Origins into Evolutionary Theory*, 1901–1921 in the pre-cited *Understanding Origins*.

OVERTURE

THE TREE OF LIFE IN AESTHETIC INSPIRATION

PATRICIA TRUTTY-COOHILL

LEONARDO'S *SALA DELLE ASSE* AND SULLIVAN'S ORGANIC ARCHITECTURE*

> He is a relentless observer. He is always active and effective in the investigation of Nature. He sees that all forms of Nature are interdependent and arise out of each other, according to the laws of Creation. In his every design a bit of Nature enters into building. His perceptions (insight) science later verifies.
>
> – Frank Lloyd Wright on Louis Sullivan[1]

INTRODUCTION

At the Congress of the American Society for Phenomenology, Aesthetics and the Fine Arts in Cambridge, Massachusetts in 1996, I was thinking about writing a paper on Leonardo da Vinci's *Sala delle Asse*, a room he decorated in 1498 as a pergola for Ludovico Sforza's Castello in Milan. He painted trees on the walls and ceiling: a powerful root system gives rise to straight trunks that branch at spandrel level to cover the ceiling with boughs trained into an interlace. The decorative scheme was meant to express Ludovico's affection for his lively young wife, Beatrice d'Este, and the joy of their union. Beatrice had died a year earlier, seven months pregnant.

Needless to say there, the story of the *Sala delle Asse* provided plentiful symbolism for a discussion of the tree of life. But as fortune would have it, my train of thought was interrupted by Richard Etlin's presentation, a phenomenological analysis of the writings of that innovator of twentieth-century architecture, that inventor of skyscraper design, Louis Sullivan (1856–1920), whose aesthetic was based on organic systems.[2] Sullivan thought that architecture completely lacking in ornament could not realize its highest possibilities. After the basic form was made expressive, the creative impulse should be carried on into ornament. Ornament, for him, is the most subtle and gracious aspect of expression. The ornament should be organic, growing out of the mass of the structure, rather than being applied. It should express the nature of the material and participate in the fundamental rhythms of the building itself.[3]

A-T. Tymieniecka (ed.), Analecta Husserliana LXVI, 19–34.

Figure 1. John Szarkowski, 1959; photograph of the main entrance of Louis Sullivan's Carson Pirie Scott Department Store (formerly Schlesinger-Meyer), Chicago, built 1899–1904. Szarkowski's photographs key into the dynamism and order of Sullivan's style and the relationship of the building design to those it shelters.

Sullivan could have been describing Leonardo's decorative scheme! I wondered whether he had seen Leonardo's room. As a Leonardista, my first inclination was that Sullivan's organic ornamentations came from Leonardo's schemes. So I set the question: Is a direct influence of Leonardo's design on Sullivan's organic theory a possibility? The answer was quick in coming. No! The *Sala delle Asse* was only recovered in the winter of 1893–1894, twenty years after Sullivan's exposure to Italian art as a student. The lower level of the walls where Leonardo had painted the tree's roots penetrating a rocky ledge had not been visible because the chief of restoration, Luca Beltrami, had had them covered over because he thought they were "strange misinterpretations" of Leonardo's intention.[4]

How to resolve the question of what I felt was a great similarity of approach? After all, Frank Lloyd Wright's description of Sullivan (above) could just as well apply to Leonardo. The aesthetics of both Sullivan's skyscrapers and the *Sala delle Asse* are premised on a tripartite division of the elevation – Sullivan's massive and plain base and Leonardo's roots; Sullivan's panels thrusting vertically through the office spaces and Leonardo's tree trunks; and Sullivan's frieze and cornice corresponding to the branches and leaves in the spandrels. In both cases the tripartite division derives from the same source, the order of nature. The problems for both artists were the same: to introduce nature into artifice and to make the built structure relate to organic structures. As we shall see, both men experienced Florence as youths, Leonardo when its buildings were being raised, and Sullivan when his consciousness was. Let us enter into this exploration so that we can grow to appreciate how both artists expressed life.

LOUIS SULLIVAN'S "GREAT ADVENTURE"

In 1924, Claude Bragdon explained Sullivan's career this way:

> Now in 1876 along comes Louis Sullivan, fresh from Europe, but unglamored by the light of its magnificent yesterdays. He held the conviction that no architectural dictum, or tradition, or superstition, or habit, should stand in the way of realizing an honest architecture. . . . To him the tallness of the skyscraper was not an embarrassment, but an inspiration. . . . It must be a proud and soaring thing. . . . Accordingly, flushed with a fine creative frenzy, he flung upward his piers. . . . Then, over it all, he wove a web of beautiful ornament – flowers and frost, delicate as lace and strong as steel.[5]

Since then the criticism of Sullivan's architecture has taken it for granted that his approach to architecture was to break with the past. The notice on him in Harold Osborne's *Oxford Companion to Art* is a representative example.

Louis Sullivan (1856–1924). One of America's greatest architects and the first to design large commercial buildings which instead of recalling the architecture of past periods were a direct expression of their metal frame structure. . . . He developed an organic theory of architecture, relying for aesthetic effect on proportion and strength of line.[6]

Such criticism dismisses Sullivan's account of his 1874–1876 travels in Europe, a trip that he called his "Great Adventure."[7] He went to Paris to study at the École des Beaux Arts. After reading Hippolyte Taine's criticism of Michelangelo's *Last Judgment*, he went to Rome to experience the Sistina himself. Sullivan describes himself in the third person:

And Louis gazed long and long, as one enthralled. And with his own eye, with his own responses, he discerned more and more. . . . From beneath the surface significance there emerged that which is timeless, that which is deathless, that which in its immensity of duration, its fecundity, its everpresent urge, we call life. . . . He saw that Imagination passes beyond reason and is a consummated act of Instinct – the primal power of Life at work.

After Rome, he "saw Florence" and was so enchanted that he "did not know how . . . to break the golden chains that bound him there, a too willing captive. It needed a full six weeks to part a net that seems but of gossamer; or was it the fragrance of Lotus Land?"[8]

In Florence Sullivan would have seen the many palazzi and must have noticed the lightness of Alberti's Palazzo Rucellai (1450–1470) compared to the heavily fortified quality of Michelozzo's Palazzo Medici (1444–1460).[9] Alberti emphasizes the vertical. Applied pilasters balance the horizontal flow of the cornices making up the floor level. The ornament expresses the potential of the multistoried structure, its height. Sullivan must have "gazed long and long" at these buildings and instinctively felt the greater quality of "life" in them, a quality that had much to do with the absorption of the vitalism of the Gothic as well as a deeper understanding, a renaissance, of the values of the classical world.

Sullivan also must have seen Verrocchio's tomb for the Giovanni and Pietro de' Medici, one of the most vital of all Florentine quattrocento works. Its elegant grillwork and verdant foliage are related to a painting often associated with Leonardo, the *Annunciation* at the Uffizi (c. 1473).

The marble base of Mary's lectern recalls the Medici tomb ornament. It is, in fact, the most energized and vital element in the virginal garden.

Such a garden was, no doubt, the dream of the quattrocento householder. The relationship of the garden with architecture is a theme Leonardo discusses, as did Alberti, and other quattrocento theorists.

THE RELATION OF ARCHITECTURE AND GARDEN IN QUATTROCENTO FLORENCE[10]

Italians built gardens near their countryside villas from the beginning of the fourteenth century. The more magnificent gardens had a fountain in the middle, enclosed by a pergola. Pietro de Crescentius, whose *Il Libro della agricultura* was published in Florence in 1478 and Venice in 1495, even speaks of a cottage made from living pines and laurels. Such gardens were made for pleasure. And pleasure was defined as the recreation of the spirit.

When, as early as 1450, Leon Battista Alberti made designs for a new Vatican garden, he suggested that the garden should follow the precepts of architecture, in the manner of the ancients. He regarded trees as architectural materials:

> laurels, cedar trees and junipers, closely interwoven, should make circular patterns according to the most highly praised architectural plans. . . . The pavements are to be stained in imitation of the bellflower with its branches twining about very beautifully, or inlaid in mosaic in imitation of garlands and branches of trees.[11]

LEONARDO DA VINCI: ARCHITECTURE AND GARDENS

Leonardo da Vinci goes one step further – he links garden design to urban architecture, as we can see, e.g., in his annotation of a c. 1488 drawing: the lower levels are devoted to service while "at the highest level will be accorded walkways for gentlemen, the *piani nobili* [will be composed] of beautiful houses and hanging gardens."[12] Indeed, Carlo Pedretti points out that Leonardo's botanical studies "started out as a subject pertaining to architecture," in keeping with his sources, e.g., Crescentio.[13]

Leonardo made this "houses and hanging gardens" drawing in Milan, where since c. 1482, he worked for Duke Ludovico Sforza in the ducal palace, the Castello Sforzesco. It was Ludovico who commissioned the *Last Supper* for the refectory of the Dominican convent of S. Maria delle Grazie that Leonardo painted between 1495 and 1497. The mural

Figure 2. Leonardo's design (restored), corner of the vault of the *Sala delle Asse*, Castello Sforzesca, Milan, 1498, as a refuge for Duke Ludovico Sfora. Sixteen trees are painted on the walls of this square room. Their branches form a leafy ("unstructured") bower growing from their elaborately structured interlaced.

was dedicated to the memory of Ludovico's young wife Beatrice d'Este, who had suffered a miscarriage and died two days after dancing and drinking the 1497 New Year night away. Although their marriage had been politically motivated, Ludovico seems to have been genuinely captivated by the vitality of his young wife. When she died, he gave up his right to their country "Paradise" at Vigevano, assigning it to S. Maria delle Grazie, the institution which became the focal point for his religious life.

To assuage his grief Ludovico urged Leonardo to complete the decorations, begun four or five years earlier, of two rooms in the square northern tower of his Milan Castello. Nothing is left of one of them, a somber retreat called the *Saletta negra*. A portion of the other, the *camera grande de le asse*, the room of the boards (a name given to the space before Leonardo designed the decoration), was recovered by Paul Müller Walde in the winter of 1893–1894.[14] It is the *Sala delle Asse*, Martin Kemp observes, that most closely reflects the courtly Leonardo. Here is a decorative system that foreshadows Louis Sullivan's theory of ornament: it expresses the life of the building.

THE *SALA DELLE ASSE*

The *Sala delle Asse* has remained intact as a space since the fifteenth century, but the walls Leonardo designed were heavily restored, in fact remade, by Ernesto Rusca under the supervision of Luca Beltrami in 1901 and 1902. Basing his work on traces of the original, Beltrami repeated them until he had covered the whole vault. No attempt was made to determine the species of the trees before the vaults were repainted.

Recent arguments that the trees should have been mulberry are persuasive. The mulberry, in Italian *moro* or *gelso*, was the emblem of Ludovico, called *Il Moro* because of his dark complexion. Gold mulberry leaves and berries were placed on crimson satin as a decorative motif when Ludovico was proclaimed Duke of Milan. Furthermore, the leaves of the mulberry reflected the economic activity Ludovico encouraged, a growing silk industry.[15] The mulberry's reputation for hardiness and resilience was most appropriate for this usurper of the Milanese throne.

The scheme of the decoration of the Sala is that of an indoor garden, surrounded by trees whose boughs make a pergola. The trunks of sixteen trees rise at regular intervals up the wall and fork at the springing of the vaults, the branches forming complex ribs. The boughs are so arranged that they intertwine, making an interlace that holds at a gold-rimmed oculus. Such intertwining and knotting are symbolic of the intricacy and order of Nature.[16]

Within the oculus is an armorial shield bearing the conjugal motif of the arms of Ludovico and Beatrice. A gold chord winds and makes knots, *vinci* in Italian, in the branches. Leonardo would borrow the *vinci* for his "Achademia Leonardi Vinci." But this golden knotted chord is also a riddle, as Carlo Pedretti explains:

> A Renaissance lady could well play with it so as not to expose her feelings and make herself vulnerable. Such must have been the meaning of the knots in the dresses of the ladies at the Sforza court in Milan in the early 1490s, when Beatrice d'Este made them

Figure 3. *Sala delle Asse*, lower wall painted as a stone cliff penetrated by the roots of the trees that grow above.

fashionable as *fantasie dei vinci* – not a pun on Leonardo's name, to be sure, but rather a reference to Dante, *Paradiso* XIV 127–120:

> Io m'innamorava tanto quinci
> che 'n fino a lì non fu altra cosa
> che mi legasse con sì dolci vinci

(I was so enamored therewith that until then there had not been anything which had fettered me with such sweet bonds).[17]

And so the golden cord is Beatrice herself, capturing Ludovico's heart,[18] love entwined in life.

The 1954 restoration brought back to view a large monochrome fragment on the east wall: a rocky ledge, through which the roots of the trees penetrate. The Sforza tree emblem includes the roots, as seen in a roundel at the piazza ducale of their pleasure palace at Vigevano. Martin Kemp suggests that these could have been accompanied by the motto "stare saldo," to stand firm. When many trees are conjoined, as in the *Sala delle Asse*, the motto could be "stare saldo e congiungersi," to stand firm and united.[19]

The roots swell up into the base of a tree trunk, perturbations protruding. The trunks are lost but a drawing by Cesare da Sesto at Windsor, RL 12417, possibly an inking of one of Leonardo's own chalk drawings, gives a sense of the vitality Leonardo would have infused into the columnar mass.

Beltrami's heavy-handed restoration was itself restored in 1954, and much of the heavy repainting was removed. At one place, above the windows of the east side, a thorough cleaning was done. Anna Maria Brizio tells us:

> This reveals a greater opening to the sky, the leaves are lighter and more wavy, studied in a much more naturalistic way, and rendered in their morphological details. The vegetal motif, although used for a decorative purpose, is animated by a charge of organic life and a naturalistic significance which is completely new.[20]

The life-poetics of the composition, and the roots especially, affirm in visual terms one of the fables Leonardo wrote to entertain the Sforza court: The Walnut and the Wall. A nut was carried to the top of a lofty campanile by a crow, who accidentally dropped it into a crevice of a wall. The nut asked the wall for help: "by that grace which God had bestowed upon it to be so exalted and great, and so rich in having bells of such beauty and of such mellow tone." The wall was moved with compassion and:

was content to give it shelter in the spot where it had fallen. And within a short space of time, the nut began to burst open and put its roots in between the crevices of the stones and push them apart and throw up shoots from its hollow; and these soon rose above the building, and as the twisted roots grew thicker they began to thrust the walls apart and force the ancient stones from their places. Then the wall too late in vain implored the cause of its destruction, and in a short time it was torn asunder and a great part fell in ruin.[21]

In the *Sala delle Asse*, Leonardo demonstrates the elegant refined artifice of courtly life in the boughs ordered according to man's taste. The roots are Life in Nature, powerful enough to break the stone walls of man's artifice, an epoche that implodes sclerifications. Leonardo moves far beyond political symbolism to express the full energy of the vital natural world.

The life force that enables man's artifice can also destroy it. "O Time! Consumer of all things; O envious age! Thou dost destroy all things and devour all things with the relentless teeth . . ." (Codex Atlanticus 70A, Richter § 1163). This outburst of Leonardo parallels a response he wrote in the 1490s when he was troubled by Ludovico's suggestion that the Sforza horse should not be made of bronze – because it could be stolen – but of stone. On the fly-leaf of Madrid MS I, he wrote that stone, no matter how heavy and clumsy, "may be turned into walls and lime. Do as it pleases you, for everything has a way to perish."[22]

LEONARDO AND THE TREE OF LIFE

The dynamic aspect of nature will become more and more apparent as Leonardo after 1508 became concerned less and less with the uses of plants and landscapes in art and instead became more interested in plant morphology, physiology, ontogeny and ecology. He sought to understand, e.g., the causes of ramification, the bending of boughs, and the origin of the bark of a tree, as can be seen in an illustration from Jean Paul Richter's translation of the notebooks of Leonardo da Vinci, which was first published in London in 1883.

In his later career, Leonardo realized that in the quattrocento "science" he depended on, the science of botany – even in Padua and Pisa – served practical ends of healing, cooking, or decoration in a garden. Emboden explains:

Botany was in its infancy at the beginning of the sixteenth century. As a science, it was

Figure 4. John Szarkowski, 1959; photograph of Sullivan's Auditorium Building, Chicago (1886–1889). The heavy rustication forms a rocky cliff above which the building springs.

still in the descriptive phase. . . . In contrast, . . . Leonardo wished to understand what forces curved boughs to bend in a particular direction, as much as he sought to understand the movement of waters in particular patterns.

For Martin Kemp, Leonardo's late style is dominated by "his reverence for the awesome intricacy of natural creation: the elaborate perfection of natural systems in their full complexity of organic design; the dynamic sense of natural processes in incessant motion. . . ."[23]

From 1503–1508 Leonardo envisioned the union of man and nature in his treatment of the theme of Leda and the Swan. We know his ideas only from his drawings and painted copies. Jupiter was enamored of Leda, the wife of Tyndareus, king of Sparta. The god came to her from the water as a swan. In his drawing, Leonardo makes a temporal elision, so that their children, the heavenly twins Castor and Pollux, Helen of Troy and Clytemnestra, hatch. Leonardo enlivened his drawing by twisting and gyrating the human forms. He adds a curving vitalism to the plant life. One of the best loved of Leonardo's studies for the painting is the Windsor *Star of Bethlehem*, RL 12424, C. 1506. Here the vortex of the waters that delivered Jupiter inhabits the foliage, turning the botanical study into a demonstration of the life-force.[24]

For Leonardo this life-force united all the natural world. In a most beautiful drawing, Windsor RL 12283 r, tree branches are implicated into a man's arm. In his demonstration against Galen's assertion that the liver generated the vascular system, Leonardo, in Windsor RL 19028 r,[25] instead compared the heart with a seed. "Just as the roots of plants arise from the earth, the reply to this analogy is that plants do not have their origins in the roots but the whole plant has its origin in its thickest parts. . . . (As in) the growth of the peach. . . . The heart is that which produces the tree of the vessels." Martin Kemp assures us that the technique Leonardo developed for illustrating "the tree of vessels," "the tree of the nerves," and other ramifying systems "was entirely novel and highly sophisticated." What he wanted to demonstrate was the organic unity of the life-forms.

ORGANIC ARCHITECTURE

Leonardo used natural analogies and similes frequently. In 1488, he used one for an architectural work, the *tiburio* of the Cathedral of Milan (emphasis added).

> Doctors, teachers, and those who nurse the sick should be aware what sort of thing man is, *what is life*, what is health, and in what manner a parity and concordance of the elements maintains it . . . ; and also one, with a good knowledge of the nature of the things mentioned above, will be better able to repair it than one who lacks knowledge of them. . . . The same is necessary for the ailing cathedral, in that the doctor-architect *understands what kind of thing a building is* and from what rules a correct building derives

Figure 5. John Szarkowski, 1959; photograph of Sullivan's Guaranty (now Prudential) Building (1894–1895), revealing the role of skyscrapers as trees of the urban landscape.

and whence these rules originate. . . . And what are the causes which hold the building together and make it permanent, and what is the nature of weight and what is the potential of force, and in what manner they may be conjoined and interrelated, and what effect they will produce combined.[26]

As Kemp maintains, even though the comparison between the doctor and the architect was an old one, "none of his predecessors took the analogy to the same lengths as Leonardo, and none of them exhibited in their designs such a highly developed sense of a building as a kind of natural organism,"[27] an idea not unlike Sullivan's "organic architecture."

And so it is not surprising that Louis Sullivan's wishes in the *Kindergarten Chats* could be read as Leonardo's:

> I would give you of our art some adequate notion of its possible beauty, its endless capacity for expression, its fluency, its lyric quality, its inexhaustible dramatic power – when it comes into kinship with Nature's rhythms. And how can you know this unless you have felt it? And how can you understand it if you have not known it? And how can you express it unless you have lived it and understood it? . . . When all is said and done, the great masterpiece, or the little masterpiece, is but the condensed *expression of such philosophy of life* as is held by the artist who creates it.[28]

CONCLUSION

My paper has relied, not on a direct "factual" relation between those two figures, but on the similarity of their consciousnesses. My attempt to "grow together" – as Leonardo explains can be done when saplings are planted close[29] – the ideas and practices of these two men in the light of the theme of the tree of life has led me to apply their understanding to the cityscape. Our cities will be much more alive and livable when our soaring skyscrapers are designed as replacements for the natural world they have supplanted. Rather than machines for living, skyscrapers should be trees of life for the urban organism.[30]

Siena College
Loudonville, New York, USA

NOTES

* Photographs of Sullivan's buildings are from John Szarkowski, *The Idea of Louis Sullivan*, Minneapolis: University of Minnesota, 1956. Most of the Leonardo images I cite can be found in the following:

KEMP, p. #: For convenience, most of the images may be found in Martin Kemp's fundamental text, *Leonardo da Vinci, The Marvellous Works of Nature and Man* (Cambridge: Harvard University Press, 1981).

WINDSOR RL # indicates that the drawing can be found in the three volumes of Kenneth Clark and Carlo Pedretti, *Leonardo da Vinci Drawings at Windsor Castle* (London: Phaidon, 1969).

[1] Frank Lloyd Wright, *Genius and the Mobocracy* (New York: Horizon Press, 1971), p. 17.
[2] Richard A. Etlin, "Louis Sullivan: The Life-Enhancing Symbiosis of Music, Language, Architecture and Ornament," The Second International Congress of the American Society for Phenomenology, Aesthetics, and the Fine Arts, *The Orchestration of the Arts*, Cambridge, Mass. (1997). Sullivan's "form follows function" is an aesthetic formulation of the rule of biology, "ontogeny recapitulates phylogeny."
[3] Hugh Morrison, *Louis Sullivan, Prophet of Modern Architecture* (New York: Museum of Modern Art and Norton, 1935), p. 254.
[4] The painting was recovered by Paul Müller-Walde; see Luca Beltrami, *Leonardo e la Sala delle "Asse" nel Castello di Milano* (Milan: [Typographer, U. Allegretti], 1902), p. 66.
[5] Claude Bragdon, Forward to Louis H. Sullivan, *The Autobiography of an Idea* (New York: Dover, 1956), p. 6.
[6] *The Oxford Companion to Art*, ed. Harold Osborne (Oxford: Oxford University Press, 1979), p. 114.
[7] Sullivan, *The Autobiography of an Idea*, *op. cit.*, p. 215.
[8] *Ibid.*, pp. 235 and 236.
[9] It is a truism of architectural criticism that his earliest large structure, the Auditorium Building in Chicago (1887–1889, now Roosevelt College) owes much to Henry Hobson Richardson's contemporary Marshal Field Warehouse Store (Chicago, 1885–1887, demolished 1931). In the Marshall Field Warehouse Store the floors were supported by metal columns and beams. The exterior walls, however, were load-bearing and this function is expressed in the mass of its red granite and sandstone rusticated walls. In Sullivan's Auditorium not only is the vertical articulation more emphatic because of the proportional system, but also the layering of the window frames into planes makes the value gradations more gradual, the loggia-echo of the colonnade recalls the fluidity of Albertian architecture. The greater "lightness" of the building is the result of the steel-cage construction, but the articulation of that quality is due to the new organization of the exterior.
[10] In explaining the relation of quattrocento architecture to the garden, I have relied on William Emboden's account in his *Leonardo da Vinci, On Plants and Gardens* (Portland: Dioscorides Press, 1987).
[11] Leon Battista Alberti, *De re aedificatoria* (1485), quoted in *ibid.*, p. 36.
[12] Leonardo da Vinci, Paris, Bibliothèque Nationale, MS. B, fol. 16 r (see Kemp, *op. cit.*, p. 118).
[13] Carlo Pedretti, *The Literary Works of Leonardo da Vinci: A Commentary to Jean Paul Richter's Edition*, Vol. 1 (Berkeley and Los Angeles: University of California, 1977), p. 323.
[14] Leonardo, 20 April–23 April 1498, mentions both rooms, noting that "time is not to be lost," and promises to finish by September. See Luca Beltrami, *Leonardo e la Sala delle "Asse," op. cit.*, for the nineteen- and early-twentieth-century work. Emboden, *op. cit.*, p. 38, suggests that the decoration embodies an idea of Francesco Colonna in the *Hypnerotomachia Poliphili* of a garden enclosed by the "curious twisting of branches and their green leaves so artificially twisted and grown together, that you might ascend by them." The *Hypnerotomachia Poliphili* was written c. 1467, and published in Venice in 1499, and was dedicated to Duke Guidobaldo of Urbino.

[15] For the most recent discussion of the iconography, see Dawson Kiang, "Gasparo Visconti's *Pasitea* and the *Sala delle Asse," Achademia Leonardi Vinci* II (1989), pp. 101–109.
[16] See Ananda Coomaraswamy, "Iconography of Dürer's Knots and Leonardo's Concatenations," *The Art Quarterly* VII (1944), pp. 109–128. For the most recent discussion of the knots in the *Sala delle Asse*, see Carlo Pedretti, *"Nec ense," Achademia Leonardi Vinci* III (1990), pp. 82–90.
[17] *Ibid.*, p. 83.
[18] It is interesting to note that when the thirty-year-old Leonardo da Vinci arrived in Milan in 1482, he inventoried his work. Of the thirty-three types he listed, the first was "many flowers copied from nature . . ."; sixth on the list was "many designs of knots." The Sala delle Asse that he would design a dozen years later combined the knots with nature. For a recent study, see Francis Ames-Lewis, "Leonardo's Botanical Drawings," *Achademia Leonardi Vinci* X (1997), pp. 117–124.
[19] *Ibid.*, p. 187.
[20] Anna Maria Brizio, "The Painter," in *The Unknown Leonardo*, ed. Ladislau Reti (Maidenhead: McGraw-Hill, 1974), p. 36.
[21] Leonardo de Vinci. Codex Atlanticus, 66 a; see Jean Paul Richter, *The Literary Works of Leonardo da Vinci* (New York: Dover, 1970), § 1277. Richter records seven fables based on plant life, see § 1275–1279.
[22] Pedretti, *Commentary*, II, *op. cit.*, p. 240, points out the thematic relationship between these two comments. Leonardo ends with the comment that "most of the time the work honors more the maker than the payer." See Ladislao Reti, *The Madrid Codices* (New York: McGraw-Hill, 1974).
[23] Kemp, *op. cit.*, p. 277.
[24] Illustrated also in Kemp, p. 274. See Francis Ames-Lewis, "Leonardo's Botanical Drawings," *Achademia Leonardi Vinci*, X (1997), pp. 117–124.
[25] "If you should say that the veins arise in the protruding perturbations of the liver. . . . Just as the roots of the plants arise from the earth, the reply to this analogy is that plants do not have their origins in the roots but the whole plant has its origin in its thickest part, and in consequence the veins have their origin in the heart where there is the greatest thickness. . . . And the example this is to be seen in is the growth of the peach which proceeds from its stone as is shown above. . . . The heart is that which produces the tree of the vessels."
[26] Leonardo da Vinci, Codex Atlanticus, 270 r c; in Richter, *op. cit.*
[27] Kemp, *op. cit.*, p. 108.
[28] Louis Sullivan, *Kindergarten Chats*, as cited in Morrison, *op. cit.*, p. 246.
[29] "Many trees planted in such a way as to touch, by the second year will have learnt how to dispense with the bark that grows between them and become grafted together; and by this method you will make the wall of the gardens continuous and in four years you will even have very wide boards." Codex Atlantics 76 r a; see Edward MacCurdy, *The Notebooks of Leonardo de Vinci*, Vol. 1 (New York: Braziller, lst ed., 1939), p. 299.
[30] This paper is dedicated to Maia Ray, whose graduate studies of Louis Sullivan provided the final impetus to this paper.

NANCY GOLDFARB

CREATIVE TIMBER: POETS AND TREES

In Appalachian lore, he who plants a cedar tree will die when the tree grows tall enough to shade his grave.[1] Because of its slow growth, the life of a tree is frequently seen to parallel human life. Trees planted by one generation often stand to shade the next, lending themselves to a sense of history and ancestral connectedness. In their representations of the tree, poets often work out the complex relationship between life, death, and the hope artistic creativity offers for transcendence.

Though the tree originates in and invokes the natural world, its representation is anything but natural. It is, in fact, a construct concealing but also revealing the aesthetic assumptions of its representer. For William Butler Yeats, the preference for art over the natural world is established in "Sailing to Byzantium." Here the gilded tree offers the elderly poet an escape from the natural world in which he is dying, supplanting nature with the world of art. When John Keats claims that "if Poetry comes not as naturally as the Leaves to a tree, it had better not come at all,"[2] he is naturalizing his romantic assumptions about poetry, that it is written effortlessly and without any need to revise, being, as William Wordsworth calls it, the "spontaneous overflow of powerful feelings."[3] As a measuring and reflective instrument in poetry, the tree reveals the poet's ideas about poetry, life, and the relationship between the two.

The three distinct poetic approaches to trees discussed in this paper suggest three different approaches to poetry and its relationship to human life. The poets who write primarily about the natural world – Wendell Berry, Jim Wayne Miller, and Mary Oliver – are humbled by the tree, seeing in it the renewal and continuity of life. The tree, for these poets, records a history of the land and of the life of the land that precedes and will survive themselves and their poetry. By contrast, Marge Piercy uses the tree for polemical ends, to make a statement about being a woman in a patriarchal culture. Finally, Wallace Stevens sees the tree, as he sees much else, as a symbol of poetry. The history it evokes in him is not so much a natural as a poetic history, invoking the trees of his poetic ancestors.

A-T. Tymieniecka (ed.), Analecta Husserliana LXVI, 35–44.

WENDELL BERRY

In "The Old Elm Tree by the River,"[4] Kentucky poet Wendell Berry identifies his own life with the life of an elm outside his window: the two grew up together, and as death stands up in the tree's trunk, he hears death's call within himself:

> Shrugging in the flight of its leaves,
> it is dying. Death is slowly
> standing up in its trunk and branches
> like a camouflaged hunter.

The great old elm appears to be solid and strong, yet it is slowly dying. When the fall of a branch in the night awakens Berry from his slumber, it becomes the poet's wake-up call, forcing him to recognize his own mortality:

> In the night
> I am wakened by one of its branches
> crashing down, heavy as a wall, and then
> lie sleepless, the world changed.

The noise of the crash awakens the poet from both an actual sleep and a metaphoric one, waking him from his denial of death. The crash changes his perception of the tree, forcing him to recognize the decay and dissolution concealed within it and within all living things, including himself. But if the world is changed as a result of the poet's changed perception of it, it has also changed independently of him, as a result of natural and socio-economic processes.[5] The ambiguity of the phrase "the world changed" suggests something of the harmony between the poet's inner and outer worlds.

This harmony is enacted in the poem by a change in pronouns. The first seven lines involve a distinction between the observer and the natural object being observed. Yet the falling of the branch and its reverberations in the poet's ear disrupt the subject-object distinction that the poem had hitherto established. The parallel construction in effect conjoins what have hitherto been two separate identities, poet and tree:

> That is a life I know the country by.
> Mine is a life I know the country by.

The remaining lines of the poem subsume "I" and "it" into the collective pronoun "we."

If the gesture of the poem is a conjoining and identification with the elm, what ultimately links man and tree is their connection to the earth, into which both plant roots and from which both derive their strength: "In us the land enacts its history." Sharing a dependence on the land, tree and man both respond to the changes within it, and their responses are recorded within their own bodies. This emphatic one-line sentence reveals a change in the attitude of the speaker. In the middle of the poem, the speaker understands his connection with the tree by attributing to it a human capacity to will its own existence:

> Willing to live and die, we stand here,
> timely and at home, neighborly as two men.

Even while trying to assert the unity of man and tree, Berry does so in human terms, personifying the tree as a neighbor. With the line "In us the land enacts its history," on the other hand, Berry begins to understand his own life in terms of the tree, whose rings in its trunk record hundreds of years of climatological information, including annual rainfall, soil quality, etc. Now, instead of attributing consciousness to the tree, Berry looks inward at the historical information recorded in his own body.

JIM WAYNE MILLER

While in "The Old Elm Tree by the River," Wendell Berry sees death in a living tree, in "The Fencepost,"[6] another Kentucky poet, Jim Wayne Miller, discovers life in what he thought was a dead tree. As he is mending an old fence he built with his deceased grandfather, Miller notices a wooden post that is unusually firm in the wet ground:

> Mending the fence we built one fall together,
> I come to the spring below the mountain field
> This stake here by the spring drain needs no bracing.

A stake in wet ground tends to be loose, requiring some support or bracing, but this one is too solid in the ground to be dead:

> It's sprouted now, you left the post unpeeled,
> And feels so firm, so solid in the ground,
> I'd say it's taken root.

Having been set in a wet place with its bark intact, the cut post managed to sprout.

The remaining lines of the poem set death against life's potential for renewal. Miller's personification of death applies the conventional image, complete with sickle, to southern Appalachian culture, and the effect is rather comic.

> Death's sickle-sweep
> Is wide: – Bone-man on a branchbank,
> In denims and an old black hat, still half-asleep
> At foggy dawn I've heard him giving his blade
> A whetstone's lick and promise by the spill.

The image of death in overalls and an old hat does personification one better: it countrifies him, as does the use of regional dialect in "bone-man" and "lick and promise." Death as an Appalachian farmer makes him all the more human, thereby limiting his devastations. In this way, agrarian culture is imagined to triumph over death, exerting a continuity that even the Grim Reaper cannot break:

> But death would have to be a new ground
> Grubber, and dig out every root, and still
> He'd not be sure roots wouldn't send down roots,
> Or sure that stubble wouldn't send up shoots.

Grubbing new ground is a process by which fields are cleared of new sprouts after the trees and brush have been cut down. But southern agrarian life, embodied here by Miller's dead grandfather, lives on. By the very act of mending the fence he built with his grandfather, Miller carries on the culture of Appalachia. Shared by grandfather and grandson, tree trunk and post, the soil is the fertile medium by which farm culture exerts its continuity, laughing in the face of death.

MARY OLIVER

With "The Black Walnut Tree,"[7] Mary Oliver introduces economic and practical considerations into her reflections on the tree beside her house. The poem opens with the poet and her mother in a dilemma: they could pay the mortgage with the timber from the tree. Besides, the old black walnut is something of a nuisance: it could fall on the house; its roots are clogging the cellar drains; its droppings are becoming increasingly cumbersome to clean up. Alone with this decision, mother and daughter endeavor to make it without the sentimentality traditionally associated with women:

> We talk
> slowly, two women trying
> in a difficult time to be wise.

The decision they come to is spiritual and aesthetic rather than a practical one: removing the tree would leave a glaring empty space in their lives:

> What my mother and I both know
> is that we'd crawl with shame
> in the emptiness we'd make
> in our own and our fathers' backyard.

For Mary Oliver, as for Wendell Berry and Jim Wayne Miller, the tree takes on a historical significance, connecting the poet with her ancestors who planted it:

> That night I dream
> of my fathers out of Bohemia
> filling the blue fields
> of fresh and generous Ohio
> with leaves and vines and orchards.

Despite their desire to see it practically, the tree means more than money:

> . . . something brighter than money
> moves in our blood – an edge
> sharp and quick as a trowel
> that wants us to dig and sow.

Mother and daughter feel the same need to plant, build and preserve that flowed in the blood of their ancestors, a need based on more than economic gain. Wise or not, Oliver celebrates life by rejecting materialism with a playful, humorous tone that mimics the lively activity of the trees:

So the black walnut tree
swings through another year
of sun and leaping winds,
of leaves and bounding fruit,
and, month after month, the whip-
crack of the mortgage.

MARGE PIERCY

The tree in Marge Piercy's "A Work of Artifice"[8] does not deal so much with facing death as with never having been allowed to live. The conceit of this poem compares a bonsai tree, whose roots are cut and contained in an attractive pot, to a woman whose growth has similarly been stunted by cultural messages that encourage her to be small and weak:

With living creatures
one must begin very early
to dwarf their growth:
the bound feet,
the crippled brain,
the hair in curlers,
the hands you
love to touch.

Through a bitter impersonation of the gardener, the dwarfed tree and woman are taught to be grateful for their dwarfing:

how lucky, little tree,
to have a pot to grow in.

The gardener is right, in a sense: if the bonsai had been allowed to grow unfettered, it would probably have encountered at one point or another a cruel or violent death, as it does in Piercy's imagination:

The bonsai tree
in the attractive pot
could have grown eighty feet tall
on the side of a mountain
till split by lightning.

By potting the tree, the gardener protects it from the elements, but he also deprives it of the chance to be the highest tree on the mountain top, where it would be vulnerable to lightning. Furthermore, being fettered deprives the bonsai of the possibility of receiving the aesthetic qualities traditionally associated with bolts of lightning: divine inspiration, sexual energy, and creativity. While for many poets artifice is a source of renewal, for Piercy it often deprives women of the opportunity to live, casting them as the object of the gardener's creative activity.

WALLACE STEVENS[9]

Though Wallace Stevens might have envied the relative simplicity of the nature poets, his poems suppress any potentially pathos-laden connections with the natural world. Trees in Stevens' late poems are at once reminders of the passage of time and of hope for renewal. The tree gives and renews life in these poems when, like the lyric poem itself, it contains metaphor. When, on the other hand, metaphoric language is suppressed, as it is in "The Course of a Particular," the tree points toward – and even, in a sense, brings about – the poet's death.

The four words that begin "The Course of a Particular" constitute the only action of the poem: "Today the leaves cry." The remaining 14½ lines reveal the consequences of this phenomenon, dramatizing the poet's inner struggle between identification with the leaves as they face winter[10] and his solipsistic refusal to associate his own pain with that of others:

> The leaves cry. . . . One holds off and merely hears the cry.
> It is a busy cry, concerning someone else.
> And though one says that one is part of everything,
>
> There is a conflict, there is a resistance involved;
> And being part is an exertion that declines.

When the impersonal third person "holds off and merely hears the cry," he chooses to restrain his initial pathetic response, as if by suppressing pathos he could hear the sound objectively. But even from the first, when rustling leaves are imagined to be crying, the sound of leaves is integrally linked with the poet's imaginative personification of them. In holding off his pathos, the speaker also holds off the metaphorical associations that potentially give them life. As a result, the cry is "the cry of leaves that do not transcend themselves," and the poem, "in the absence of fantasia," closes with a vision of the absence not only of meaning but also of an observer for whom the cry's meaning is even relevant. In a sense, the leaves in "The Course of a Particular" are simultaneously mourning the poem's absence of metaphor and the anticipated absence of the poet.

While the tree in "The Course of a Particular" is represented synecdochically as a collection of leaves that cry, "Prologues to What Is Possible" and "Of Mere Being" bring into view an entire tree, complete with bird. Both poems use the tree to suggest the excitement associated with discovery and renewal. For Stevens, "one of the motives for writing is renewal,"[11] and this arboreal revitalization occurs in both poems as a result of metaphor. The metaphor that opens Part I of "Prologues to What Is Possible" has the effect in Part II of frightening the man involved:

The metaphor stirred his fear. The object with which he was compared
Was beyond his recognizing.

But while the metaphor is initially frightening, it becomes in the course of Part II the small element that, added to a commonplace, "creates a fresh universe out of nothingness by adding itself." This is figured in the last stanza of the poem as a bird in early spring newly arrived from the South whose addition illuminates and renews the trees upon which it alights:

> A flick which added to what was real and its vocabulary,
> The way some first thing coming into Northern trees
> Adds to them the whole vocabulary of the South,
> The way the earliest single light in the evening sky, in Spring,
> Creates a fresh universe out of nothingness by adding itself,
> The way a look or a touch reveals its unexpected magnitudes.

The tree's job in "Prologues to What Is Possible" is to contain and shelter the metaphoric bird, the undiscovered possibilities within the self.

Likewise, the tree that opens "Of Mere Being" is a palm whose existence begins and ends not just in the mind but on the farthest reaches of it: "The palm at the end of the mind." This purely imaginary tree embodies for Stevens the very center of being. The palm, however, like the poetic structures that contain and facilitate metaphor, is only part of the picture. The other is the bird whose gold feathers, together with the "bronze decor" of the background, recall Yeats' vision of a golden bough, upon which a gilded bird sings in Byzantium, the "city of perfected art."[12] Revising Yeats' artistic vision, Stevens makes the scene more realistic: the tree is a real tree – a palm – and the bird is gilded only in a natural sense: it has gold-colored feathers. Though stylized, the scene is a natural one, illuminated by a setting sun:

> The bird sings. Its feathers shine.
>
> The palm stands on the edge of space.
> The wind moves slowly in the branches.
> The bird's fire-fangled feathers dangle down.

In this last poem of Stevens, written, like Yeats' "Sailing to Byzantium," by an old poet facing his own death, the natural and the aesthetic converge at a point beyond the human: "without human meaning/Without human feeling." Life goes on without us in these lines, whose short, periodic sentences accentuate the verbs, suggesting activity even in the absence of humans to set things into motion. With this final vision of life and art, Stevens suggests that poetry – the bird in the tree – will simultaneously outlive and illuminate us all.

CONCLUSION

As you can see, poets use trees to various ends. What most of them have in common is the tree's relationship to life, death, and renewal. For Stevens, the tree is the image by which life can be renewed and redeemed through metaphor. For Marge Piercy, the pruning of the bonsai tree represents life denied, particularly through denying women the opportunity for growth. For the nature poets – Wendell Berry, Jim Wayne Miller, and Mary Oliver – the tree is not so much a metaphor of their own lives, but it stands like themselves as a member of a greater context of nature; even though they die, the community of life of which they were a part lives on. In the face of this knowledge, death loses its sting.

ACKNOWLEDGEMENTS

The author would like to thank Professor Dorsey Grice for his insightful and suggestive readings of several of the poems discussed in this article.

Western Kentucky University

NOTES

[1] Newman Ivey White, gen. ed., *The Frank C. Brown Collection of North Carolina Folklore*, Vol. 7 (Durham: Duke University Press, 1964), p. 74.
[2] John Keats, letter to John Taylor, 27 Feb. 1818, *The Letters of John Keats, 1814–1821*, ed. Hyder Edward Rollins, Vol. 1 (Cambridge: Harvard University Press, 1958), pp. 238–239.
[3] William Wordsworth, "Preface to the Second Edition of the Lyrical Ballads, 1800," in *Lyrical Ballads*, ed. R. L. Brett and A. R. Jones (London: Methuen and Co. Ltd., 1963), p. 260.
[4] Wendell Berry, "The Old Elm Tree by the River," in *The Country of Marriage* (New York: Harcourt Brace Jovanovich, 1973), p. 3.
[5] Much of Berry's work, particularly his essays, emphasizes the impact of the changing economics of farming on the relationship of humankind to the earth and to nature. With increased mechanization, we are further removed from the soil.
[6] Jim Wayne Miller, "The Fencepost I," in *Copperhead Cane. Poems by Jim Wayne Miller* (Louisville: Green River Writers/Grex Press, 1995), p. 29.
[7] Mary Oliver, "The Black Walnut Tree," in *Twelve Moons* (Boston: Little, Brown & Company, 1972), pp. 56–57.
[8] Marge Piercy, "A Work of Artifice," in *Circles on the Water* (New York: Alfred A. Knopf, 1982), p. 75.
[9] Poems cited are from *The Palm at the End of the Mind*, ed. Holly Stevens (New York: Vintage Books, 1972).
[10] Harold Bloom, *Wallace Stevens: The Poems of Our Climate* (Ithaca: Cornell University Press, 1977), p. 357.
[11] Wallace Stevens, "The Irrational Element in Poetry," in *Opus Posthumous*, ed. Milton J. Bates (New York: Vintage Books, 1990), pp. 220–221.
[12] D. J. Gordon and Ian Fletcher, "Byzantium," in *Yeats: A Collection of Critical Essays*, ed. John Unterecker (Englewood Cliffs: Prentice-Hall Inc., 1963), p. 136.

SUZANNA B. SIMOR

THE TREE OF THE CREDO

Symbolism of the Tree in Medieval Images of the Christian Creed

The tree is a powerful and near universal symbol of life in ancient myths of many peoples around the world. The *arbor vitae* linked their human and profane condition to the realm of the divine and the sacred. At an early date, this motif was adopted by theologians to illuminate some of Christianity's chief mysteries. Salvation through the Cross extended the ancient idea of cosmic *renovatio* symbolized by the Tree of the World, given the early Christian association of the Tree in Eden with Christ's Cross. As the Tree of Christ's Sacrifice, i.e., of Suffering and Salvation, the Cross was homologized with the Cosmic Tree at the center of the world.

In the twelfth century this link reappeared in a legend of the Tree-Cross, the *arbor crucis*, devoted to the cycle of sin and Redemption, beginning with the Original Sin and its punishment. Multiple legends tell that the "true" Cross was made out of wood descended symbolically from the Tree of Life, and materially from the Tree of the Knowledge of Good and Evil, both of which grew in the Garden of Eden, with the Tree of Life growing at its center (Gen. 2.9). The two trees were frequently contrasted in art, one fruitful, the other withered. In the paradise to come, at the end of time, this Cross-Tree was seen to be again at the center "with its twelve kinds of fruit, yielding its fruit each month [i.e., continuously]" (Rev. 22.2; the description derives from Ezek. 47.12).

This interpretation, made familiar through the liturgy, *The Golden Legend* of Jacobus de Voragine, and the *Biblia pauperum* tradition, was connected with the typology of father of mankind, Adam. By eating from the Tree of Knowledge, Adam introduced sin into the world for which he was exiled from Paradise, and eventually buried in the hill of Golgotha – i.e., again at the omphalos – the *Mons salvationis* where the Tree of Life then grew from three seeds (or, in another version, from a twig) of the Tree of Knowledge that had occasioned the Fall, and from the wood of which was made the Cross later planted there. Adam was then baptized, i.e., reborn and redeemed, by the blood of the sacrificial Death of the new Adam, Christ. This connection between the Fall and the

A-T. Tymieniecka (ed.), Analecta Husserliana LXVI, 45–54.

Redemption was drawn in the early observance of March 25, before it became the Feast of the Annunciation, as both the day of Christ's Death and a day dedicated to the memory of Adam. The bloodied Cross-Tree gives forth, for all times, the oil, wheat, grapes, and herbs used to prepare the matter of the sacraments administered by the Church. Thus the Living Tree of the Cross, symbolic at once of sin and of redemption, binding earth and heaven, God and His creation, was paradoxically conceived as both occasion of sin and instrument of salvation, and became a most precious object, the holiest relic of Christendom.

The identification of the Cross with the Tree of Life was advanced in the thirteenth century by St. Bonaventure's *Lignum vitae*, which developed the metaphoric parallelism here as it recounted twelve aspects of Christ. Its chapters covered the twelve mysteries of Redemption, seen as the trunk, branches, and fruit of a tree simultaneously identified with the Tree of Life and the Cross of the Crucifixion, after a vision of the crucified Christ as the Tree of Life, i.e., the Cross in leaf once again with the Savior's sacrifice. This influential guide to meditation, composed by the General of the Franciscan Order (1221–1274), was widely disseminated, and provided inspiration to over a century of Christian mystics.

In visual representations of the Death of Christ, the Cross had been interpreted as the Tree of Life from the sixth century on. The subject of the Tree-Cross occurs in art sporadically from the twelfth century, concurrently with the spread of the legend of the Cross, but became common only in the fourteenth century, under the influence of Bonaventure and his followers. The institution, in 1376, of the Feast of the Finding of the True Cross, served to popularize the theme.

Theologically, the Crucifixion is the culmination of the Christian creed, the declaration of fundamentals of the Christian doctrine, the summary and core of the Faith, which, for those who believe, is the means of personal salvation made possible by Christ's Death on the Cross. Linking the Crucifixion and the *Credo* thus has the support of theology, and, occasionally, images of Tree-of-Life *Crucifixions* presented the *Credo*. In each case, the so-called Apostles' Creed was employed, i.e., the oldest creed, which is, with the Nicene Creed, the principal formulation of Christian dogma, the basic creed of the Universal Church. In some of these representations, the *Credo* is rendered by the mediation of those apostles believed to have been not only its disseminators but also its authors, they having compiled it under the inspiration of the Holy Ghost before they dispersed to preach the gospel. In this schematic type of *Credo* representation, the *collegium apostolorum*, for which models were plentiful, proclaimed the "revealed truth" of the Church. In images rooted in the speech scrolls of the prophets in Early Christian art of both East and West, the Credo now became an attribute of its bearers, who expressed, transmitted, and embodied their creed.

We have depictions linking the apostles with articles of the Creed and the Crucifixion. None predate the thirteenth or, rather, the fourteenth century, when lay literacy was becoming more widespread. The Credo was always a basic tool of catechesis, and interest in its representation expanded rapidly under the influence of the pastoral initiatives that followed the Fourth Lateran Council (1215). Images of the Apostles' Creed, with its text, served as a visual aid to communicate the formulated synthesis of Christian dogma, possibly also to insure its uncorrupted transmission.

Because Christ called himself the true vine, and His disciples its branches [John 15.5], representations of the Tree-of-Life Cross could allude to the Apostles by the number of their branches, and then make those correspond to the traditional twelve articles of the Apostles' Creed; see the concept in, e.g., Rabanus Maurus' *De laudibus Sanctae Crucis*. *The Tree of the Cross* in the central *Crucifixion* of a small Sienese triptych of c. 1400 now in Hanover (Niedersächsische Landesgalerie) includes evangelists and apostles with articles of the creed. In this instance, apostles displaying Credo scrolls are placed *in* the twelve branches surrounding Christ. All figures from the *arbor crucis* legends and from the liturgy, i.e., the Original Sin, its punishment and Redemption, Adam and Christ, Eve and Mary, are included. The Tree-Cross grows directly from Adam, who personifies the Fall. Its punishment, the Expulsion that Adam and Eve suffered, are recalled in the fur coverings of the first parents and the branches from the Tree of Knowledge in their hands. Their presence at the foot of the Tree of Life in itself points back antithetically to the Tree of the Temptation, and Adam proclaims *MULIER DECEPIT ME ET COMEDI*, "The woman tempted me and I ate." Eve, at Adam's feet, points to the salvation to come through Mary (*FUNDA NOS IN PACE. MULIERIS EVE NOMEN MARIA*, "Keep us in peace. The name of the woman Eve is Mary"). Mary, kneeling to Adam's right, confirms salvation (*EVA LACRIMIS MARIA GAUDIENS IN VE[N]TRE PORTAVIT*, "Eve bore tears, Mary joy, in the womb"). Opposite is Mary Magdalene (*DE MISSA SU[N]T MICHI PECCATA MULTA*

QU[O]N[IAM] DILEXI MULTU[M], "My sins which are many are forgiven for I have loved much"), completing the group. All the four figures are nimbed, because they have been redeemed by Christ's Sacrifice. Juxtaposed under the Cross, Eve, mother of mankind, and Mary, Mother of God and second Eve, restate the Fall-Redemption antithesis.

As seen in an illustration of Dante attributed to Giovanni di Paolo, Adam and Eve turn from the Tree of Sin toward the Cross; depicted in between is the Annunciation to the humble Mary that made restoration possible (c. 1438–1447; London, Brit. Lib., ms. Yates Thompson 36, fol. 141r). Depiction of both Eve and the Crucifixion is rare in Italian art. The two motifs were brought together in a Tree-Crucifixion in Giovanni da Modena's complex program in S. Petronio, Bologna (1421) which confronts and unites both trees of Paradise in an allegory of Sin and Redemption. Under a Crucifixion scene in a barren Tree of Knowledge to the left of the Cross (the Tree that brought about man's Fall also brings him Redemption), Adam and Eve are in the act of Fall, with Old Testament figures. To the right, Mary-Ecclesia catches the blood of Christ in a chalice. Mary's and Eve's inscriptions are contrasted: For Eve, "The race is destroyed by an idle morsel, because I have closed the gate of heaven you will die"; for Mary, "I now open heaven, which, you, Eve, had closed; through my Son I will redeem every sinner." On the opposite wall, a second Crucifixion fresco shows a "Living Cross," i.e., a crucifix from whose four ends hands emerge in action. Below, are mottoes of the Synagogue and Ecclesia linked to those of Eve and Mary: "The blood of the sacrificial goats seduced me like the snake . . . ," says the Synagogue, while Ecclesia proclaims, "Provided with (His) blood, I am called the Bride of Christ. . . ."

The poetic force of the image of Bonaventure's Tree assured it a ready acceptance in the visual exegeses of the thirteenth- and fourteenth-century schemata as found in paintings and in *Speculum theologiae* manuscripts. Many of the latter included diagrams of the Apostles' Creed, as do the English Arundel Psalters (e.g. Brit. Lib., ms. Arundel 83 I, fol. 13; Arundel 83 II, fol. 125v). Fritz Saxl's observation that from the thirteenth century on, *Lignum vitae* is frequently found in collectanea together with texts dealing with death and the final things not only illuminates the nature of the composition's attraction but suggests how the transference of its iconography to pious, moralistic depictions of the Crucifixion may have been facilitated.

An early Bonaventurian example is found in the French Psalter of

Yolande de Soissons (illuminated at Amiens, c. 1280–1290; New York, Pierpont Morgan Lib., ms. 729, fol. 345v). Early on, twelve prophets, justified by their sayings though unexplained by Bonaventure, were included in the branches and present in numerous schemata. In Italy, Bonaventure's Tree was visualized in Pacino di Bonaguida's monumental *Arbor vitae*, now in the Accademia in Florence (c. 1320–1325). There roundels in the Tree contain scenes from Bonaventure on the Origin and Life, Passion, and Glorification of Christ. In Pacino's reduced, more profoundly abstract version in an illuminated Bible (Milan, Bibl. Trivulziana, ms. 2139, fol. 435), the titles of the twelve sections (*fructi*) of Bonaventure's book are displayed on the scrolls of prophets in the branches. Many fourteenth-century paintings derive from this scheme, including Taddeo Gaddi's extraordinarily "verbal" fresco for the main refectory of S. Croce in Florence where the fruits contain Bonaventure's chapter headings and the twelve branches end in Old Testament prophets with relevant prophecies.

Bonaventure's Tree developed visually from representations of the Tree of Jesse, identified with that of *Life*. Busts of apostles with articles of the Creed were grouped around a Throne-of-Grace Trinity in an initial of a fourteenth-century manuscript of *La Somme le roi*, a compendium of doctrine and morals complied in 1279 for Philip III, "the Bold," of France, son and successor of St. Louis (Lille, Bibl. Municipale, ms. 116, fol. 1). Like Christ's royal ancestors, the Credo apostles became the branches of Trees in works still to be seen in churches in France and Germany. One such Tree, issuing from Christ and crowned by God the Father, was painted in the vault of the church of Vielle-Louron (Hautes-Pyrenées). Inscriptions below identify the subject as *LBRE DE NOTRA FOY E LA SINBOLA/DES APOTRES E CREDO IN VNVM DEVM*. Another Tree, in the choir of the parish church of St. Leonhard in Frankfurt (1536), is abloom with busts of the apostles presenting the Creed while the enthroned Redeemer, at the top, sends them on their mission (Matt. 28. 18–19): *DATA EST MIHI OMNIS POTESTAS IN COELO ET IN TERRA. EUNTES ERGO DOCETE OMNES GENTES BAPTIZANTES EOS IN NOMINE PATRIS ET FILII ET SPIRITU S[ANCTI]*.

This tradition provided the iconographic and formal lineage for the painting in Hanover by an artist close to Paolo di Giovanni Fei (fl. 1369–1411), a work that came at the end of the theme's evolution. The triptych's presentation of the total work of salvation is complete and clear, its combination of subjects unique. And Apostles with the Credo and prophets with their sayings frame, on the right and left respectively, several didactic fifteenth-century Venetian Crucifixions, even more crowded with biblical, allegorical, and dogmatic themes. These examples are from the Museo Correr (early fifteenth century) and from Prague (c. 1450). Within Tree of Life compositions, however, apostles are rare. Their heads stand for the twelve crops of fruit borne by the Tree of Life in the Apocalypse of Queen Isabella (c. 1313; Paris, Bibl. Nat., ms. fr. 13096, fol. 83).

Treating his model loosely, the artist of the Hanover painting compacted the tree into roundel branches that recall the traditional medallions. He omitted *Lignum vitae* inscriptions, and, instead of the prophets and saints sometimes present in earlier paintings to testify to Christ's Life, introduced the Credo, i.e., the totality of faith. The triptych was likely created for the private devotion of an as yet unidentified patron. Devotions proceeding from Bonaventure's *Lignum vitae*, the source for the main scene, while characteristic of the spirituality of the mendicant orders in general, were especially embraced by the author's Order, both its official branch and the Spirituals (the Cross, of course, was an attribute of St. Francis). The commission might thus be assumed to be Franciscan. Gertrude Coor, on the other hand, who forty years ago discussed the triptych in an article, concluded that it was probably Augustinian, on the evidence of passages in Augustine and Pseudo-Augustine that deal with Adam and Eve as types of Christ and Mary, and in the context of Redemption. Augustinians, however, were not the only religious order that paid respectful attention to Augustine's teaching, and sources other than his writings, patristic to Late Medieval, could have provided these themes. Nonetheless, the recent discovery that Paolo di Giovanni may have painted in the church of S. Agostino in Siena is of interest, for S. Agostino then still had a representation of the Credo, in Ambrogio Lorenzetti's once famous frescoes in the chapter house, and the Order commissioned other visualizations of the Creed. The provenance of the Hanover triptych is simply unknown. But those Italian fourteenth-century Tree-of-Life Crucifixions of known provenance were mostly Franciscan, including both the most important, Taddeo Gaddi's, and the most influential, Pacino's.

The composition of the Hanover painting is simple in form but spiritually rich. The small size and the private character of the work doubtless dictated its simplification and abbreviation. The forms are those of a miniature and the text is tiny, both to be contemplated only in proximity. The central placement of the apostles and the creed they teach leads the beholder's eye repeatedly across the painted field in a serpentine sequence enfolding the crucifix, as does the sequence of the Bonaventurian narratives in the wings (the Annunciation, the Nativity, the Adoration of the Magi, the Carrying of the Cross, the Harrowing of Hell). It is interesting to note that some other fifteenth-century representations of the Apostles' Creed, in sets of apostle figures, are similarly composed around an altar: the sculptures at St. Martin in Landshut (1424), stained glass windows of the Duke of Berry's former Sainte-Chapelle at Riom, and the choir sculptures of Albi (both of the mid-century). This disposition underscores the Creed's [the Tree's] centrality to God's plan for salvation: Faith, in the words of the divinely inspired apostles (like the crucified Christ, they are shown as the fruit of the Tree of Salvation), is the Tree's harvest, offered by the apostles as redemptive food. This fruit confers everlasting life. (An example of the idea of the fruits of the Tree being redemptive food as well as the redeemed being themselves the fruits of the Tree is also found in the corner figure of Christ next to the Crucifixion relief on Nicola Pisano's Siena pulpit, a figure also adopted by his son Giovanni in his later pulpits.) The teaching of the Church is the Hanover painting's central message, developed through multiple allegories of redemption from sin through faith. Within a cosmological and eschatological framework, it contrasts and connects Adam and Christ, Eve and Mary, including the archetypal redeemed sinner Magdalene. Pointing out mankind's perilous earthly journey, the painting introduces the theme of man's choice. By presenting the path for Christians seeking salvation, the road from sin to redemption, the image offers warning and guidance: Contemplating the allegory, the viewer, as Adam's heir, could spiritually ascend to Christ, and advance toward his own salvation and eternal life, Christ's grace descending upon him. Such notions of ascent and descent belong to the multiple allegories of Tree-of-Life Crucifixions. The inclusion of the Credo apostles transformed this Tree into a symbol of the Church.

Because belief centered upon the dogma of Redemption, Fides was closely connected to the theme of the Death of Christ on the Cross, and the Tree of Life carries visualizations of the Creed also in some medieval representations of the personification of Faith, the first theological Virtue. With Fides, of course, the Credo functioned as the identifying, emblematic attribute of a theme that itself symbolized it.

Giotto's Fides, in the Arena Chapel in Padua, from the first decade of the fourteenth century, holds a scroll with the beginning of the Apostles' Creed. In the *Belleville Breviary*, illuminated in Paris about twenty years later, Fides in the *bas-de-page* of Psalm 1 is enveloped in an aureole radiating miniature symbols of the Creed (Paris, Bibl. Nat., ms. lat. 10483, fol. 7).

In a group of fourteenth-century Italian illustrations, most of them miniatures pregnant with Bonaventurian imagery, Fides is a tree whose fruit are the articles of the creed. A Spanish poem of 1396, one inspired by Dante, by the Genoese Francisco Imperial, active in Seville, makes use of the same imagery. The poem places Fides after Caritas (in translation):

> The second dame a tree entwined secure:
> Its roots sprang forth from rock of crystal pure,
> Twelve branches from its trunk did thrust amain;
> To plead the Credo's dozen their domain.

These personifications appear in cycles of the Virtues and the Liberal Arts based on Augustine. Most are found in manuscripts with eulogies of secular rulers: Robert of Anjou (King of Naples, 1309–1343) in manuscripts in Rome (Galleria Nazionale d'Arte Antica, Gabinetto delle Stampe, *Canzone delle Virtù e delle Scienze*, fol. 6v), Florence (Bibl. Naz., ms. B. R. 39 [formerly II. 1, 27; Magliabecchiano Cl. VII. 17], fol. 32v), and Vienna (Österreichische Nationalbibl., Ser. Nov. 2639, from Castle Ambras, fol. 2v), the panegyric poem attributed to the Pratese humanist and Petrarch's tutor Convenevole; and Bruzio, son of Luchino Visconti of Milan (who was banished in 1355) in a manuscript in Chantilly (Musée Condé, ms. 599 [formerly Cod. XX 1426], fol. 4v), whose eulogy is an adaptation of Convenevole's *Cantica* by Bartolomeo de' Bartoli of Bologna. Another personification belonging to this group is found in a manuscript of homilies and prayers in Paris (Bibl. Nat., ms. ital. 112, *Homélies et oraisons latines*).

Two reduced versions signed by Niccolò di Giacomo da Bologna (c. 1330-d. by 1402) are in Milan (Bibl. Ambros., ms. B. 42 inf., fol. 1, with Giovanni d'Andrea's *Novella in libros Decretalium*, dated 1354) and Madrid (Bibl. Nac., ms. 197 [D. I. 2], a *Triumph of St. Augustine*, with the *Commentari* by Bartolo di Sassoferrato). Two frescoes painted for Augustinian churches in northern Italy and related to Niccolò's miniatures belong with this group: Giusto Menabuoi's in the Eremitani in Padua (1370), and a *Triumph of St. Augustine* from St. Andrea at Ferrara, today in the Pinacoteca (c. 1378).

In the Rome manuscript, whose silverpoint drawings outlined in pen are accompanied with the most text, each personification is centered between a Latin laudatory poem and a prose definition of the virtue, with vernacular commentary, all extracted from Augustine. Marginal glosses provide bibliographic references to his relevant works. Fides Catholica is a crowned and veiled woman embracing a tree whose feathery branches end in the Apostles' Creed. At the apex, the middle of the fifteen roundels bears the words *TRINUS ET UNUS DEUS*. The Creed's text is given in the remaining fourteen roundels. (This division of the Credo into fourteen articles, found in this group, was the simplest of variants employed in the Middle Ages, one preferred by theologians, canonists, and synods, and dear to thirteenth-century scholastics. In visual representations, it is relatively unusual since these, largely involving apostles, typically employ the more common twelvefold division of the text.) The tree grows out of an altar in a church, glossed *PETRA AUTEM ERAT CHRISTUS* (I Cor. 10. 4) and *ET SUPER HANC PETRAM EDIFICABO ECCLESIAM MEAM* (Matt. 16.18), and Fides triumphs over the heresiarch Arius.

This very full representation of Faith, instructive and in context as it is, records the nature, function, and significance of the virtue in image and word, complete with a representative historical archtransgressor trampled in a graphic demonstration of his defeat as an embodiment of the opposite vice. The choice of Arius, the heretic of the fourth-century theological debates over the relationship of the Second and the First Persons of the Trinity from which the orthodox creed was formulated, underlines the identification of Fides with the Creed. Arius and his adherents questioned the divinity of Christ and were anathematized by the council of Nicaea in 325, the first ecumenical council, convoked at the behest of Emperor Constantine. (According to tradition, in the following year, 326, the emperor's mother, Helena, prompted by a dream, located the place where the True Cross of Christ's Crucifixion lay buried and had the wood unearthed.) The Arians were again condemned in 381 by the second ecumenical council at Constantinople. Both councils promulgated important creeds that stated their doctrinal positions. To the Middle Ages, Arius, for disputing consubstantiality of the Son and the Father, represented a particularly fundamental heresy.

All the representations of this group are similarly developed, and the illustrations vary in form but not in iconography. For example, in the Vienna and Florence codices, where the Virtues are paired with the Liberal Arts, the body of Fides Katholi[ca] is merged with the trunk of the lush, leafy tree that bears the twelve fruits of the Credo. Most complete is the Chantilly codex. In the reduced variant in juridical codices, on the other hand, Niccolò da Bologna presented all the figures in a single scene. In the Madrid Triumph, the Virtues/Arts iconography was employed to project a complex theological message. Here Fides is at the center, directly below Augustine, and the creed reads from top to bottom, across Fides, like in the Hanover Tree of Life. The frescoes share this iconography. In Padua, a vestige of the tree trunk is still visible in Fides's hands though no branches remain, and of the original crown of Credo fruit, only the first and the last roundels are preserved (at Faith's right is the first article, *CREDO IN DEUM PATREM OMNIPOTENTEM CREATOREM CELI ET TERRE*; its opposite is the last, *CREDO IN VITAM AETERNAM*). The Ferrara fresco is a clarified counterpart to the Madrid illumination.

There is little agreement on the origin of these illustrations. They seem to derive from the colored pen drawings in the Chantilly manuscript which itself belongs to the school of Bologna. Many of these Augustinian cycles refer to Bologna. However, their relationship is complex and they probably derive from a common prototype.

That both Fides and the Crucifixion could be shown with a twelve-branched (twelve-fruited) tree, standing for the Apostles' Creed, reflects the presence of the theme of the Tree of Life at the very heart of Christian dogma.

Queens College, City University of New York
USA

SECTION I

THE DIALOGUE BETWEEN LIFE SCIENCES AND PHILOSOPHY

VIJAY SHANKAR RAI

THE ORIGIN OF LIFE

Individuation and Evolutionism

The problems of the metaphysics, phenomenology, ontology or origin of life, or of the origin of universe or of any aspect of philosophy are always related to the problem of ultimate reality, whether it be matter or mind or both. The ultimate reality is the main source of energy of life in the universe. So every phenomenon is always related to ultimate reality. Therefore we have first of all to find out what ultimate reality is, viz., matter or mind or both. Without revealing ultimate reality, our struggle to find out the origin of life and evolution will be in vain.

In chess, the victory of the Blue Thought computer over world champion Garry Kasparov, has started a new debate regarding human intelligence vs. artificial intelligence. The superiority of the Blue Thought has been accepted worldwide. Even Kasparov himself has accepted it: "Something truly unbelievable happened." Deep Blue can think millions of positions in a second; last year Kasparov alleged that "it was no less than God."

But this victory of the Blue Thought is not a victory of matter over mind, but is simply a victory of more information over less information in a particular field. It is information, which corresponds to consciousness, that plays an important role in the functioning of individuals, as well as that of the whole Universe. The astonishing fact is that energy and information always go together. So first of all we have to search out ultimate reality on a scientific basis.

The origin of life, individuation, diversity and evolution are only different processes and different patterns of ultimate reality, viz., consciousness. While physicists hold that individuality is nothing but different patterns of energy, one of the great physicists of this twentieth century, Stephen Hawking, has stated that the total energy of this universe is zero. Not only this but another great physicist, Roger Penrose, has stated that in the 'big crunch,'

Not just all matter becomes destroyed in this way, but even the very space-time must find its end! Such an ultimate catastrophe is referred to as a space-time singularity. The reader may well be asking how it is that we know such catastrophes must occur, and under what circumstances matter and space-time are destined to suffer this fate. These

A-T. Tymieniecka (ed.), Analecta Husserliana LXVI, 57–71.

are conclusions that follow from the classical equations of general relativity, in any circumstances, when a black hole is formed.

Now there is a fundamental question, that of how did from nothingness this universe, matter and space-time come into existence? The great physicist Stephen Hawking bewildered after deep observation of this mysterious universe answered, "However, the laws do not tell us what the universe should have looked like when it started – it would still be up to God to wind up the clock and choose how to start it off."

Stephen Hawking puts a certain question before the philosophers of the world: "Even if there is only one possible unified theory, it is just a set of rules and equations. What is it that breathes fire into the equations and makes a universe for them to describe?" The usual approach of science in constructing a mathematical model cannot answer the questions of why there should be a universe for the model to describe. Why does a universe go to all the bother of existing? Or does it need a creator? Up to now, most scientists have been too occupied with the development of new theories that describe what the universe is to ask the question why?

On the other hand, the people whose business it is to ask why, the philosophers, have not been able to keep up with the advance of scientific theory. Philosophers have reduced the scope of their inquiries so much that Wittgenstein, the most famous philosopher of this century, said, "The sole remaining task for philosophy is the analysis of language. What a come down from the great tradition of philosophy from Aristotle to Kant."

This is an age when science has crossed the boundary of the physical world and tried to enter into the quantum world or the metaphysical world. Therefore, now is the proper time for rethinking logical positivism.

Now I assure Mr. Stephen Hawking and my friends that I have tried to answer the above questions raised by Mr. Hawking and some more fundamental questions in this paper and in my book, *Contemporary Science and the Absolute*.

Now we come to the problem of the origin of life, evolutionism and individuation.

The century-old problem is that of whether consciousness and life are by-products of matter or matter is a product of consciousness. Some contemporary physicists and materialists held that all phenomena are the result of matter in motion. According to the Dialectic Materialism of Karl Marx, "Our consciousness and thinking, however supra-sensuous they may seem, are the highest product of a material, bodily organ, the brain. Matter is not a product of mind, but mind itself is surely the highest product of matter."

Thus, for the materialist, consciousness or mind is not eternal but is only a by-product of matter. This materialistic world-view has made headway in the world over the centuries. Materialists believe that everything can be explained by matter and motion as the effect of necessary laws. They believe that this universe is made of matter only, that all living beings and every thing in the world are made of matter and differentiated through the laws of motion.

To examine the truth of materialism and in particular, the deterministic evolutionism of Darwin, it is necessary to pose certain questions. Thus we can investigate how far the materialists are correct, because they hold that they can explain each and every thing through matter and motion.

Therefore, I hereby put the following questions before the materialists and scientists of the world. I very humbly request of them their answers to the following questions, so that we may acknowledge the ultimate truth of reality:

Q. 1. All contemporary scientists including Albert Einstein, Stephen Hawking and Roger Penrose, endorse the expansion model of this universe.

Why did the universe start out with so nearly critical a rate of expansion, not one greater or lesser?

This rate of expansion is so astonishingly critical that if the rate of expansion one second after the Big Bang had been smaller by even one part in a hundred thousand million millions, this universe would have recollapsed into a black hole.

Q. 2. Why are spin and motion found in tiny quarks, in electrons, in planets, in stars and in farflung galaxies?

If spin and motion were taken away, what would be the fate of this universe? Would life survive in this universe?

Q. 3. Do you find any of God's maxims and God's design in this universe? Please tell us the evidences of them.

Q. 4. Why are there exact numbers, sizes and measurements for electrons, protons, neutrons, Hadrons and sub-particles and atoms? If slight changes were made in electrons, protons, neutrons and in the atom, what would be the fate of this universe and life?

Q. 5. Why is there a critical proportion in the strength of nuclear forces? Why is it not greater or lesser? If these forces were lessened by even one part in a hundred thousand millions what would be the effect on the universe and on life?

Q. 6. Why is there a critical proportion in the strength of the force of gravity? Why is it not greater or lesser? If it were less by even one part in a thousand in its strength, what would be the effect on the universe and life?

Q. 7. Why is the universe so uniform on a large scale?

Q. 8. Why do information and energy go together?

Q. 9. Why are there nearly about 200 subparticles in atoms? Why not more or less? If there were no proton or no electron or no neutron, what would be effect of that on this universe and on life?

Q. 10. What is the cause of the physical law of causation itself?

The theory of causation is fundamental to science. As Albert Einstein stated, the law of causation itself is the ultimate basic postulate of all natural sciences (Einstein p. 261). But scientists avoid finding out the cause behind the above fundamental questions on phenomena. Perhaps they do not know the cause of the above phenomena. Without knowing the cause of the above phenomena, we cannot understand reality as a whole. But I praise Stephen Hawking, who has accepted scientists' inability to answer the above fundamental questions and has asked philosophers to answer them.

This limitation is imposed on scientists because their knowledge is limited to physical law. Since those physical laws related to the origin of life and the functioning of the universe are not fundamental laws of nature, these laws have slipped away or collapsed. Einstein clearly stated that many physicists maintain that in the face of these facts not merely differential law but the law of causation itself, hitherto the ultimate basic postulate of all natural science, has "collapsed".

Einstein's finding shows clearly that the known physical laws are not fundamental to nature because through these physical laws all the phenomena of the universe cannot be explained.

These physical laws are governed by certain divine universal laws. Scientists and physicists have limited themselves to physical laws only. They have failed to appreciate the all-pervading divinc universal laws. Only through these divine universal laws, can the mystery of the origin of the universe and of life be understood properly. It is through these divine universal laws that the very mysterious processes of this universe are revealed.

All of the above questions are related to the origin of life and the evolution of the universe.

EVOLUTION AND ORIGIN OF LIFE

Shifting scientific knowledge cannot be granted the status of true and ultimate knowledge. Only the universalized multi-angle tested knowledge of philosophy is to be accepted. Therefore, a rethinking of Logical Positivism is necessary.

Biologists believe in Darwin's theory of natural selection, which would be the sole mechanism responsible for the evolution of species and the origin of life. The great difficulty is that no evidence of such is found in the early universe. If natural selection is a universal law, evidences of it must be found in the early universe.

The evolution of the universe is an accepted fact. Physicists have accepted that the Big Bang is the boundary for the origin of space and time. Albert Einstein and Roger Penrose accepted the standard model of the Big Bang as the boundary for the origin of matter and space-time.

Now we have to examine whether the mechanism of natural selection is found in the first three minutes of time and the early universe. A well-known description of the cosmic fireball is found in Steven Weinberg's book *The First Three Minutes.* I will enumerate here only a few of Weinberg's important findings.

The Big Bang actually started at $T = 10^{-2}$ second, the first hundredth of a second. In the first hundredth of a second the temperature of the universe was 100 billion kelvin. It was dominated by radiation. The electron-positron pairs were produced by radiation. These annihilated to produce radiation and the massless neutrons and their anti-neutrons. The nucleons were being constantly bombarded by the electrons, positrons and neutrons. Particles changed repeatedly from neutrino to proton and back again. No mechanism of natural selection can be found in the above events.

As the temperature of the universe declined, many reactions took place in the universe of particles. About a third of a second after Time = 0, a great change occurred in the universe. Neutrons ceased to occur in the universe. The neutrons ceased to interact with electrons, positrons or anything else. These neutrons remained in the background as cosmic background radiation.

At last, the reactions described in detail by Gamow and his colleagues, nucleo-synthesis, rapidly built up nuclei of helium 4. But matters stopped there, for physicists have calculated that there are no stable nuclei with masses 5 or 8. Therefore helium 4 developed, there being no guarantee in helium 5 or helium 8 for stable nuclei. This clearly shows intelligent planning, a design or programme behind the evolution of the universe. From time zero, and for millions of years, however, no mechanism of natural selection is to be found.

Thus, the theory of natural selection fails to explain the evolution of the universe and the origin of life in the early universe. The origin of life is related not only to the earth, but as scientists have found, its sequence extends to deep space also. It needs planned evolution from the Big Bang onwards. Only then could life flourish in this universe. All the building blocks of nature, like particles, nuclei, atoms, all the natural forces and environment are responsible for the origin of life in this universe.

Now we have to search out that Divine universal law which is the fundamental law of nature governing even the mechanisms of natural selection.

DIVINE UNIVERSAL LAWS AND EVOLUTIONISM

We have to go beyond singularity and the Big Bang origin of the universe and space-time, because these are temporal.

Physicists have revealed that in this universe two kinds of substances are found. First there is Information and secondly quantum energy or radiation or the matter of physical appearance. Stephen Hawking and Roger Penrose have come to the conclusion that "not just all matter becomes destroyed in this way but even the very space-time must find its end. Only Information which corresponds to consciousness remains after the final crunch. Consciousness is eternal and Pure Consciousness is the cause of all physical appearance."

When there was no matter, no energy, no space and no time, how did singularity come into existence and why had it created the Big Bang? We have to find out where energy came from in the first place. This zero state surely requires a super-natural explanation.

Contemporary physicists have inferred that before the Big Bang, there was a singularity. This is an edge to space-time. A singularity is the nearest thing that science has found to a supernatural agent.

Contemporary physicists have come to the conclusion that there is some supernatural agent (singularity) or Divine agent to which no physical law applies. When scientists give witness to a Divine agent, there must be His universal laws also. Stephen Hawking agrees in this matter: "He (God) would, of course still have had the freedom to choose the laws that the universe obeyed." Now we have to mark out the function of these Divine universal laws in the universe. If we should succeed, certainly it would be a great triumph over the deterministic and materialistic view of nature.

The absolute supernatural agent or Divine agent operates in this universe through following mysterious Divine universal laws whose function is precisely testified to in this universe. Observation attests to the existence of these Divine laws.

1. The Divine universal law of transformation.
2. The Divine universal law of self-replication.
3. The Divine universal law of the photo-copy system.
4. The Divine universal law of programme.
5. The Divine universal law of code.
6. The Divine universal law of self-organization.

These laws are the main cause of the laws of physics and programmed matter, radiation, the Big Bang, gravity, particles, etc. nicely and mathematically, so that the laws of physics may function properly. The divine universal law of programme has programmed this universe, its building blocks, particles, atoms, energy and all natural forces.

Physical laws are not sufficient to explain each and every event that occurs in space-time. They have certain limits in their applications.

Now I will describe the Divine universal law of programme because the mechanism of natural selection itself is one of the programmes of the Divine law of programme.

Due to the Divine universal law of transformation and that of the photo-copy system, singularly the Big Bang occurred and the universe came into existence. Roger Penrose and Stephen Hawking describe it thusly, "God abhors a naked singularity."

Therefore, we have this singularity, be it called supernatural agent, Golden Egg, Brahma or God's maxim or Divine law. It is difficult for the human mind to imagine a thing which is not in space-time.

God has not created this universe or singularity, because He has no unfulfilled desire. He is far away from all these events. That there was the initial singularity (supernatural agent) and then this universe is due to the operation of His Divine universal laws.

Divine universal laws throw enough light on the mysterious functioning of this universe. Many centuries-old problems relating to metaphysics, the new physics, biology, philosophy and religion have been solved through these Divine universal laws.

Roger Penrose and Stephen Hawking are eager to know the ultimate reality lying behind this universe. Hawking is not satisfied with physical truth, and therefore he is asking questions:

> Even if there is only one possible unified theory, it is just a set of rules and equations. What is it that breathes fire into the equations and makes a universe for them to describe? Is the unified theory so compelling that it brings about its own existence? Does it need a creator, and if so does He have any other effect on the universe? And who created Him?

This is the problem of evolutionism. The evolutionary model of universe follows from the Divine universal laws of programme and code. Many physicists have asked the question of why this evolutionary model of the universe has been choosen at all? The above given Divine universal laws have evolved the evolutionary model of the universe and programmed all processes, fundamental buildings blocks and the environment to reveal intelligent life in this universe.

Due to the mysterious functioning of these Divine universal laws, philosophers and scientists of the world always have been bewildered. I quote here Stephen Hawking's statement: "These laws have originally been decreed by God, but it appears that He has since left the universe to evolve according to them and does not now intervene in it" (Hawking). Due to the Divine universal laws of programme and code, this universe functions like a supernatural automatic plant. Therefore scientists and philosophers have been divided into two groups. Some believe in determinism and others in indeterminism. Albert Einstein was also trapped by this choice and gave a statement, "God cannot play dice with this universe."

Mr. Stephen Hawking is somewhat correct in this matter. God has nothing to do with this universe. It is the photocopy of His ideas, namely, the universal Mind or supernatural agent, who governs this universe through His Divine laws. Due to the operation of these Divine universal laws, this universe seems somewhat deterministic and like a digital computer in its operation. But this universe and reality are actually non-deterministic.

I have come to the conclusion that this is a programmed and planned universe. There is order and uniformity in it. The expanding model of the universe, the Big Bang, particles, atoms, molecules, radiation, energy, natural forces have been programmed very nicely to make this universe more fit for intelligent life and for expressing Divinity's sublimation.

Deterministic physicists are unable to demarcate the difference between a stone and a watch. But any sensible person would conclude that the latter mechanism was devised by some intelligent being for a purpose. In the same way the intricate and delicate organization of this universe and joint functioning of its components and individuals is overwhelming evidence for its design, purpose and designer.

In the same way a super design and super programme are found in the atom. A tiny atom is wonderfully programmed with nearly 200 programmed particles. Take away some components from a watch and it would stop functioning. In the same way take away the electron, protons or quarks from an atom and it would stop their functioning to create different elements, and the whole universe would change dramatically and there would be no life at all. So, the origin of life is related to the Big Bang, the origin of particles, a motion and spinning programme, etc.

Many volumes of books would not be enough to cover such programmes as are found in this universe. These programmes are necessary for this universe and for the origin of life. Numerous programmes can be regarded as evidence of this programmed universe. This very universe has a deceptive nature like a particle which sometimes behaves like a wave and sometimes like a particle. So materialists and idealists, and determinists, are caught in a trap.

THE ORIGIN OF LIFE

The origin of life is not an accidental event. Ours is a planned and programmed universe meant for the revelation of intelligent life. None can see the evolution of the species apart from evolution of the universe. The universe functions as an organized whole, and the living species are only an integral part of the whole. These species have no separate history apart from the history of the universe in the space and time. The part is governed by the whole.

THE DIVINE UNIVERSAL LAW OF SELF-ORGANIZATION AND REVELATION OF LIFE

It is by the Divine universal law of self-organization that the evolution of the universe and the living species became possible. Due to the self-organizing quality of 200 particles and subparticles, atoms emerged; and due to the self-organizing quality of atoms, helium and hydrogen gases and various elements came into existence. In essence, Dalton's theory is nothing but the theory of the self-organization of atoms.

The magnitude of the Divine universal law of self-organization covers group combination, viz., genes, chromosomes, molecules, cells, the brain and the human body, etc.

The self-organized brain system is even more remarkable. In the brain some 10^{10} to 10^{12} neurons and nearly 10^{12} to 10^{14} synapses come together and develop a self-organized system of mental activity and of memory. If there were no Divine universal law of self-organization systems, there could be no particle, no atom, no molecule, no cell, no gene, no chromosome and finally no living being.

Due to this law, every particle, subparticle, atom and object of the universe, shows individuality, through their individual functioning and these particles, atoms, and objects coordinate within the organism, and each and every individual organism coordinates with universal organisms. Every individual is a child of this universe, because he also is made of the same atoms and particles found there and needs a certain environment in the universe. As quantum mechanics has revealed, the part is governed by the whole.

THE THEORY OF NATURAL SELECTION, LIFE AND INFORMATION

According to Johannes Müller (*Lebensweisheit* [Berlin: 1929]), life depends on self-reproduction and it has the unique ability to replicate self and mutate so as to multiply variants. Life has the ability to self-copy and mutate with the mutant types then being able to multiply themselves. This does not mean that the earliest forms of life were able to copy themselves precisely. It is likely that in the beginning there might have been many more mistakes in the copying process. It is conceivable that mutation might have occurred earlier in the history of life than self-reproduction.

Natural selection would have selected those with the greatest ratio of correct vs. wrong copies. In other words, it would cause a lowering of the mutation rate. The gene through its ability to self-reproduce and to mutate has served as the primary basis of life. Natural selection can become effective to a certain extent only when these two properties are present.

That means that there must be some conscious agent who determined which is the correct copy in the primaeval self-copying process or in the formation process of the primaeval gene. This conscious agent, nature (God) or universal mind had to select only the correct copy of genetic material because then no such mechanism was available. In the primaeval pond, there was no genetic material as such, gene, DNA or loose string of DNA. So the conscious agent must have evolved the raw genetic material Himself. In that genetic material mutation occurred. So before mutation there must have been some genetic material which was doing self-copying. There must have been some genetic material with a self-copying system before mutation occurred in it. That means genetic material was first programmed with a self-copying system. Without a codifying programme and information, a self-copying system could not have occurred in this universe. It is only according to coded programmes that the functioning of the universe occurs. Due to the universal law of the programme, no system could occur in this universe without a coded programme. Thus first information which corresponds to mind was codified in raw genetic material to accomplish self-copying. Because of coded information, the mysterious self-copying system appeared in genetic material. This also is authentic proof for the existence of a universal mind, or a supernatural agent. There must have been some supernatural agent codifying the information for the self-copying system in raw genetic material. First there is information, then there is a gene, then there is life. There should first be bits of information, then that information is transferred as a chemical code to form a gene, and from the programmed and coded gene with information there evolves life in this universe.

Now the question arises of just how it can be judged that a certain mutation was right or wrong and is then to be or not to be selected. To judge a thing to be right or wrong, there must be certain criteria. To judge mutations to be right or wrong implies some plan or design to be implemented, for which a genetic plan, a particular genetic mutation is then selected. Only being useful then, will the mutant be right. For what would that mutation be useful? It must be useful in carrying out the genetic plan or genetic design of this universe. So there is a genetic design or genetic plan for this universe. Thus there must be a genetic plan or genetic design of the universe. Therefore evolution from the beginning should move towards the fulfillment of the genetic plan, for without a genetic plan nothing can be said to be right or wrong.

We can say that this universe is a genetically planned universe. Thus this universe has been evolved by a supernatural agent, viz. a universal mind or master mind, according to Divine universal laws.

A. G. Cairns Smith, the writer of *Seven Clues to the Origin of Life*, declares as I do that

> Genetic information is the only thing that can evolve through natural selection because it is the only thing that passes between generations over the long term. Although held in a genetic material, genetic information is not itself a substance. It is form . . . evolution can only begin once there is this a kind of form when the conditions exist for the replication of genetic information.

Thus a clue to the origin of life is first information, then there is the gene, and then life. The gene contains coded information. Hence the gene is called a programmed and coded gene. It is not worth pointing out that this is an authentic fact: information or consciousness or mind first is not matter. For the evolution of life, information or mind is a prior condition. Therefore, ultimate reality is mind or consciousness, not matter.

Hence in the primordial pond, when no genetic material was available and no replication system was to be found, information was needed first, namely, the information (form) of the primordial genetic material and the information of the replication system. With that, life in the universe originated. First information, then life, and information corresponds to consciousness or mind. Thus consciousness first, then genetic material, then gene, then replication system. Thus consciousness or information or mind plays an important role in the origin of life and the evolution of a universe.

INFORMATION, ENERGY, INDIVIDUATION AND COMPUTERS

Energy has been programmed so that the fundamental particles might evolve, then atoms and the forces of nature, then the emergence of intelligent life in this universe that carries information through a Divine universal law setting a programme and code. Prior information is a precondition for the origin of life. Therefore the main function of energy is to evolve fundamental particles, atoms and natural forces and to carry information.

It is found in the sequence of the purine and pyrimidine bases. A single gene probably carries considerable information. For example, one gene (cistron) possesses about 1,000 to 1,500 nucleotides. So the number of permutations is 4^{1000} (since there are A bases) or 4^{1500}. This number is much more than the number of elementary particles in the known universe. When we consider the number of genes present in a cell, we can well imagine the enormous amount of information contained in the base sequence in DNA.

This universe and its energy exist only for the expression of information. Therefore energy and information go together. It is an astonishing fact that energy itself is a photocopy of information which corresponds to a consciousness. So energy is a very mysterious reality. A single man has such enormous information in him, bits in numbers millions and millions times greater than the number of particles in the whole universe. It is scientifically proven that an individual man is a bundle of information and his body is a container of information. The individual is made of information. He is not made of blind matter. So he is the product of information, and he is not a product of matter.

This information or consciousness evolves individuality and diversity in this universe.

The person or thing or a computer which had more and more information, would meet with more and more success. This is an empire of information.

A tiny atom contains huge energy in the same way that a little gene contains enormous information, more possible permutations than the total number of particles in the universe. When a man or a species has millions and millions of genes, it is difficult even to imagine the huge amount of information before us. Then there are the countless viruses, bacteria and the varied species, plant and animal. It becomes even more difficult to imagine the huge amount of information here. We can say that all the particles and matter of the whole universe are actually like a tiny ball in a vast ocean of information or consciousness.

So information or consciousness is the ultimate substance of this reality. The little matter of the universe is only the photocopy of information and nothing else.

Once there was nothing, that is, there was only consciousness. This whole cosmos is only the photocopy of the idea of consciousness or mind. So there is Advait.

The brain is a photocopy of 'mind' or 'self' or information or consciousness, and mind or self is a localized wave of consciousness. Individuality or self is a wave in the ocean of consciousness. This wave of consciousness takes the shape of 'individuality' with the encouragement of Divine universal law or self-organization. Consciousness is the only ultimate reality. Therefore there is Advait only. All chemical components or physical appearances are a photocopy of information. Therefore, energy and information go together.

Our emotions and information take the form of a mass, viz., neurons and out of these neurons' molecules there emerge our emotions and feelings again. This is the very mysterious process of our mind. The functioning of the Divine universal law of transformation and the photocopy system is very explicit.

The Nobel Prize winners of 1994 in Medical Science, Dr. Alfred Gillman and Dr. Martin Rodbell of America have concluded that "it has been revealed that as soon as our sensation crosses the surface of the cells of our body, it transforms into a chemical code and it increases the chemicals of the cell also."

ESSENCE

The Divine universal law of transformation, of the photocopy system, of the programme and the code system, and of self-organization that I have explored here is authentic and based on scientific findings of the twentieth century, even the work of Nobel Prize winning scientists. These Divine universal laws have explored a very astonishing universe and its mysterious processes.

Divine universal laws play a very important role in the origin and evolution of life, to promote emerging individuality and to demarcate the 'self'. The Divine universal law of transformation is the main cause of singularity and the universe, while the universe and its individual entities are photocopies of information or mind or consciousness. Thus, prior information plays a very important role in the origin of life. We are like a tiny dot in the ocean of information.

There is diversity and individuality amid the organic unity of the universe. Particles, atoms, genes, chromosomes, cells, bioorganisms, and diversity reflect the Divine universal law of self-organization. There is no difference between particles and waves, information and chemical code, physical appearance and information, body and mind, universe and consciousness. Particles transform into a wave of information and vice-versa. All of these pairs transform into one another. This is very mysterious reality. So pure consciousness is only reality. There is Advait, no plurality.

'Self' is a wave of consciousness. Pure consciousness is the only ultimate reality. All physical entities and this universe are photocopies of information or consciousness. Pure consciousness is the only ultimate reality. Thus there is Advait and no duality.

The Nobel Prize winners of 1994 in Medical Science, Dr. Alfred Gillman and Dr. Rodbell of America have concluded that "it has been revealed that as soon as our sensation crosses the surface of the cells of our body, it transforms into a chemical code and it increases the chemicals of the cell also." Thus mind or information transfers into a chemical code or into matter. Therefore matter (chemical code) is only a photocopy of consciousness or information. Consciousness or mind is the eternal and ultimate reality.

Jabalpur

BOGDAN OGRODNIK

ON THE METAPHYSICAL FOUNDATIONS OF LIFE

1.

Life in its actual form and aspects must have its foundation in the deepest level of the real world. This simple truth was already known by ancient Greek philosophers. For instance Plato believed in a world soul which was the source of all regular movements present in the universe. According to his thesis each living being had a soul which participated in the world's soul. This was especially expounded on by Plotinus. Solutions to the problem of genesis of life on that level were sufficient up till the moment when mankind came to know that Earth was only a speck of dust in the universe. Life lost its special position in the known world to the benefit of inanimate matter. This was the triumph of mechanicism in the sciences and philosophy. The almost empty universe cannot be alive but must generate life. So the whole is poorer than a part. At first sight it seems to be a paradox, but it will no longer be so if we have a good theory of the object in which we can frame the problem.

In my opinion the best theory of the object was proposed by Polish phenomenologist Roman Ingarden. In his theory he described an object, especially a real object, from different points of view: as an assemblage of parts of a whole, as a composition of properties and their subject and as a composition of actions and their subject. The three main aspects build a full object, and many complicated relations between them are discovered. What is more, each mentioned aspect of an object and the object itself have three ontological levels, i.e., some mode of existence, some form and some pure qualities. All of these aspects and levels of an object are joined together by its essential structure based on its constitutive nature. The ontology of an object, merely mentioned here, gradually turns into a theory of partly isolated systems. In order to construct a good idea of such a system, it is necessary to elaborate a typology of real objects. On the grounds of Ingarden's ontology there are three types of real objects: things, processes and events. Generally, if things are joined together by processes and events, then a partly isolated system

A-T. Tymieniecka (ed.), Analecta Husserliana LXVI, 73–81.

may be constituted, one which possesses its own essence. Of course, there are many conditions that must be fulfilled. Ingarden showed that an organism is a partly isolated system, but of a particular type.

The ontological language proposed by Ingarden allows us to describe the very complicated structure of a living being, but the fundamental question of the metaphysical conditions of the possibility of life in general is still unsolved. One of the reasons for this situation is the fact that ontology and metaphysics are separated. According to Ingarden's thesis an ontology searches ideas and pure possibilities which are being perceived in their content. A theory of real objects is an example of such research. Reality is one of the possible modes of existence, while factually real or indeed real objects are studied by metaphysics. Metaphysics studies the essential structure of the factually real world and it contains existential sentences, too. However, not every existential sentence is a metaphysical one. An existential sentence which is also metaphysical must show the necessity of the existence of a being or the connection of its existence with another existing necessary being.

This conception of metaphysics originated directly from the tradition of rationalism present in European philosophy, but there is a hidden assumption in it of the identity of ratio and nature. This assumption limits metaphysics to a small area of ratio. But the world of facts is infinitely rich. In this situation there is only one solution: to allow that metaphysics must be open to all experiences and, in keeping with their contents, that it must have the possibility of modifying its thesis or, to express it more strictly, that the methodological status of metaphysical theses is that of hypothesis. This solution was very strongly propagated by Nicolai Hartmann, but unfortunately, hardly anyone has dealt with the problem. To do so, some assumptions must be made.

The first is to accept the analogy between the relation of mathematics to physics and the relation of ontology to metaphysics. Mathematical structures approximate physical notions in the same manner as ontological structures of pure possibilities approximate metaphysical notions. An indirect conclusion to be derived from the first assumption is: the stronger or richer ontology is, the better the metaphysics that can be achieved.

The second assumption is the hypothetical status of all metaphysical theses. An indirect conclusion derived from the second assumption is that metaphysical theses gain increased adequacy through the necessity of their perpetual modification.

After this introduction, some reflections on the metaphysical foundations of life can be presented.

2.

Nothing (i.e., no being or its level) can come from nothing – *ex nihilo nihil fit*. This archaic thesis may be understood in many ways, but the main idea is very simple. If something exists, there must be some metaphysical reason or foundation for this fact. If not, the world must be absurd. This statement is superior to the distinction between determinism and indeterminism because both determinism and indeterminism must possess their own reason or foundation. For instance, a metaphysical foundation of total determinism may be *Logos* (written with a capital letter) and the foundation of total indeterminism may be the absolute freedom of God's will. Both, Logos and God are the required metaphysical foundations. Then, the questions arise: What is the metaphysical cause of life? And/or what is life's metaphysical foundation? The answer must indicate either something in the real world or something transcending it. The something must possess such a nature that allows us to understand the possibility or necessity of the existence of life. Biology, chemistry and recently physics have searched for conditions and laws in the light of which every scientific aspect of life can be understood. But life as such, the life from which consciousness emerges, is a mystery in the same way as the existence of each being is a mystery. The same laws are present in each part of the universe, but conditions differ. This means that not every being can participate in every law, but only some beings in some laws. The degree of complexity of a being seems to play a crucial role.

3.

The following rule can be formulated: the particular degree of complexity of a given level of being can arise only from a foundation of a higher degree of complexity. Of course, neither the term "complexity" nor the expression "level of being" have a strict meaning. This rule is one possible version of the principle: *ex nihilo nihil fit*. The major levels of being: inanimate being, living being, man and the sphere of divinity have been known since the oldest antiquity. Nowadays, besides these levels of being, many levels in being itself, especially in real being, are known. This multilevel or, more strictly, hierarchical structure of real being must be taken into account in any contemporary metaphysical system. It is very important not to forget the various aspects which build a real object as a whole. A good example for this rule may be a comparison of the degree of complexity of a crystal taken as a whole object with the degree of complexity produced by its parts, i.e., atoms and the relations between them. We obtain a similar result confronting a cell taken as a whole object with the cell's compartments and all the connections between them. One can relatively easily imagine a complexity made up of these compartments or parts. What does the complexity of something taken as a whole mean? A distinct ontology must be used here. In fact, the degree of complexity does not refer to a whole built by parts but refers to an object constituted above the whole. Ingarden proved that to identify the form of wholeness and the form of the object is a fundamental mistake which leads to reductionism and, as a consequence, to mechanicism. This mistake is not so dangerous for the sciences as it is for philosophy. The degree of complexity of an object is fixed by all its properties and other compounds of its ontological structure like form, modes of existence, etc.

4.

The highest degree of complexity we know is indicated by life. The evolutionary development of life without a doubt is directed towards an increasing degree of complexity. However, one cannot forget that every living organism is a dissipative structure, i.e., such a structure which can exist only in a continuously flowing stream of matter, energy and information. In other words, a process is possible which has the property of self-organization. Such a process produces an unimaginable complexity which, using Whitehead's words, "is apprehended in the unity of the organism". Complexity as such is a derivative of both multiplicity and unity. The most general question is: what is the source of multiplicity and unity in the real world? In each epoch the attempt has been made to solve this ancient problem; so it is in our epoch.

5.

From the principle *ex nihilo nihil fit* and from the thesis that the highest degree of complexity is realised in life, the degree of complexity of its real world foundation can be deduced. A good example of how we might imagine this foundation may be the idea of a quantum vacuum in contemporary physics. This theoretical notion postulates an undetermined being which fulfills the Heraclitean *pantha rhei* thesis. It is important to realize that this vacuum generates fluctuations which are called 'virtual elementary particles'. These particles are of all possible types. But no virtual particle can exist as a real particle unless we specify some additional conditions. Contemporary physics shows that a very deep level of the real world may be an exemplification of the dynamic principle of multiplicity. David Bohm went further when he proclaimed that a reality has an infinite number of levels. Each level determines a higher level only in part. The infinite complexity of the world is divided into particular levels. It seems, however, that the concept of an infinitely mutlileveled reality is too sophisticated and that Ockham's razor must be used. The problem is: how many levels does the real world have and does the last level exist? In fact, Bohm's answer is an extrapolation which does not leave the realm of scientific notions.

A quite different solution was presented by Alfred North Whitehead. His answer was fully metaphysical when he said in his *Process and Reality*:

> "Creativity" is the universal of universals characterising ultimate matter of fact. This is that ultimate principle by which the many, which are the universe disjunctively, become the one actual occasion, which is the universe conjunctively. It lies in the nature of things that the many enter into complex unity.

In his earlier book *Science in the Modern World*, Whitehead described this foundation, i.e., creativity, as a substantial activity of individualisation or a substrate activity, etc. 'Pure Activity' seems to be the most suitable name for this foundation. This Pure Activity is a power which may create and destroy each being and which is absolutely undetermined in its action. Any other aspect of the real world is, in fact, only a mode of this foundation. This allows us to formulate a thesis that the quantum vacuum is a possible manifestation of Pure Activity. Of course, the transition between Pure Activity and the real world is not simple, but the explanation of this transition is the main purpose of modern metaphysics. It is necessary to adopt a very general heuristic principle when constructing modern metaphysics: the system must be open to the facts that point out basic metaphysical categories.

The nature of a quantum vacuum and the nature of life are such facts. These two facts indicate the same Pure Activity, but each of them in a different way. In a vacuum there is a balance between creativity and destruction. Forms become and immediately decay. In life creativity dominates destruction. Forms develop in the direction of complexity, which implies the heightened unity of the organisms of some species. Intensified unity gives these organisms more self-dependency and allows them to became more impervious to environmental conditions.

Both the quantum vacuum and life are sources of multiplicity but in different ways. This means that there are other categories which modify the same Pure Activity and lead to different effects.

6.

As was mentioned above, besides a particular degree of complexity, the unity of animate being is also of the essence of life. Nowadays the laws of Nature are regarded as the source of unity. Scientists, especially physicists, look for one fundamental law which would describe the whole Universe. This type of unity is widely accepted but incomprehensible in itself if lacking a certain metaphysical foundation. For Whitehead, the unity of the actual occasion appears as the effect of creativity. However, creativity expresses mainly the synthesis of some multiplicity into a new unity. Both, earlier actual occasions and the hierarchy of eternal objects are unified there. The eternal objects determine the novelty of form yielded in the becoming event. God has an essential influence on the process of the event's actualisation. God harmonises the process of actualisation with a sphere of pure possibilities, i.e., a sphere of eternal objects, through an aesthetic ordering. God is also a mode of creativity.

This metaphysical solution to the problem of the origin of unity described by Whitehead lays particular stress on the role of synthesis as the main aspect of creativity. This was necessary because for Whitehead – as a logician – the sphere of eternal objects presented a multiplicity of an essence's units albeit strongly connected with one another. There are other possibilities. Plato wrote about one of them. According to him each unity emanates from the highest idea – the Idea of Goodness-Beauty. Plotinus developed this doctrine into the system of henology. In this system the second hypostasis is *nous*, which generates the whole sphere of ideas. As mentioned above, nowadays a very strong conviction of the unity of all the laws of nature prevails in physics.

The posited unity consists in the common origin of all laws in only one universal law. This universal law, which in fact has not been discovered so far, is held to be diversified by boundary conditions which change throughout cosmic or global time. Moreover, any concrete thing is, then, subject only to these derivatives of the universal law, which can be perceived through it.

These different plots may be joined in the notion of "logos". *Logos* is a mode of the Pure Activity. If it is the source of definiteness, then it must possess a few properties. Each of these properties is also a mode of Pure Activity. The first is creativity. *Logos* must be continuously created in a way analogous to that in which a notion is created in a man's consciousness and its existence sustained by some effort. In a similar way a kind of "effort" produces some heterogeneity in Pure Activity. In this case the heterogeneity has a particular degree of self-reference. The self-reference is an immediate effect of the constant maintenance of the heterogeneity's existence, which is supported by the creativity of Pure Activity. Both the heterogeneity and the self-reference are comprehended as being dynamic, but the degree of the dynamism is moderated. Self-reference is something weaker than identity, but it is a necessary condition for any definiteness to come about. The definiteness created in this way is a primordial essence. All these stages from constant creation to an essence are going on simultaneously and jointly build the whole *logos*. *Logos* is the highest possible unity generated by Pure Activity, but it also has an elementary metaphysical structure. All of the elements of this structure are necessary because each of them is a necessary condition for the next. The primordial essence has a monadic nature and, to a certain degree, it is creative. For these reasons we can try to compare this comprehensive *logos* with Leibniz's concept of the monad.

In this way the origin of definiteness and unity and the origin of multiplicity are the same, i.e., Pure Activity. It is very important to notice that although the primary source of multiplicity was described, some knowledge about its immediate source is necessary in order to understand the real world completely. One must come back to the intuition that follows from the idea of a quantum vacuum. Is a metaphysical vacuum possible? Of course, it cannot be understood as absolute nonbeing. The appropriate description of the nature of such a metaphysical vacuum must be made using only notions which are derived from the nature of Pure Activity and its modes. At this point a lecture on the entire metaphysics implied by the whole situation should be given. That is impossible in this short article, so only a few final theses on the nature of a metaphysical vacuum will be presented.

First of all some remarks about the various modes of Pure Activity should be made. These modes are pure existence and pure act. Pure existence is apprehended as the maximal homogeneity of Pure Activity. Pure act is apprehended as the maximally directed heterogeneity of Pure Activity. So there are three modes of Pure Activity: *logos*, existence and act. From these three modes all beings are derived. Unfortunately, this cannot be done directly. An act of creation is necessary which makes these modes refer to one another in a particular way. The boon of this creation are three hypostases: vacuum, eternal being and light. All the names of these hypostases designate purely metaphysical notions because each of the notions is explained using terms derived from the nature of Pure Activity and its modes. As was mentioned above, the vacuum is the source of the multiplicity and internal dynamics of beings. Correspondingly, eternal being is the source of the essences of real beings, and finally light is the source of the unity of the space-time and of the movement of real beings.

Every real being participates in these three hypostases in a particular way. This can be said even more strongly: the immediate metaphysical causes of essence and of all the particularities which belong to a real existing being are the three hypostases. Of course, the physical cause of a real being is another real being or beings. The range and depth of the participation of a given being in the hypostases is crucial to the metaphysical status of a given real being, i.e., for the degree of its perfection.

Now the essence of life may be explained. An animate being realises an increasing range of participation in the immediate foundations of the real world. Besides, the process of participation goes deeper and deeper. These two processes converge in the unity of the hypostases. This means a regression to the primary foundation of the real world, i.e., to Pure Activity. Then again, life becomes a source of *logos*, existence and act which are the three modes of life here. Life generates these modes only to the degree to which it realises the unity of the hypostases in itself. This fact is fully manifested in a human being. The state of man's metaphysical freedom consists in the unification of the three modalities: a necessity emanating from eternal being, an indeterminacy issuing from the vacuum and a self-dependence coming from light.

7.

The question then arises: does Pure Activity realise the unity of its being? This is a very difficult question, but one can try to give an answer to it indirectly. In accord with the nature of Pure Activity, heterogeneities are generated. From one side the complexity of all the various heterogeneities may be of any degree. From the other side the support of the existence of the real world is evidence of steadiness in the act of creation. This steadiness leads to the assertion that a kind of unity is present in Pure Activity, a unity which lies "above" or "below" the level of the heterogeneities.

If this unity lies "above" that level, then it joins the heterogeneities in some way. In this case the unity leads to something like an act of will that subordinates many heterogeneities to one aim.

If this unity lies "below" that level, than it must generate heterogeneities. This situation resembles Plotinus' first hypostasis, that while remaining absolute, oneness is creative.

8.

If that is so, then the degree of complexity of the heterogeneities generated by the primary foundation may be greater than that of animate being, and the unity of that foundation may be greater than the greatest unity realised in nature, i.e., the unity of man's consciousness or man's soul. This means that something analogous to consciousness may be one of the aspects of Pure Activity. This implication is only probable and it depends on the validity of the metaphysics which is taken into account. Apart from this, a derived thesis should be understood only in an analogical sense. The theory of metaphysical analogy is widely used in Neo-Thomism, and its systematic use in the metaphysics proposed here must be studied anew. The question of the essence of life is strictly connected with the question of the essence of the real world and its foundation. The answer may be given only in such metaphysics as is structurally open to new findings of science. In general, every type of experience must have an adequate consequence in the metaphysical system. A metaphysician must continuously look for these facts, which lead to new metaphysical categories or the modification of old ones.

Silesian University
Poland

A. J. ANTONITES

CREATIVE EMERGENCE AND COMPLEXITY THEORY

Phenomenological philosophy intends to find a vantage point from where everything has its proper place. In this paper it is philosophy of science for which this vantage point becomes actual and relevant in a special way. The context of the theme is the problem of an untenable dualism which ought to be addressed. I mean that the synthesising activity of the philosopher is just what is at stake here. In its contemporary phase, phenomenology to a large extent can be conceived of as an interdisciplinary philosophy: "This fact speaks for itself. Philosophy can reach to its source in the human condition only by taking into account from the start the enormous number of new insights, approaches, and results that contemporary science/scholarship continues to bring forth; it has to pursue its investigation in an interchange with scientific research" (Tymieniecka, A-T. 1986: 386). This is exactly what this paper intends to do.

Since Husserl stated intentionality, it has become one of the basic tenets which has been accepted by all scientists and especially philosophers who are associated with phenomenological thinking. What is more, even outside the ambit of phenomenological thinking, intentionality is in one or another way integrated into diverse thinking. I think the reason is that intentionality is one of the certain insights philosophy succeeded in reaching. It confirms itself as an answer to the problem of the dualist gap between consciousness and the world. The Cartesian substance dualism has in principle been set aside. I would describe this as a "horizontal" dualism. It can be called a certainty because it avoids all the problems involved in a dualist approach. However, it is important to note that the term "certainty" used here, does not imply eternal verities. It is a stable insight, but like all other knowledge in science, this certainty is also provisional in nature.

There is, however, another dualism which is very seldom, if at all, stated in phenomenological philosophy and discussed: It is dualism of another kind which in the past was not so much experienced as a problem with the concomitant absence of attention. I henceforth would describe this as a "vertical" dualism. I call it vertical because it stands clearly

A-T. Tymieniecka (ed.), Analecta Husserliana LXVI, 83–95.

illuminated by the insights of empirical science, *viz.* biological evolution. For this reason I do experience it as a problem. This is the dualism between consciousness, person, on the one hand and biological life, body, flesh and organic matter on the other hand.

Als Begrundung fur diese Behauptung wird auf drei punkte hingewiesen: erstens auf den Sachverhalt, dass das Denken unkorperlich est und das mechanistische Modell daher nicht auf dasselbe anwendbar ist, zweitens auf die Tatsache, dass auch das Wesen des Korpers nur vom Geist erkannt werden kann und drittens wird behauptet, die Eigenschaften des Korpers seien defiziente Modi der res cogitans (Hutteman, A. 1996: 139).

There would therefore still be a conflict between the qualitative distinct first-person reality of the creative human being as a spontaneous meaning-bestowing agent, a marker of his life world, and his natural condition. Why? Because the natural condition is accepted as processlike and causally-deterministically understood and explained. It is one of the insights and facts of science.

This brings us to the second reason why this dualism has not been addressed. It was and is still argued that the vertical dualism need not exist for Philosophy: It is an issue of the natural and biological sciences; it lies outside the scope of Philosophy. Philosophy could comfortably ignore the vertical dualism and keep itself busy with other issues. I, however, disagree completely: Firstly the horizontal dualism also involved the Cartesian-Newtonian and mechanist paradigm of science. Husserl took it into account and addressed it. There is no reason why this would not also apply to the vertical dualism. Secondly, there is always a fruitful exchange between phenomenology and the sphere of knowledge, the sciences and the arts: "Phenomenology has proven itself to be enlightening beyond the strict humanities extending to biology, all the branches of sociology, technical studies and architecture, and the phenomenology of life has much to contribute to ecology and environmental studies" (Tymieniecka, A-T. 1991: xix). Although differing amongst themselves, the various branches of knowledge have sought a common ground in philosophy. Even if it is the case that philosophy could not match the precision of the scientific observation of facts, philosophy is not relieved of its task to supply answers to the most puzzling questions concerning the ultimate status of life and the meaning of existence – it would leave the human ship without a helm on an uncharted ocean:

No piecemeal or roundabout approaches to the universe, nature, and human life would do. . . . And yet Edmund Husserl made a monumental effort to revamp the entire philosophical field. He proposed laying foundations for a new "holistic" philosophy, which under the name of "phenomenology" would lay bare the ground for the primeval philosophical intuitions of mankind . . . (Tymieniecka, A-T. 1986: 388).

I would argue further that because of the synthesizing activity of the philosopher and especially here the philosopher of science, he cannot leave the situation intact. Why? There are two indicators: One is from the immediate experience and *immediate insights* from our life world. It refers to intuitive and primitive evidence. The second is from science and especially from biological sciences, which adopt a natural attitude and qualify themselves as empirical sciences. This refers to the theory of evolution. Even though the philosopher is not a biologist, and even though it is not for the philosopher to decide on empirical issues of other sciences, the philosopher of science and the philosophical anthropologist are continuously invited to address the reality of the insights from contemporary science. The theory of evolution today is a well confirmed theory. Even though the philosopher argues on the second-order level, he cannot leave this discontinuity, this dualism, intact. It just cannot satisfy his philosophical curious mind. I think that Husserl was right here when he stated that there is a continuity between the various sciences and philosophy, in spite of the difference mentioned in orders or levels. Phenomenology must not be misunderstood as a closed sphere of investigation without openness to real objects and consequently to real problems – intentionality as the directedness of consciousness towards objects is not in contradiction with the phenomenological reduction as a return to consciousness. In his *Crisis* Husserl pushed this further, even to the reverse situation, *viz.* bringing philosophical reflection back into the activity of scientists. This work has partly been done by philosophers such as Popper, Kuhn, Lakatos, Feyerabend and others (Molchanov, V. 1991: 370, 371).

I would argue that a solution to the problem of vertical dualism can be found in the advent of the human condition, where life take a special circuit: an evolution of life. In the past some phenomenological thinkers like Max Scheler did acknowledge biological evolution, but did not pursue the vital connection. The problem of dualism was not explicitly present. Another phenomenological philosopher who was also a medical physiologist, F. J. J. Buytendijk, went further than Scheler: He surprisingly came to remarkable conclusions about similarities in primates in comparison with human beings. Thus, before the results of recent and ongoing research on the cognitive capabilities and rationality of primates, dolphins and others became known today, Buytendijk foresaw some of these results! Yet he also did not explicitly address the problem of discontinuity at the vertical level.

How, then, could the problem of vertical dualism and discontinuity be solved? I would like to argue, firstly, that it can be abridged in terms of the theory of evolution, but with certain qualifications. Not the theory of evolution alone, but how we as philosophers make sense of it from our vantage point. This "but" is very important, however! The solution to the problem can be of such a nature that the two components are not just waived away, but taken up to make sense within a philosophical coherence and totality. The mentioned qualification is that the classical Darwinian evolution alone would not bridge the discontinuity. It operates solely with a classical mechanical explanatory principle. And this, exactly, is part of the problem of vertical dualism: A mechanical-determinist explanation with its materialist supposition by implication, is necessarily and logically in conflict with everything basically that phenomenologically orientated philosophical anthropology states about the human condition. I argue that the coming of emergence and complexity/chaos theory as a rather radical modification of the theory of evolution, strongly invites the philosopher to rethink in terms of synthesis and coherence.

One could, of course, argue that the discontinuity could be retained as something "good," as a paradox. And certainly I would be the first to admit that there are many paradoxes in life and science. Many are, to the contrary, quite enriching to science, including philosophy. Admitting paradoxes in certain cases would also not be in conflict with the synthesizing activity of Philosophy as a total science. In this case (radical dualism) however, the immediate awareness of our concrete unity as body-soul-spirit as well as the unity implied by the theory of evolution, forces us to go beyond this dualism.

Chaos or complexity theory is already operating as a paradigm, even though, like quantum theory in its initial years, it was not accepted by all. Some still are of the opinion that it is too early to say whether this research program will flourish or even to tell whether its notion of emergence is well defined (Lange, M. 1996: 240). However, firstly, it is well established empirically as a new paradigm and more than one success has been recorded. Secondly, Nobel prize winners like Murray Gell-Mann and Ilya Prirogine, both biochemists, are some of the few who apply this paradigm. It involves evolutionary biologists, but also scientists from physics, chemistry, biochemistry, psychology, archeology, computer science, other areas of biology, economics, linguistics, sociology, political science, and of course philosophers. Sometimes the terms "complexity" and "chaos" are used interchangeably. Yet both are involved, but can be distinguished. It is rather the edge of chaos, i.e. between total order and total randomness or change, that became the focus of scientific interest. It is an area which is, ontologically speaking, highly non-linear, and lies beyond standard mathematical analysis. When graphically pictured on the computer screen, chaos shows infinite patterns, qualitatively rich, where novelties are continously created – you never know what to expect next.

In the *classical approach* but which so many still think is the standard approach, the *Newtonian and Leibnizian clockwork model* holds sway. It is characterized by repetitiveness, and therefore natural laws which allow for predictability. The launching of the recent Galileo spacecraft to Jupiter depends on predictability. Alter the trajectory just a little bit, and its new path, which deviates slightly from the original, can again be predicted using the equations of the motion. This is a linear ontology which was accepted. In a sense it must still be accepted today – but under certain conditions and with certain qualifications.

Non-linear ontology comes with non-reductive accounts, including supervention and emergence. They operate with an unpredictable world. *Sensitive dependence on initial conditions*: small inputs can lead to dramatically large consequences, and this is called the "butterfly effect." Very slight differences in initial conditions produce very different outcomes. That is one of the bases of its unpredictability. It is sometimes connected with strange attractors. It was and is still argued (an argument which does make sense in light of the "ingrained" classical way of thinking) that the chaotic/complex phenomena of complexity theory are negligible: eventually they would be explained away by the deterministic linear model of explanation. A parallel kind of argument was also used by classical realists against the official Copenhagen interpretation of quantum theory. Chaos/complexity theory has indicated that this argument is incorrect. And as an alternative to reductionist theories, complex/chaotic nonlinear systems have non-reducible emergent properties.

There are many complicated things which interact, and so produce quite divergent behaviour. It looks random, but is not. This is called "deterministic chaos." It is not yet the "real" thing of complexity theory.

"Complicated" is not the complexity of chaotic phenomena – even though the term "complexity" is also used here. It could still be specified by equations. In "real" non-linear random dynamic systems there are interactions between component parts of a system at the "bottom." From interaction, something new then emerges suddenly, a global reality at the "top." It is something that you cannot predict from what you know from the component parts. The *emergent global property* then "feeds back" to influence the activities of the individual components below. In dynamical systems stability can sometimes develop into total chaos and instability. Just before it turns into chaos, i.e. at the edge of chaos, a remarkable interaction occurs – and a novel global entity emerges. The creative principle of emergence comes to the fore intermediately between chaotic, and fixed or periodic behaviour. It is a narrow region, where chaos and stability pull in opposite directions, the area of the onset of total chaos. A power emerges here were chaos and order meet. At this balance point there is a constantly shifting battle zone between anarchy and stagnation (Waldrop, M. M. 1992: 12). Chaos is creative because it forms an infinity of patterns, and unpredictable as it is, you never know what will happen next. Like a rapid pendulum, the system moves through activity states, stable, then chaotic. It thus evolves to a critical state, hovering above equilibrium, but eventually comes to rest, with optimized fitness. It is then poised at the edge of chaos with maximum evolvability. The edge of chaos shows an analogy with phase transition in the inorganic physical world. It appears that emergence can occur everywhere, not as something mystical (or is it?!), but as a result of local interaction.

An important creative characteristic of the new emergent reality is the impossibility of explaining creative emergence in terms of the properties of the "bottom" physical constituents of the system. It is thus qualitatively different. An emergent quality belongs to a whole and not to its parts. Its behaviour cannot be deduced even from complete knowledge of its parts, or components. It can be contrasted with the classical mechanical views, because there is no general law of composition that dictates the relevant properties of the whole on the basis of the properties of its components:

For an emergent evolutionist, a property of a system is emergent if its existence is novel at the level of evolutionary or physical complexity in which the system is found. In this context, a property of a system is novel if its existence cannot be predicted or explained, given full knowledge of the constituents of the system (Newman, D. V. 1996: 246–247).

Such an explanatory principle would account for the change from single-celled organisms to multi-cellular organisms. This occurred 600 million years ago, i.e. in the *Cambrian era.* For 3 billion years, soon after the earth cooled sufficiently, the highest form of life was the single cell. A degree of complexity then emerged a little more than a billion years ago, when cells developed packaged nuclei and included mitochondria. It was eon upon eon of mind-numbing sameness. Then suddenly, with spectacular effect, the trick of cellular differentiation and aggregation into multicellular organisms evolved. An explosion of new forms occurred – a bewildering variety of complexity (Lewin, R. 1993: 18). The change and innovation were not only in terms of quantity, but were a remarkably qualitative change. This implies a dramatic, big evolutionary leap, a shift from single-celled to multi-cellular organisms. . . . This pattern of the Cambrian evolutionary explosion is regarded as normative and basic to all emergent innovation. Therefore it could now be argued that billions of individual component parts of the brain, i.e. neurons, interacted non-linearly to yield a higher global and new property, consciousness. While no single neuron is conscious, the whole human brain is. A statement is now put forward and explained by complexity theory – but remarkably, it had been stated by Husserl and other phenomenologists long before! The statement is that to have an experience of the color blue is quite different from knowing the brain mechanisms for that experience. Husserl cautioned against a reduction of life world experience, e.g. of blue, to the thematized scientific blue with its vibrations and frequencies of light. Legitimate as the thematized world is, the life world is the basis of that scientific world. Its "reality" is undiminished. I would conclude that this is a surprising convergence of statements by two ways of thinking which developed independently of each other!

On the one hand, thus, we see that the emergence in evolution is not predictable from its properties, not in terms of a deterministic approach. On the other hand, laws of nature, uniformities and equivalence are not done away with. There is *convergence.* The highly complex, uniquely new property, the eye, evolved 40 times among different species and in different times – independently of each other. Then there is a strikingly similar morphology in widely divergent species. Similar adaptations shape similar anatomies and behaviour. A rerun of the Cambrian explosion is even possible.

In a *classical Darwinian paradigm* the explanation was one in terms of *causal law*. The causal mechanism was *natural selection*. This would explain the marvellous order in nature. Natural selection involves gradualism and continuity. Nature is seen as determined where events occur per change, random without any purpose. In Neo-Darwinian thinking causal determinism and change also play a leading role. Accidental mutations and recombinations are the most important causes of the origin and maintaining of genetic variability. By recombination, the newly originated genetic properties are continously connected (by mutations) to new combinations of other hereditary properties – both processes are as accidental as the mechanism of natural selection (Soontiens, F. 1993: 74).

From this it can be deduced that modifications of classical evolutionary theory occurred. "Darwinian evolution can be conceptually distinguished from non-Darwinian evolution which does not involve selection as a causal force of genetic changes" (Tuomi, J. 1992: 435). But it then involves other causal forces. This development according to the new paradigm of chaos/complexity theory is something positive, but insufficient and one-sided. Now there is *spontaneous order, self-organization* as natural properties of complex genetic systems. "There is order for free out there, a spontaneous crystallization of order out of complex systems, with no dominant need for natural selection or any other external force" (Kauffman, S. in Lewin: 25). It does not exclude natural selection, but combines natural selection and self-organization. The self-organization also lies at the edge of chaos. Organisms are the cause and effect of themselves. Neither natural selection nor genes are the *causes* of organisms. Organisms are self-causing agencies. Natural selection alone cannot account for marvellous intricate structures like the eye or brain.

In the classical mechanist Darwinian view, organisms are causal effects of random mutation and natural selection, mindlessly following adaptation in one direction, and then in another. On the other extreme vitalists saw an outside force directing life. Where a mechanist in natural selection only sees a movement from bottom to top, vitalists mostly only see a movement from top to bottom. Creative emergentism has both movements included, by accepting intrinsic self-organizing principles. Self-organization involves a measure of purposiveness. Although this view thus is neither reductionistic to a mechanist-materialist approach, nor to a classical vitalism, I would conclude that there is something of an Aristotelian scientific world view returning. In contrast to a Cartesian dualism, the idea of emergence within a supervenient dualism can in a sense also be traced back to Aristotle. The soul is something immaterial, but Aristotle insists on the substantial unity of body and soul (Leiber, T. 1995: 381, 382).

With this goes a *holistic approach*. Brian Goodwin, together with his tutor C. H. Waddington, a major British biologist, maintain that organisms must be studied as *wholes*. The principal challenge of biology was to understand the genesis of form. Brian integrated this with molecular biology (Lewin: 28). According to him the intrinsic logic of a complex system, such as feedback, repression, control of enzyme activity – gave rise naturally and spontaneously to oscillatory behaviour and global patterns. It is the generation of order as an inevitable product of the dynamics of the system. The concept of form, the unity of form and so the whole of an organism, become fundamental. Species become attractors in a dynamical system. In linear systems there are also wholes, but these wholes are always equal to the sum total of their parts. In non-linear systems the whole is always more and greater over and above its parts. This also, I would say, has a distinct Aristotelian flavour.

In a sense, this new development in complexity theory is a continuation, in quite a radical way, of the decisive issues which Husserl has opened up, e.g. realism/idealism, objectivity/subjectivity, and life/reason.

To come towards a conclusion: We have noticed that important and decisive issues in phenomenological thinking: non-reduction, non-predictability in several areas, a non-mechanistic perspective on life and even spontaneity, are convergent and compatible with complexity/chaos theory's spontaneous emergence, qualitative difference change, holism, non-linearity with its non-predictability. Looking from the vantage point of philosophy, we notice especially a compatibility and coherence with regard to *teleology*. In the fact of our life world, teleology can be experienced. We ourselves are telic beings: "Moreover, if I follow a Husserlian train of thought I find that the world is pregiven, even at the level of instincts, as being goal directed and, correlatively, that my consciousness is instinctively oriented towards a telos. Thus, I find that teleology is implicit in the *life world*, prior to any attempts to reflectively thematize it" (Allen, J. 1979: 214). We have indicated that purposiveness, and thus a form of teleology, has reoccurred in complexity theory.

Granted that it is not an outside force as with vitalism, it is indeed comparable to Aristotelian teleology – although less metaphysical.

The mechanical approach since Galileo has eschewed teleology. Logically it is the opposite of intentionality, plan design and directedness. Changes in an event are the result of the coincidence of many independent series of causally determined events. Both complexity theory and Phenomenology with Husserl developed quite independent of each other (!) and both moved beyond the reduction, also with quite similar arguments. Is this not a bridging of the problematic dualism? Yes, I think so indeed. In the case of the human person, it is the determining of oneself, the interior dynamic of the structure of one self who constitutes the act of willing, in the subjective "I". It is a teleology which does not involve a solipsism, but a living contact with the world, the whole reality. It brings with it the opening of the self:

> The self-teleology of man indicates only that this life-giving exchange is accomplished on the level and by the standards of the personal "I"; in him it finds its starting point and its goal; to a certain extent it begins with him and is ultimately founded in him; it is from that that it takes its shape, and also shapes him (Wojtyla, K. 1979: 207).

Granted that organisms do not have the same development of teleology, not having a transcendent reference to truth-fulfilling aims, and not connected to intersubjectivity or a telic process of self-understanding (or so it seems at this stage!) it is quite clear that there is an inherent form of teleology. From this it follows that both humans and non-human organisms partake in teleology, which is intrinsic. It also involves the physical-material biological brain. It no longer stands dualistic and discontinuous over and against consciousness and personhood!

At this closing stage, I think it is also worthwhile to notice that the change and randomness within the mechanical-Darwinian context is also relativized by the fact that what we mean by change and accident is itself less simplistic than one may think. This refers to the stressing of accident as change and randomness in both the classical Darwinian paradigm and some neo-Darwinian variants. On the one hand they, if natural, are called random, not so much because they are arbitrary or undirected, but because they act opportunistic, depending on whether the environment changes or not. On the other hand evolution is called random not so much in the sense that it is not causally determined or completely arbitrary, but in the sense that there is no design or plan involved. Kimura (a molecular biologist) refers to the random drift of evolution on the molecular level. What accident and random mean is minimized by the view of other biologists like Rench and Schoffeniels as well as the system of theoretical evolution of Riedl and the thermodynamical evolution of Brooks, Wiley and Eicken. In all these cases the role of natural laws determines the process of evolution: the laws of physics, chemistry and thermodynamics. Embryologists like Goodwin also concluded that an accident is limited by several mechanisms in developing organisms (Soontiens: 75). According to Dobzhansky, mutation can be the result of "change" but is not in the least random! Natural selection is random in that it is not determined by a telic directedness, a plan or design, and so is opportunistic. But it is at the same time "anti-random" in that it does create order and directedness in its activities. From this it can be gathered that the philosopher has an enlightening role to play! It thus appears that "change" has more than one meaning, *viz.* the following: unpredictable, not foreseen, not determined by known causes, undetermined, chaotic, not lawlike, not correlated, not deliberate, not designed, not directed, arbitrary, improbable, not essential, not necessary (Soontiens: 77).

In contradistinction to classical reductive materialism and contemporary eliminative materialism, but also in contradistinction to vitalism, emergence in complexity theory, with its non-linearity, claims that these systems are composed of physical entities. It does not introduce any special non-physical properties: "it is plausible that the properties of that theory strongly supervene on that of physics" (Newman: 260). Yet, saying that, I think a large difference is ensuing here. That matter and the physical that is being spoken of here is not the same matter of classical physics and the mentioned materialism. It is seen from the philosophic vantage point, quite considerably moved away from materialism. It did move nearer to the human sciences – strange as it may sound.

We have, with this analysis, entered into the immense complexities that life offers. In this, the synthesizing powers of the philosopher's mind sought to seek interrelations, reasons and causes, with the purpose of introducing some order. "No doubt, to philosophize means to seek the origins of order or disorder, explanations, interrelations, reasons, etc. and to seek unity in what appears dissipated. . . . It is a challenge to the analytic/synthetic/conjecturing mind to seek to dissociate reality and recompose it by means of rational articulations, symmetries/asymmetries, harmonies, principles, and reasons" (Tymieniecka, A-T. 1991: 290). To understand the human being solely from itself, in the sense of singling him out from the universe and nature, would amount to indulging in a speculative, rational isolation of humanness. It would render an anthropocentrism, ignoring that we are embedded in material-physical nature. Being so we would then try to find links with the bridges to the empirical dimension of the human being so defined by our isolating operations, and to nature in which it is embedded. We saw nature as more universal with the insights of contemporary science, without thereby reducing philosophy to pseudo-science. "In short all philosophical anthropologies which start by focussing on the specificity of man as he is not in himself and alone inevitably run into anthropomorphism and into dichotomies of various kinds" (Tymieniecka: 293, 294).

Our approach on the one hand indicated that being human is not merely a product of mute nature, thereby reducing man to a thing amongst things. He is much more than the result of a causally determined process. In fact he is much more than processlike entities. Nature is no longer so mute and inane. His relationship with other creatures and amongst other thinking non-human living beings becomes more interwined and benevolent. In fact nature itself no longer stands over and against himself as an exploitable dead entity. Without falling back to animism, he can now respect nature with greater awe. On the other hand, with the bridging of the vertical dualism, he, as a self-creative, self-realizing and co-realizing being does not stand alienated from a *res extensa* world.

I think that a new itinerary of research for philosophy has been initiated.

University of Pretoria

REFERENCES

Allen, J., "Teleology and Intersubjectivity," *Analecta Husserliana* **IX**.

Blasco, P., "Science and Dialectics in a Phenomenological Anthropology," Analecta Husserliana **XXXV** (1991).

Hutteman, A., "Geist und Mechanik," *Philosophische Rundschau* (1996).

Lange, M., "Life, Artificial Life and Scientific Explanation," *Philosophy of Science* **63** (June 1996).

Leiber, T., "Funktionalismus und Emergenz in Aristoteles' naturphilosophischer Psychologie," *Philosophische Jahrbuch* **102**(2) (1995).

Molchanov, V., "Strict Science and *Lebenswelt* in Husserl's Phenomenology," *Analecta Husserliana* **XXXV** (1991).

Newman, D. V., "Emergence and Strange Attractors," *Philosophy of Science* **63** (June 1996).

Rizzacasa, A., "Epistemological and Phenomenological Considerations about the Natural Sciences in the Thought of E. Husserl," *Analecta Husserliana* **XI** (1981).

Soontiens, F., "Evolutie en Toeval," *Wijsgerig Perspectief* **33** (1992/3).

Tuomi, J., "Evolutionary Synthesis: A Search for the Strategy," *Philosophy of Science* **59** (1992).

Tymieniecka, A-T., "The Human Condition within the Unity-of-Everything-There-Is-Alive: A Challenge to Philosophical Anthropologies," *Analecta Husserliana* **XXV** (1991a).

Tymieniecka, A-T., "Worldwide Phenomenology Fulfilling Husserl's Project," *Analecta Husserliana* **XXV** (1991b).

Tymieniecka, A-T., "The Interdisciplinary Phenomenology of Man and the Human Condition," Analecta Husserliana **XXXI** (1995).

Wojtyla, K., "The Transcendence of the Person in Action and Man's Self-teleology," Analecta Husserliana **IX** (1979).

SPAS SPASSOV

CONTEMPORARY LIFE SCIENCES AND THE SCIENTIFIC WORLDVIEW

One of the most characteristic features of our time is doubtless the crisis of values provoked by modern scientific culture and the destructive impact of the modern scientific worldview on the traditional moral systems. This issue has been widely discussed in the literature, and there is a huge variety of definitions, points of view, and solutions proposed by authors of various disciplines and theoretical orientations. We have to admit, however, that despite the impressive quantity and variety of these discussions, they are still far from achieving a kind of consensus or from proposing a realistic solution to the problem. Instead, they only reinforce our feeling for the profound character of this crisis and reveal the real difficulties for any attempt to find a way out of it. In this paper, I would like to propose some reflections on the question of why it is so difficult to deal with the issue of scientific nihilism and, accordingly, so unrealistic to follow the widespread trend in confronting modern science, considered to be the main source of the crisis. For that purpose, I will first make explicit the very core of the problem, and then discuss the general approach common to many authors who are trying to propose an alternative to the nihilism of modern science by rejecting, or at least by reducing, its significance and its value.

It is not in my intentions here to argue with one or another theory that expresses a hostile attitude toward modern scientific culture. It would be very difficult, in a few pages, even to give a precise definition of some basic notions, such as the notion of scientific nihilism itself. Indeed, it is not easy to briefly define something common to so many different theories, such as the nineteenth century scientism, the various schools of existentialism and their precursors like Nietzsche and Dostoyevsky, and contemporary sociobiology, to mention only these examples. This would become even more discouraging if we tried to overview the existing variety of definitions coming from authors with different backgrounds. Not all sociobiologists, for instance, would admit that their discipline professes a nihilist attitude. It seems to me, then, that a more general approach to this problem is justified, since my main purpose here

A-T. Tymieniecka (ed.), Analecta Husserliana LXVI, 97–106.

is to make explicit the real questions that underlie the contemporary discussions on scientific nihilism.

Contemporary scientific nihilism is a result of the whole scientific progress in modern time, but it is especially related to contemporary biological sciences, that is, to Darwinian and post-Darwinian biology. Contemporary biology has significantly affected not only our knowledge of human biological nature, but first of all our self-image, the idea which man had created for himself and for his place in the universe. In his famous book *On Human Nature*, Edward Wilson gives a concise description of two major dilemmas created by modern biology. The first one follows from one of the main implications of Darwinism, namely the idea that there are no transcendental ends in nature. In considering the human beings and the other living creatures as emerging through a simple interplay of chance and necessity, Darwin's theory advances the idea that there are no other ends in nature beyond the instinct for survival and reproduction. Thus, even the human mind that makes man the only rational creature, has to be considered as a mere physical means, or a functional adaptation, for survival and reproduction. But if this is so, if human life has no transcendental meaning, if man does not have to do anything else on Earth than to live and reproduce, than he is losing the dignity that gives value to his life, and even worse, he is losing his motivation to maintain the civilization. The second dilemma created by modern biology, according to Wilson, comes from the fact that man has been given the power to control the genetic mechanism. The danger this fact brings about is that we can now decide how "human" we want to remain. The power of influencing the mental inclinations that we inherit as elements of our biological nature, allows us to make choices between them (Wilson 1978). The underlying idea here is that such a perspective and the possible results are not very attractive.

The fundamental issue that underlies the discussions on scientific nihilism refers to the question of the meaning of human life, or to the lack of such meaning, that has some transcendental and necessary foundations and possesses as such a universal status. The view denying a universal meaning to human life, a meaning that refers to something called "human nature" (whatever that could be) and that is subject to some "natural" or transcendental laws, different from the laws of a purely cultural character, looks dangerous and destructive from the ethical point of view. Another recent author, Larry Arnhart, defines two aspects of nihilism provoked by modern biology and professed openly, according to him, by one of its secondary products – sociobiology: 1) the idea that man is nothing but one form of life among the others, and that his existence has no other natural goal than to survive and reproduce; 2) the idea that there is no rational norm in nature which confers any meaning to human life and which can therefore serve for him as a reference (Arnhart 1988). The idea of depriving human life of a universal meaning makes it impossible to found the moral system on a kind of necessary transcendental or natural basis. This idea looks very destructive for any theory seeking universality and the necessity of human values, and for any attempt to put moral values beyond the arbitrariness and temporality of man. If there are no transcendental laws underlying the moral principles, if we cannot make reference to anything other than human culture as a source of ethics, we cannot escape the cultural relativism and the arbitrary nature of moral values. This conclusion appears completely unacceptable to those who are trying to save ethics from nihilism.

These examples show well the basic issue underlying the discussions on scientific nihilism. the ultimate question in these discussions refers to the process of demystification of nature, the nature of the universe and the nature of man, brought about in one or another way by modern science. We can describe this idea by another concept, the concept of the objectivity of nature. In his book *Chance and Necessity*, Jacques Monod defines the principle of the objectivity of nature as a systematic refusal to accept that any interpretation of natural phenomena in terms of final causes or a "goal" that can lead to a "true" knowledge. This principle is a pure postulate, as it is impossible to imagine an experience proving the non-existence of a project or a goal somewhere in the universe. Nevertheless, it is "consubstantial to science itself", it is the "cornerstone of the scientific method" and has led the scientific process for three centuries (Monod 1970: 32–33). This scientific method and the worldview of the objectivity of nature it implies, form the metaphysical foundations of modern science. The epistemological root of the principle of objectivity can be found in the law of inertia formulated by Galileo and Descartes, a law that has destroyed the ancient and medieval teleological view of the universe. This principle is far more destructive in regard to the old worldview than the new cosmological system of Copernicus (Monod 1988: 143). The Copernican principles are anti-anthropocentric. The principle of objectivity goes far further – it denies any transcendental or natural teleology in the universe.

Remarkably, the scandalous implications of this new worldview have been fully understood much later because, for a long period of time, the principle of objectivity has been limited to the physical sciences and applied only to the non-living world. The principle of objectivity, which lies in the basis of modern physics, has been kept far away from biology for a long time. Living beings, with their complexity and the intentionality they show in their structures, evolution and behaviour, resisted any interpretation in non-teleological terms. The first biological theories of evolution have even strengthened the old conviction that there exists some design in the universe. The evolution of the living world could be interpreted as proof of the existence of a guiding force in nature, leading by necessity the evolution from the inanimate to the living world and, finally, to man.

The principle of the objectivity of nature was implemented in the field of biology when all the philosophical implications of the theory of natural selection were fully understood and widely accepted. This became possible with the foundation of genetics and the coming of molecular biology. The most important implication of contemporary biology in this regard refers to the aleatory character of the origin and the existence of the living world in its actual form. The emergence of life and even of man himself is considered a cosmic accident, resulting from the blind interplay of chance and necessity. Life could never appear on Earth, and in any case the universe remains equally indifferent. Thus, the modern scientific approach enforces the idea that man is an accident or a stranger in the universe, and threatens to destroy the "old link" between man and the rest of the world (Monod 1988: 145). As one of the founders of molecular biology describes this idea, contemporary biological science is particularly disturbing, because it has proven that we can easily affect the very substance of life on Earth, and especially because it means that we have to consider life and ourselves as the mere result of a cosmic tinkering (Jacob 1981: 93).

What modern science and the principle of objectivity lying in its foundations have destroyed, is ultimately the old idea attributing to man a necessary place in the universe. In other words, they have destroyed the idea of a mutual link, based on necessity, between man and nature, a link which is the necessary condition for finding in nature the meaning of human life. The natural teleology underlying the system of values proper to every society is based on this idea. Any moral structure, mythological, religious, or simply based on some "natural laws", presupposes the idea that there exists a necessary foundation for the system of values, a foundation which transcends man's power of choice. Reason and faith can only identify and recognize moral values, but man should not have the power to define or modify them. Again using the words of J. Monod, man knows today that he is alone in the universe where he has emerged by chance; not only his destiny, but his duty as well is written nowhere (Monod 1970: 195). In destroying this natural teleology, contemporary science has deprived the moral system of its necessary foundations and thus of its universal character. The principle of objectivity has threatened the whole ancient tradition that assured stability to the moral and social structures.

It is not difficult to understand why the moral authorities have never accepted the principle of objectivity. In fact, the idea of the objectivity of nature is hardly widespread beyond the scientific community and even there it is far from being unanimously accepted. One of the reasons, probably the main one, for the feeling of fear and rejection with regard to scientific culture, a feeling that is constantly manifesting itself under various forms, can be seen in the way some authors describe the history of science as a history of progressive alienation (see Prigogine 1989: 400). The objectivity of nature, the absence of natural ends and necessary foundations of moral values, make it impossible for any transcendental and universal teleology to underlie the moral system. There is but one step from this view to the establishment of a bridge between the principle of objectivity and the well-known view of Ivan Karamazov, opposed to any moral teleology, that "everything is allowed". The principle of "everything is allowed" is the pure form, but also *l'enfant terrible* of the objectivist worldview. It places all human behaviour on the same ontological level. Violence and altruism, for example, would be equally possible because equally deprived of universal meaning. The fear of establishing a link between this destructive idea and the principle of objectivity is, in my view, one of the main reasons for the opposition to scientific culture.

All the ideas briefly broached here, though rarely discussed directly and explicitly in the literature, describe an important aspect of the problem that underlies the discussions on scientific nihilism. The conflict between ethics and science, the opposition to scientific nihilism, are not inspired by a kind of fundamental doubt about the "truth" of a scientific explanation of nature, and do not question the scientific correctness of this worldview. The opposition is basically nourished by the fear of the implications and conclusions that the scientific view on nature, the nature of the universe and the nature of man, contains. The refusal to recognize for science the right to deal with anything belonging to the field of ethics and human values, is ultimately inspired by the refusal to accept these conclusions. The objectivist worldview imposed by contemporary science confronts the most natural and solid basis of any moral system aspiring to universality. The acceptance of the scientific explanation of nature and the contemporary scientific worldview can lead to a denial of the necessary and transcendental foundations of moral values. It is indeed difficult to abandon this idea that has served quite well, until the modern age, as an alibi and justification for various moral and social orders. The reason why man has always looked for natural and transcendental foundations of moral values is simple – they give these values a stable and unquestionable character. This is a necessary condition for assuring stability for the ethical and social structures.

We have to ask, however, whether it is possible to save these foundations while denying the worldview of contemporary science. It is clear that the fundamental issue in these discussions is not a matter of some isolated scientific theories that we can oppose or simply ignore. It is the scientific method itself that is involved in the very core of the conflict. There are, of course, many abuses with this method, producing statements or even theories that can easily be denied. These are, however, only secondary products that are not representative of contemporary science and its methods. In order to assure the universal and necessary character of the moral system, many authors try to deny the very foundations of scientific knowledge, of scientific culture itself. Defending such an attitude, however, hardly looks realistic, especially in a society based on science and taking advantage of its power.

On the other hand, we cannot substitute the attempts to find some necessary and transcendental foundations of the system of values, with the search for such necessary foundations in man's biological nature. There are authors who, in order to escape the cultural relativism, are trying to define human nature by making reference to some biological imperatives. These authors propose, as an alternative to biological nihilism, to found the moral system on a kind of biological teleology. As a highly organized system, man, like the other living beings, possesses a kind of internal teleology that coincides with the principles of its organization. This internal teleology is the reason for the goal-oriented character of his functioning and behaviour. The necessity of this teleology, however, does not transcend the principles of organization of the system itself. As a consequence, it cannot define the existence of a transcendental human nature, able to serve as the necessary foundation for the system of values, and putting them beyond man's power of choice. It is well known that we cannot derive what should be the case from what is the case. In other words, it would be unacceptable to derive moral principles from biological facts, to prescribe universal norms of behaviour or to define our moral choices through biological causes. It is true that biological sciences enlighten and deepen our understanding of human nature and, as a consequence, biological facts can at least inform, though not determine, our moral choices. This is not to say, however, as proposed by some authors, that we should found the system of values on this biological teleology, arguing that what is good for man as a human being is the full accomplishment of his natural inclinations (Arnhart 1988a: 224). This idea looks to me completely unacceptable not only for theoretical, but for practical reasons as well. It is difficult to imagine something more terrifying than a society where all the natural inclinations of people could really be actualized without limits or constraints. If it is true that there exist some natural inclinations that man receives as elements of his genetic inheritance and that we can define as a kind of biological or "human nature", I do not think that it is a good idea to found the ethical system on this biological teleology.

One of the implications of scientific nihilism that raises much of the opposition toward the principle of objectivity, as I have mentioned, refers to the relativism of moral values. If it were true that Darwinian biology had shown the impossibility of grounding moral principles on some intangible laws, then we would have to admit that our moral and political preferences are nothing but preferences, and that we cannot deny the arbitrary character of our values (Arnhart 1988: 174). Moral and cultural relativism, however, though far from being absolute, is nevertheless a reality that we do not see only if we do not want to. In order to find their limits by looking for some kind of transcendental human nature or necessary teleology, we have to confront the whole modern scientific worldview.

If confronting science in what is the very essence of modern scientific culture – the objectivist worldview, looks unacceptable or hopeless, it would be perhaps more realistic not only to admit it, but to use it, in order to create and found the moral system on it. The universal character of such a moral system would not be based on some kind of necessary transcendental or "natural" laws, but on a kind of deliberate and thoughtful consent. The acceptance of contemporary scientific explanation and its objectivist worldview does not necessarily lead to a denial of any possibility of a universal ethical structure. Finding some necessary natural and/or transcendental foundations of moral values is not the only possible way to build up a sustainable ethical system.

Without denying some indirect biological interference in the realm of values, it seems to me that moral values belong to the world of human culture and have direct cultural determinants. If this is so, however, we also have to admit their quasi-arbitrary character. Then, instead of trying to escape moral and cultural relativism in looking for necessary foundations of moral values in some intangible human nature, it would be more realistic and preferable to concentrate our efforts on resolving the moral controversies by way of deliberate and considerate discussions. Such discussions need a real and constructive dialogue between science and the other branches of human culture. If there is a real threat to the stability of the social and ethical order, it does not come from the cultural relativism which only bears witness to the flexibility of man as a cultural and intelligent creature. On the contrary, much more fearful is the insufficiency, or sometimes even the lack, of reasoning capable of guiding human behaviour. If there is my hope for our survival as a society, it does not lie in our submission to some blind necessity. Rather, it lies in our will to use our reason as much as possible – this instrument that makes man the only intelligent being on Earth. Unless, of course, this last statement is one more example, among many others, of human vanity.

Institute for Philosophical Research
Bulgarian Academy of Sciences

REFERENCES

Arnhart, L., "Aristotle's Biopolitics: A Defense of Biological Teleology against Biological Nihilism," *Politics and the Life Science* **6** (1988).

Arnhart, L., "Author's Response," *Politics and the Life Sciences* **6** (1988a).

Jacob, F., *Le Jeu des possibles* (Paris: Fayard, 1981).

Monod, J., *Le Hasard et la nécessité* (Paris: Seuil, 1970).

Monod, J., "La Science et ses valeurs," *Pour une éthique de la connaissance*. Text chosen and presented by Bernardino Fantini (Paris: Editions La Découverte, 1988).

Prigogine, I., "The Philosophy of Instability," *Futures* (August 1989).

Wilson, E., *On Human Nature* (Cambridge: Harvard University Press, 1978).

A session of the conference.

PIOTR LENARTOWICZ, S. J. AND JOLANTA KOSZTEYN

ON SOME PROBLEMS CONCERNING OBSERVATION OF BIOLOGICAL SYSTEMS

INTRODUCTION

We propose to discuss some basic descriptive concepts used in biological observation. We believe that the current descriptive methods give an inadequate insight into the essential properties of living things. Our objections are numerous.

Our first objection concerns the current language of biology. The words "whole", "part", "system", "integration", "disintegration" are used without any definite illustration and without any unambiguous definition of their meaning. This produces an illusion that a "biological system" can be represented, for instance, by a beautiful, symmetric snowflake. In addition, it is difficult to grasp whether those words refer to an empirical reality or just to mental structures constructed by a learned observer. Those words, moreover, are usually missing among the entries of philosophical encyclopedias, dictionaries and textbooks – their meaning being taken for granted.

Our second objection is that – according to a current, tacit, but clearly distinguishable methodological "principle" – every descriptive abstract idea has to exclude any irreducible difference between living beings and dead, physical matter. In this way the question of the supposed reducibility is "solved" *a priori*, independently of the empirical data. Sometimes this "principle of reducibility" is treated as a meta-methodological assumption, inherent in the very definition of "scientific" cognition. Consequently, the meaning of the term "scientific cognition" is monopolized by a rather limited metaphysical belief and, at the same time, it becomes immune from any criticism based upon observational data.

Moreover, the above "principle" is detrimental to the selection of objects for study. Suppose one intends, for instance, to study phenomena of locomotion. According to a common, and legitimate practice one has to select the "correct model". The "principle" we are opposing allows one to select the movement of the clouds in the sky instead of the movements of a bird in the sky.

A-T. Tymieniecka (ed.), Analecta Husserliana LXVI, 107–119.

Our third objection is raised against another quite widespread assumption, namely, that any real (empirically detectable) activity has to be reduced to the pattern of physico-chemical causality (influence/effect relation). In other words only one, single pattern of activity is recognized, with no other form of dynamism being seen as (*a priori*) conceptually, scientifically permissible. This assumption does not prevent the empirical sciences from collecting new, relevant data. It merely acts as a sieve, restricting our cognitive capacity to a narrow, fragmentary sector of biological phenomena. The main catastrophic consequence appears to be the "physical theory of cognition". "Physical" means "causal", i.e., reduced to the influence/effect phenomena. Whoever takes this strange theory for granted must, consequently, accept several skeptical arguments, tenets that during the last centuries have dug an abyss between empirical and theoretical knowledge.[1]

Our fourth objection against the current "rules of scientific description" concerns a bias in favor of assemblies rather than units. "Assemblies" of identical or almost identical units are selected and treated more attentively than the units themselves. For instance, the dynamism of a multitude of chemical compounds, or the dynamism of a "whole" *Drosophila* population is examined and pondered with care, while the complex dynamism of the integrative epigenesis within a single life cycle is usually skipped, disregarded, hardly mentioned.

Primary Errors

Two fundamental cognitive errors have to be mentioned in this context.

The first primary cognitive error (the "error of negation") consists in a refusal to see what is obvious, clear and objective. In the past there were observers who did not see what was and still is clearly visible to everybody. Who, for instance, could surpass David Hume! He did not notice any causal influence between two colliding billiard balls! In our times the disputable unity of the Earth ("Gaia"), or the secondary relations within the ecological systems seems to many absolutely evident (which is good), while the primary complex integrative dynamisms and dynamic relations within a single organism are often hardly noticed and seldom mentioned – and that is wrong. Because of this error the description of some key question raising biological phenomena is seriously inadequate, incomplete.

The second error (the "error of affirmation") consists in pretending one sees, detects, observes something which in fact is nonexistent. In the history of the biological sciences some observers saw – supposedly with their own eyes – what nobody else ever saw before or after.

Medieval anatomists, for instance, described the nonexistent channel between the spleen and the stomach. In the seventeenth and eighteenth centuries followers of the preformation theory "saw" minute homunculi in human eggs (or in human spermatozoa). In our times many scientists, for instance, keep telling us how DNA shapes protein molecules, cell organelle, tissues, organs and even human behavior. From the empirical, molecular, biochemical point of view this complex, hierarchical, causal activity of the DNA polymer molecule is absolutely undetectable, both practically and theoretically, even within a single cell – just as the homunculus in the sperm remains undetectable until now. It is true, of course, that the enciphered messages found in the cellular DNA structure greatly facilitate selection of the proper sequence of the amino acids (the primary structure) of cellular proteins. Those messages, however, are not sufficient to determine properly the final, functional shape (tertiary structure) of even such a simple protein molecule as insulin.

Summing up, essential and quite evident biological data are often rejected (because of the first error), and pseudo-explanations are accepted (because of the second error).

Our essay may be considered as an exercise in ostensive definition. Ostensive definition consists of an object to be observed and a set of linguistic indicators which guide the process of observation. If there is nothing to be seen, no indicator could help. If the indicators are misleading, the essential elements of the object may remain unregistered. Our main aim is to visualize some fundamental dynamic relations characteristic of living beings. We are not attempting to interpret, or to solve the questions that those phenomena open. Such an interpretation would be too long, too complex and probably too difficult. We intend only to make some statements concerning "matters of fact", meaning the description of some fundamental question provoking elements of biological phenomena.

In the first part we are going to give a sketchy picture of our object, and in the second we will reflect on some guidelines, mental indicators. In the third part, the last, we will venture to formulate some epistemological conclusions.

I. THE OBJECT

The object is an Australian bird, which looks like a fowl. Actually it is called a jungle-fowl (*Megapodius freycinet*). Ordinary birds incubate their eggs using their own body heat – but not our bird. Who would believe that the bird collects branches to make a fire in order to warm its eggs? Nobody. However, the jungle-fowl is even more intelligent. It constructs and then operates a "fermentation mound", which might be rightly labelled a "fermentation oven".

Incubation in the Fermentation Oven

During the four winter months the bird prepares a big heap of wet leaves and branches. The heap may be up to 5 meters high and some 10 meters broad. After a time the organic matter in the mound begins to ferment, which produces quite a lot of heat. Then the bird prepares channels in the mound. The channels are dug in a strictly determined order, following the gradual spread of the fermentation wave which starts in the center of the mound. Every 2–3 days, the female lays an egg in a warm, newly built channel.

But that is just half of our story. During the incubation period the male constantly monitors the temperature in the individual channels (there are usually twenty to thirty channels in a single mound). The male ventilates the channel if the temperature rises above 34 centigrade. The incubation of the whole set of eggs may take up to 7 months, which means that the male is rather busy 11 months a year and has just one month of holidays. The hatched chicks are fully formed and desert their parents immediately after emerging.

If we now look at the schematic, crude representation of the temperature level in the bush of Australia (see Figure 1A), we may easily notice that there is a rather obvious contrast between the constantly changing temperature of the surroundings and the rather narrow temperature range necessary for the proper development of eggs. That difference might ruin the prospects of young jungle-fowls. Fortunately the deviations are compensated for almost perfectly by the activity of the vigilant father (see Figure 1B).[2]

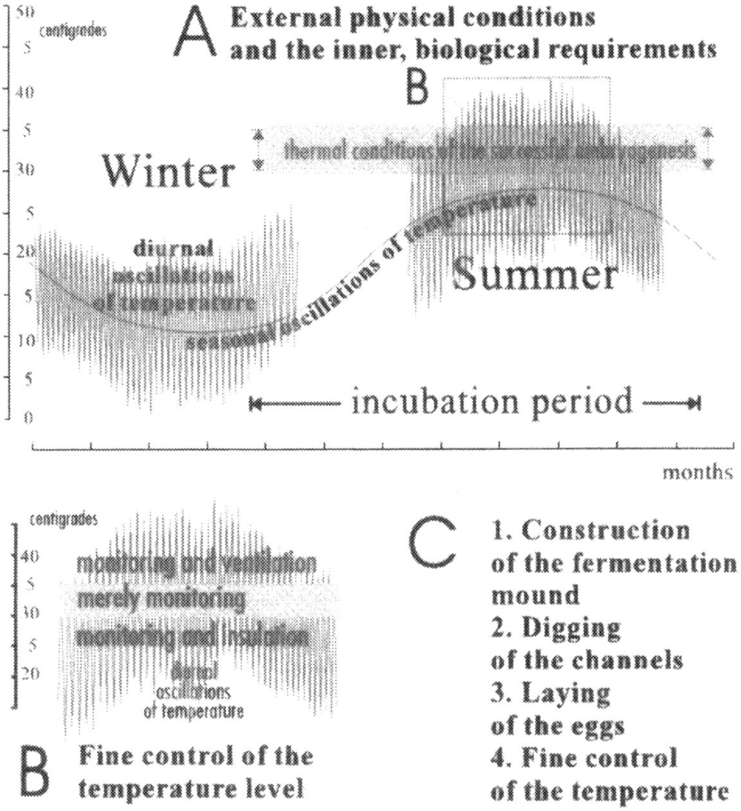

Figure 1.

Potential Forms of Incubation in Megapodius freycinet

The amazingly intelligent behavior of this bird is just a small part of its possible forms of activity. The jungle-fowl manifests a whole range of different forms of incubatory behavior:

A. On the hot beaches of northern Australia the bird buries its eggs in the deep channels made in the sand. Then the bird leaves them there without any further supervision. The sand maintains its warmth during the night, and the vicinity of the ocean lessens diurnal fluctuations of temperature. The eggs can develop safely without any further intervention from the parents.

B. Where the seashore is rocky, the bird fills the crevices of the warm rock with clay and leaves. Then the female lays its eggs there.

C. The jungle-fowl has also discovered the advantage offered by Savo Island in the Solomon Archipelago. There, warm volcanic steam filters through the vast areas of sand. Many birds of the jungle-fowl species fly there to put their eggs in that sand. They then leave them there unprotected. The bird is obviously lazy.

D. Only in the bush area where the sun (or a volcano) does not provide enough warmth, does the jungle-fowl build the fermentation ovens.[3]

II. THE GUIDELINES – CONCEPTUAL INDICATORS

We have already mentioned the "indicators" which may guide our understanding of the object. Now we will try to present those guidelines and with their help demonstrate the existence of some important inner relations within the dynamism of this bird. Here is the list of those guidelines:

Repetition	Physical causality
Selection	Correlation
Potential	Integration
Cognition	Motion

Repetition, Selection and Potentiality

The first conceptual guideline is "selective repetition". In our "model" the different forms of incubatory behavior are highly repetitive and selectively linked with specific kinds of surroundings.

The bird obviously selects among different forms of protective behavior. "Selection" denotes an abstract, mental idea. But this idea is molded in our mind by the repetitive elements of the objective, empirical data. In order to "select", one has to have a real choice. The choice, in our case, is made not between some external objects, but between the different, inner, complex behavioral patterns. Those patterns must have existed before selection was made, before (in concrete surroundings) a particular behavioral pattern was manifested. Such hidden, inner, unaccomplished existence is commonly called "potential". Where there is no potential, no selection is possible.

In the case of Megapodius we have to do with a biological and behavioral potential. The elements of this potential exist in distinct, modular, complexes that seem to be of a whole.

Cognition

The next conceptual guideline is "cognition". The proper kind of behavior depends on the proper evaluation of both stable and temporary properties of the environment. Selection of a given strategy of incubation depends upon the cognition of the relatively *stable* properties of surroundings. Selection between ventilating or insulating behavior depends on the cognition of the relatively *unstable*, transient quality of the surroundings.

An act of cognition may be selective, and the act of selection may be guided by an act of *cognition*. This does not mean, however, that the concept of *selection* could be replaced by the concept of *cognition* – or vice versa – without doing serious damage to our sense of reality. Neither cognition, nor selection as such produce changes in the environs. They do not, therefore, seem reducible to causal acts of an influence/effect type.

Motor Activity

The third conceptual guideline is "motor activity". The birds actively modify the environment, digging holes or channels, preparing mounds, collecting leaves to cover their eggs. This activity is selective and obviously guided by cognitive acts. *Motor activity* however cannot be reduced to the selective activity or to the cognitive activity. The term "motor" is used here in a very broad sense. Any transfer of a physical body or energy means "motion".

Summing up we may state that in the life-cycle of a given, concrete bird living in relatively stable surroundings three descriptive guidelines are of crucial importance.

1. Cognition

2. Selection 3. Motion

The above three abstract concepts constitute a sort of *indivisible set.* The motor activity of our birds is complex but selective, no doubt about it. The element of cognition is also hardly separable from the elements of selection and motion. Without cognition the capacity to act selectively would be hardly understandable. The very existence of *Megapodius'* brain and sense organs would look extravagant and ridiculous.

Now, in order to visualize a deeper layer of our object of study we have to explain the difference between *causal relation* and *correlation.*

Causal (physical) relation

A causal relation links an *influence* with its *effect*. For instance, the mound exemplifies *effect*, while the building behavior of the bird provides an example of *influence.*

Similarly the channel is an example of an *effect*, while the process of digging exemplifies *influence.*

A physical causal relation embraces some concrete material, an amount of concrete energy in some form, and a transfer of this energy called an "influence" that results in a modification of the material.

The causal relations of the bird's activity can be broken down into a complex series of successive movements. The energy is spent in a selective way, and the repetitive pattern of the effect is obtained.

Grasping the causal relation between influences and effects is absolutely necessary in the cognition of our object. But this concept is far from being *sufficient* in attaining a grasp of the whole reality of this object. Another, irreducible concept is necessary, namely, that of *correlation.*

Correlation

What is the biological meaning of the word *correlation*? We cannot describe it with words.[4] We have to show it.

Let us have a look at Figure 2. Do you see a certain "relationship" between the structure of the mound on the one hand and the structure of the channels? Is this "relationship" a sort of mental illusion, or, to the contrary, does this "relationship" exist independently of our cognitive acts?

Do you think this relationship can be identified with a causal, physical relation or "influence/effect"? Do you believe, for instance, that the structure of the mound influences the structure of a channel, or that the structure of the channel is an effect of the structure of the mound? Nonsense.

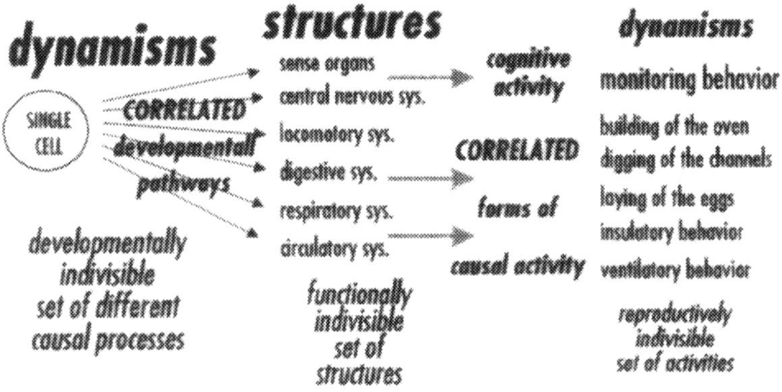

Figure 2.

Do you see any causal (influence/effect) link between the structure of the leg, for instance, and the structure of the eye, or no link at all? Is there any link between the male's reproductive organs on one hand and the female's reproductive organs on the other? Is this a causal relation? We do not think so. We see an obvious link, but we cannot reduce it to a causal relation. Therefore we shall call it a *correlation*, because the actual meaning of the word in common, ordinary language fits our purposes well.

The correlation may be detected a) in the context of a *functional dynamism* (e.g., the structure of a cylinder and the structure of the piston in the engine are correlated), or b) in the context of a *development dynamism* (e.g., the development of the bones and the development of the muscles are correlated).

The usual, and the most obvious way of detecting "correlation" consists in the attempt to separate the correlated structures. The dynamism of a piston is obviously restricted by the structure of the cylinder, the dynamism of the female reproductive organ is impossible without the dynamism of the male reproductive organ. One can "see" it, although this correlation can be detected only on the condition that the field of observation (both in space and in time) be broad enough.

Piotr Lenartowicz S. J., and Jolanta Koszteyn.

Integration

In order to obtain a new generation, the bird has to correlate many different forms of dynamism in a set which is actually *indivisible*. If any element of the set is missing, the end result cannot be reached. That is what *indivisibility* means. We call it "integration". In other words, biological integration refers to a complex set of elements which are physically necessary and sufficient to produce the end result.

This is a thoroughly *dynamic* concept of indivisibility. It does not apply to any singly structure, or to any single, atomic event, but just to some complex processes which are repetitive, and which operate close to the physically maximal economy of energy and material. Because of this economy the dynamism is exceedingly fragile, and this fragility is known in biology as the law of "*all or none*".[5]

Indivisibility

The essence of the biological idea of integration is rooted in the idea of the indivisibility of the end result. Were the end result divisible, the idea of integration could be replaced by the much simpler and much more general idea of *proportion*. In the process of incubation the indivisibility of the end result means that the alternatives are a *dead* embryo or a *developing* embryo. The life of progeny depends on the indivisibility of the complex incubatory behavior of its parents. This behavior, therefore, can be called *integrated* in the described sense.

III. SOME EPISTEMOLOGICAL REMARKS

What are the epistemological conditions of the proper method of biological description? What are the conditions of discovery of such empirical traits as correlation and integration?

The first condition is to recognize the dramatic chasm between a dead embryo and a living embryo. If one questions the dramatic chasm between the processes of decay and the processes of embryogenesis, all the rest of our story is pointless. The continuity of the species, of any single lineage depends on the survival of the embryo.

The second condition is to realize how the narrowly determined limits of temperature parameters are crucial for the survival of an embryo.

The third condition is to realize how different activities of the bird (cognition, selection, locomotion) cooperate in maintaining the proper and constant temperature for incubation (see Figure 1C). Any fault in this complex dynamism must lead to the death of the embryo. In some cases of mutilation it has been found that the survival of the system depends on a very accurate system of regeneration which only adds a new dimension of complexity and poses new questions concerning the origins of biological integration.

The fourth condition is to realize that the set of the above-mentioned activities is dependent on the integration (read indivisibility) of a determined set of body organs.

The fifth condition is to realize that the whole set of those body organs is produced within the egg. We call it *embryogenesis*.

The sixth condition is to realize that the complex set of morphogenetic dynamisms within the egg is indivisible, i.e., integrated with reference to the chances for reproduction. Any imperfection in the morphogenetic dynamism jeopardizes the chances for successful reproduction. The repetitivity of the life cycles in a line of generations is intrinsically dependent upon the integration of the morphogenetic dynamisms.

This is the end of our essay. We have simply tried to show what biological integration means. This concept – in our opinion – is not *theoretical* in the modern, subjective, or constructive sense of the word. It is a concept which can be visualized using a proper observational approach. A fragmentary, purely analytical approach has to destroy any chance of detecting, and observing the inner, genuine, essential properties of living bodies.

We believe that in biology, contrary to many methodological and/or philosophical accounts, descriptive concepts such as "integration", "whole", "part", "fragmentary" are just corollaries of the concept of "dynamic indivisibility" that we here tried to illustrate. The concept of successive, repetitive "individual life cycles" (the line of generation) and the concept of the dynamic indivisibility (the law of "all or none") constitute the primary basis for the cognition of life. We also believe that these concepts are "prescientific", i.e., that they are shaped, almost spontaneously, during the early phase of human cognitive activity, probably during childhood. The prescientific and cruel experiments carried out by children on unfortunate insects or small animals constitute, therefore, the earliest phase of the biological sciences. In those experiments the crucial question of the distinction between structural divisibility and dynamic indivisibility comes to the surface.

A mental, conceptual attempt to break the fragile but evident indivisibility of living being into fragments turns our cognitive efforts from the contemplation of living objects towards meditation on the properties of dead, inanimate matter. Like children we are left with dead remains, and like children we are unable to put life back into them.

Jesuit Faculty of Philosophy, Krakow
Institute of Oceanology, Sopot

NOTES

[1] Cf. P. Lenartowicz, S. J., "The Body-Mind Dichotomy – A Problem or Artifact?" *Forum Philosophicum*, Vol. 1 (1996), pp. 9–42.

[2] In laboratory conditions the bird was able to keep the temperature of the mound within the limits of a single centigrade. The bird achieved this is spite of the mischievous scientist who tried to trick it with the help of an electric heater hidden in the mound. See also J. Koszteyn, and P. Lenartowicz, S. J., "Biological Adaptation Dependence or Independence from Environment?" *Forum Philosophicum*, Vol. 2 (1997), pp. 71–102.

[3] Cfr. N. E. Collias and E. C. Collias, *Nest Building and Bird Behavior* (Princeton: Princeton University Press, 1984); M. Cuisin, *La Vie Secrète des Bêtes. Les Animaux des Îles* (Paris: Hachette, 1989); R. Hill, *Australian Birds* (Nelson, 1967); M. H. Hansell, *Animal Architecture and Building Behaviour* (London and New York: Longman, 1984); and D. Lack, *Ecological Adaptations for Breeding in Birds* (London: Methuen & Co. Ltd., 1968).

[4] The mathematical term "correlation" is limited to the quantitative aspects of phenomena. It is, therefore, inadequate for conveying the biologically meaningful aspect of reality. "*In correlation . . . we are concerned largely whether two variables are interdependent or covary – that is, vary together. We do not express one as a function of the other. There is no distinction between independent and dependent variables. It may well be that of a pair of variables whose correlation is studied, one is the cause of the other, but we neither know nor assume this. A more typical (but not essential) assumption is that the two variables are both effects of a common cause. What we wish to estimate is the degree to which these variables vary together*"; R. R. Sokal and F. J. Rohlf, *Biometry. The Principles and Practice of Statistics in Biological Research* (New York: W.H. Freeman & Co., 1969), p. 563.

[5] P. Lenartowicz, S. J., "Fundamental Patterns of Biochemical Integration," *Ann. Fac. Philosophicae SJ, Cracoviae* (1993), pp. 203–217.

RINALDS ZEMBAHS

LIFE-SPACE AND LIFE-WORLD

Merleau-Ponty on Situations

For me, the main question is this: can a common thematic ground be found for both phenomenology and those philosophical discourses that rest on sciences of life? Or, rather, has the fundamentality of life matters perhaps put an end to phenomenology as a primary grounding discipline? What if the "direct givenness", proposed by Husserl, cannot give us insight into *that-which-lives*, thus marking the impossibility of phenomenology as a universal and strict science? Could it really be that consciousness, with its constitutive activities, is exactly that-which-lives?

Remember Husserl, who speaks of "absolute consciousness" that remains intact even after the annihilation of the world (see *Ideen* I, § 49; the same also in Husserl 1950: 18–20). For him, that-which-lives becomes that-which-survives during a phenomenologically guided thought experiment: consciousness receives its utmost reality because this possibility is principally thinkable.

As Derrida points out in his introduction to Husserl's *The Origin of Geometry*, the main excuse for this assumption is the phenomenological evidence for that *force of negative acts* which characterizes phenomenological reduction:

> Therefore, being able anytime and freely to change or to arrest the positing of *every* . . . accidental existence, of *every* . . . transcendence, consciousness in its sense is principally independent from the totality of the world (my translation; Derrida 1962: 95).

Derrida here says: "consciousness in its sense", pointing out the achieved *eidetic* independence of consciousness. But more radical reductions are still possible, the transcendental regression has not yet come to an end (see *Ideen* I, § 81, p. 182; elsewhere Husserl adds that the eidetic reduction is only the "threshold of phenomenology", *op. cit.*, p. 202). If so, then we must proceed further in the search for that-which-lives in phenomenology. There is still some promise of reaching that aspired-to *proto-region* which can be called transcendental in the proper sense: one that is not only world-constituting, like the region of pure consciousness, but also self-constituting (Derrida 1962: 96). But what could be more "pure" than pure consciousness? What could this tran-

A-T. Tymieniecka (ed.), Analecta Husserliana LXVI, 121–128.

scendentality-beyond-purity be? Maybe some of Derrida's suggestions can give us a clue: he notes that pure consciousness still contains eidetic sediments whose structural irreversibility makes them very similar to some kind of "facticity", factual structures which cannot be thrown off anymore. He asks:

As forms of sense, would they not be marked by some order of the factual world to which *past* consciousness is tied up together with its own connections, structurally engaged in any present consciousness? (my translation; Derrida 1962: 96).

And here, somewhat retrospectively, by putting Derrida's questions as if anterior to Husserl's later works, we can trace the way back to Husserl's concept of the *life-world* ("Lebenswelt"). Of course, there is some leap from one field of problems to another: the problem of the life-world is something other; it arises from questions concerning the grounding of science and its results, seeking to reactivate the sciences' lost sense and moving away from an all too technical attitude (the pleasure of counting). In contrast, Derrida's theme of the factual structures of eidetic sediments, mentioned above, belongs to egology and, consequently, to the special problem of overcoming solipsist implications.

But, to my mind, there is some point where these two specialised analyses coincide. This is to be found in a search for the proto-region (the phenomenological that-which-lives), for the transcendental in its proper sense. For now let us set aside the task of determining the specific aspects of this possible coincidence. Here, it would be enough to say that the life-world's proto-regionality is defined as a "primary evidence" ("Urevidenz"), as a source of evidence for objective statements in science (Husserl 1954: 131).

Even more, Husserl says that the evidence obtained on merging with the life-world, is an *experience* in the strict sense which is assumed by the objective sciences as inherent to their own working (*ibid.*). When objective scientists say that they are following experience, for Husserl this means only that their achievements can be based only on primary evidence obtained in the life-world, in the ever changing relative contexts of their own surroundings (including laboratory procedure). Therefore, "the objective cannot ever be experienced" (*ibid.*), only *endless approximations* to it "in itself" can be achieved.

Speaking about the mode in which the "objective" could be experienced, Husserl compares it with an infinity of "ideas", for example, as in the case of numbers (*ibid.*). Trying to master the gap between the evidence and the "in itself", the latter being given in the idea of endless approximations, the sciences have developed the method of *idealization* as an endless approximation to essential structures; but, unhappily, this new infinity has moved them away from the primary evidence, from the life-world itself.

What then is the life-world if recovering it helps us to criticize the "objective"? if it is not actually the "objectivity" itself? If the life-world is not the "in itself", could it be that it is simply the noematic correlate of the constitutive activity of an Ego which "signifies" transcendent, "noumenal" objects? Could it be only an eidetic structure *in* consciousness?

We do not have the answer, nor did Husserl himself. Some clue is given by him when he talks about the "subjectively-relative" and names some things as belonging to the life-world ("lebens-weltliche Dinge", *op. cit.*, p. 132): stones, houses, trees. The life-world seems to be what we encounter with the "naked eye", before any "idealization" has begun: things within it are given simply, are optimally distinct for the guidance of practical action. But, when Husserl speaks of the life-world as a source of primary evidence, then we are left with an impression that this inaugurates the opening of an indefinite horizon ("Seinsweise in der Schwebe", Husserl 1954: 358) as a transcendental phenomenon of the world. The life-world, taken as a noematic correlate for the givenness of everyday objects, cannot present fundamental difficulties for Husserl's project of phenomenological analysis.

But there is another reason for calling it "enigmatic" (Husserl 1954: 134): the life-world as a source of evidence for the sciences appears to be influenced by the sciences themselves. The sciences show their *applicability*; applicability to what? Here we return to the problem of "transcendent" objects and their configurations, where this applicability shows itself as real, therefore calling for the "objectivization" of the life-world. Husserl himself could not go that far: the premises of his philosophy did not allow him to look beyond the "enigma" of the life-world's paradoxical position (at once noematic and practical); the transcendent sphere of praxis and the possible locus of that-which-lives were killed by the transcendental or, rather, the eidetic reduction. But at least, his bumping against this borderline gave impetus for reformulating the tasks of phenomenology. Among these revisions, that of Merleau-Ponty is one of the most daring.

I would like to explore some implications of his use of the concept of *life-space*, which followed on his shifting his focus from absolute consciousness to the body. The concept of life-space is coherently bound to what Husserl called the "somatological" level (see Husserl 1986: 9–12), so that there is no necessity to boost analysis to the purely noematic level. Experience of spatiality, perception within ever changing spatial configurations, is that primary locus where our *fixedness* in the world manifests itself. It is in the body's fixedness in the world that Merleau-Ponty tries to find the ultimate transcendental structures. Here there is at first no wide open horizon correlative to the constitutive workings of consciousness, but rather the obstinate arrow of fixedness which nails us to the world.

Perception is bodily participation in the world, and only after its realization are we led to the self and the world as distinct entities; Merleau-Ponty holds that there is a point at which there is simultaneous awareness of the two realms, that they have the same primordiality (see Merleau-Ponty 1962: 298). In his posthumously published work *The Visible and the Invisible*, he calls the primordial "the flesh of the world"; some further clues to this later concept are to be found in his *Phenomenology of Perception*. Here we find efforts to state the conditions of *intentionality* as such, which cannot be viewed as being purely solipsistic, but as a kind of mundane turbulence. Phenomenology tries to find these conditions out, following the phenomenally given.

Merleau-Ponty seems to be repeating Husserl when he says that the phenomenal field is a transcendental field (*op. cit.*, p. 63), that is – that the constitution of objects depends on this phenomenal field. No doubt, this assumption is the central problem of phenomenology, begging the question of its applicability: here lies the danger of an idealistic turn. Merleau-Ponty criticized both idealism and empiricism throughout his work, and yet he holds a view which can easily be misunderstood. For me, the phenomenal field is that-which-lives or, less radically, the phenomenological counterpart of it: phenomena as givens are the presencing of that-which-lives, its turbulence, torsion, circulation; after we have seen or intuited them, they pass by, go through us; maybe a phenomenon is a *superposition of an event*? Thus far, my bold hypothesis allows me to understand why Merleau-Ponty says:

> An initial perception independent of any background is inconceivable. Every perception presupposes, on the perceiving subject's part, a certain past, and the abstract function of perception, as a coming together of objects, implies some more occult act by which we elaborate our environment (Merleau-Ponty 1962: 281).

Speaking of a more occult act, Merleau-Ponty seems to point out that here we have a theme other than a description of givens: phenomenology must turn its attention to the problem of the constitution of these givens and, in the last analysis, of the life-world itself. And the first locus where the problem of constitution can be cleared up is precisely the perception of space, of spatial configurations.

As we noted earlier, this perception of space is *an expression* of our fixedness in the world. Here we have the field of phenomena referring back to a transcendental field. Our perception is commanded by a kind of *imperative*: we cannot help perceiving, even when we close our eyes, darkness appears as something perceived. Not only are all of the senses constantly sensitized by the flow of events, of their presencing – that being the "input" – but, the body sends sensation back too. The body's response to the imperatives of the world (for more on these imperatives, see Lingis 1994) consists in adjusting, in building its own space or *body-scheme* (see Merleau-Ponty 1962: 98–102). As long as something is experienced as meaningful, that means that the body had already situated itself in the world through the medium of inner coordination. The internally coordinated, schematized body is the body in a situation. The body-scheme is an emitter of situations, on this side of the *life-space*.

The concept of life-space is of a narrower scope than Husserl's concept of the life-world. Within the specific analysis of our fixedness in the world, it is not enough to acknowledge, as Husserl did (Husserl 1954: 133), that "the scientist is a human being as well"; we must move away from this somewhat abstract definition of what subject is, narrowing the scope down to the body. Merleau-Ponty is convinced that the body could reveal more about the transcendental or that-which-lives; the conscious human being is portrayed as too eidetic to help in this: whenever he comes to acknowledge his being-in-the-world, he must necessarily look at what is really incarnated here. And, correspondingly, the concepts of life-space and life-world are simply aids for phenomenologists in outlining the horizon within which that-which-lives unfolds itself. Their difference marks a shift in paradigm in phenomenology. Merleau-Ponty remains a phenomenologist even though he turned to the situated body so drastically.

The situation implies that there exists some unobjectivated point where the phenomenal ordering (patterning) of the world is made. The body-scheme has been constituted as a centre of the world, it posits the body itself as that centre, but, paradoxically, it can manifest itself only phenomenally, never becoming objectivated for itself. Maybe we have a *limit* here: that-which-lives – and we know somehow that there must be something that speeds up the living and makes it irreversible – can be grazed by intuition ("Anschauung") in just this way of expressing things where the schematized body is sunk into spatiality and unfolds the life-space out of this relation (and, simultaneously, where the pure relationality becomes a *distance*, where the possible breadth of that-which-lives is felt):

> Besides the physical and geometrical distance . . . a "lived" distance binds me to things which count and exist for me, and links them to each other (Merleau-Ponty 1962: 286).

And my life flourishes and fails, together with the way in which this lived distance supports my ability to move freely in the face of things, events and other people: the givenness of the world might cease being a total front pressing me back to the physical limits of my body.

Therefore, when the sciences of life reduce their analysis to the anatomical and physiological levels, that-which-lives is locked into the solipsistic body, but Merleau-Ponty's phenomenology believes in expression and life's unfolding in ever changing situations, unfolding into lived anthropological space; for him, this phenomenal field of the life-space is the locus where that-which-lives is what it is. That is why the formula of Merleau-Ponty's transcendental reduction tries to take the very form of lived distance:

> Reflection does not withdraw from the world toward the unity of consciousness as the world's basis; it steps back to watch the forms of transcendence fly up like sparks from a fire; it slackens the intentional threads which attach us to the world and thus brings them to our notice . . . (Merleau-Ponty 1962: XIII).

This means that transcendental reduction is repeating our factual relation to the world, simply taking another register: the phenomenal. In everyday life, we lose ourselves in objects and actions which are all too common, simply existing and happening there; involved in this blind becoming of that-which-lives, our everyday attitude (including scientific routine) has completely lost sight of the movement of presencing and appearing that constitutes the lived and the seen and gives them meaning.

Looking deeper into the phenomenal field of lived distance, the phenomenology of Merleau-Ponty seems to be close to that "total intention" (Merleau-Ponty 1962: xvii), to that unique way of Being that expresses itself in the endless qualities of things and the diverse facticities of historical situations, and, at the end, in the lonesome thoughts of philosophers. It may be the case that this somatologically oriented phenomenology is able to just simply *touch* the total intention of that-which-lives, but is not capable of ensuring that the further course of thinking would not seek sources of the phenomenal other than intuition.

What if this total intention leaves much of its movement outside of intuition, though this minimality of its intuited presence results in a meaningful outlook on the world and history, as Merleau-Ponty views them? What if we do discern intention and intuition, assuming that intentionality constitutes its own phenomenality as well? Then we could explain how the question about that-which-lives can be asked and answered at all. When we ask about the appearing of appearances, we are already fixed by that total intention that brings the given to intuition: its *approach, movement* and *turbulence* are somehow felt at the level of our fixedness in the world; these are strictly intentional "phenomena". That-which-lives as a total intention has its own sphere in bodily movements, affectivity, void spaces where leaps between positions unfold, and finally, in the enormous circulation and turbulence of materiality. If we were able to develop the "protonization" of the life-world and body from the models given by the physical sciences and biology, it would mean that the intentional had its own phenomenality, something not accepted by Husserl but pointed to by Merleau-Ponty's concept of the flesh of the world. That's why Merleau-Ponty says that

> this facticity of the world is what constitutes the *Weltlichkeit der Welt*, what causes the world to be the world (Merleau-Ponty 1962: XVII).

University of Latvia

REFERENCES

Derrida, J., *Edmund Husserl. L'Origine de la Géométrie*, trans. and introduction by Jacques Derrida (Paris: Presses Universitaires de France, 1962).

Husserl, E., *Gesammelte Werke*. Bd. I: *Cartesianische Meditationen und Pariser Vorträge*. Hrsg. von S. Strasser (The Hague: Martinus Nijhoff Publishers, 1950).

Husserl, E., *Gesammelte Werke*. Bd. VI: *Die Krisis der europäischen Wissenschaften und die transzendentale Phänomenologie*. Hrsg. von Walter Biemel (The Hague: Martinus Nijhoff Publishers, 1954).

Husserl, E., *Die Phänomenologie und die Fundamente der Wissenschaften*. Introduction by Karl-Heinz Lembeck (ed.) (Hamburg: Felix Meiner Verlag, 1986) (Philosophische Bibliothek, Bd. 393).

Lingis, A., "Intentionality and the Imperative," *International Philosophical Quarterly* **XXXIV**(3), Issue No. 135 (September 1994).

Merleau-Ponty, M., *Phenomenology of Perception*. Trans. Colin Smith (London – New York: Routledge, 1962).

SECTION II

PRIMAL ORIGIN, INDIVIDUATION, INTERPLAY

BENGT TJELLANDER

THE CONSTRUCTION OF THE CONCEPT "THE OMNIVIDUAL"

Distinguished colleagues, I want to tell you a humanistic science and fiction story. I will tell it in the Husserlian egological way.[1] But my ego is more the body-mind of Merleau-Ponty than the onlooker ego of Husserl. I am a pedagogue, a lecturer in education.

INTRODUCTION

As a student in the mid-seventies I, together with many other students, had a hang-up with the concepts of "subjectivity" and "objectivity". We were trained in the positivistic way of thinking, and we learned that research was to be "objective". Research was numbers; you should be able to calculate variables. The concept of "subjectivity" was equated with "personal opinion", and therefore it contained no "truth".

To make a personal opinion true, you had to test it in different ways, or you had to let it be judged in different ways to see if it was intersubjectively valid.

We also learned that when doing research, you should have a theory. If you could connect your findings to an accepted theory, you were doing high-status research. A description of events, actions etc. was questionable – "it is just a description" – and was dangerously close to a personal opinion.

We were a group of students who discussed these concepts a lot, and we were left with a feeling of confusion and uncertainty. We could not accept only the perspective of science or that to apprehend rightly is only a reality to calculate and possibly predict.

A birch tree is also reality. . . .

As single individuals we were not trustworthy, we were "just subjective" in our thinking. If we agreed upon a question, and started to form a "student-opinion group", we became more intersubjectively valid in our opinions.

Almost an objective fact to take into consideration and calculate with. . . .

When you had objectivity on your side, you had power on your side.

A-T. Tymieniecka (ed.), Analecta Husserliana LXVI, 131–155.

I was very irritated; something was wrong here. There was too much magic in these two concepts – what was the meaning? And so when I happened on a certain fragment of text from Hegel in which he talks about aesthetics, I really got upset. I do not remember it exactly now, but it was something along the lines that when an artist can show – make it stand out for everyone to see – his whole subjectivity, that is an objective phenomenon.

That piece of text opened my eyes to a new perspective on the dichotomy of "subject" and "object". Objectivity in the sense of "making visible" . . . that was a new meaning for me.

By the way, "making visible" are the words "phanta zein" in Greek, fantasy. . . . This led me to a point where the concept "objectivity" and the concept "phantasy" had a connection, in a most unbelievable way.

ARE THE CONCEPTS OF "SUBJECT" AND "OBJECT" REALLY CONTRADICTORY CONCEPTS?

Influenced as I was by Heidegger's search for meaning in the etymology of words and concepts, I looked for the Swedish meaning of the words "object" and "subject". I had read some of Heidegger's work in Swedish[2] and I was very impressed – for two reasons.

Pro primo: I had never ever read anyone who wrote so tightly, who elaborated the words he had just used to squeeze more meaning out of them.

Pro secundo: In the foreword to *Teknikens väsen*, the translator Richard Matz wrote that in order to understand Heidegger we had to turn to "fiction and literature, which is a source of knowledge that is not accepted by the spokesmen of science" (My translation, *ibid.*, p. 12). This last reason was revolutionary to me. Was fiction science? I had come across a scientific text, that told me that fiction and poetry could be looked upon as sources of knowledge, as science. Unbelievable! (This discovery of mine was some years ago, when I was a student. But I want to tell you this because I now have students who react in the same way I did! They are very surprised, when they hear either my or my colleague Christer Bjurwill's perspective on doing science and making sense.)

In above-mentioned book, Heidegger elaborates the concept of objectivity. His starting point was the concept of "presence – of being here". And Heidegger meant that the New Time's conceptualization of "here and now being – presence" was characterized by "Gegenständigkeit" as "objectivity". "Gegenstand" = that which stands before you to be seen. But he said nothing about "subjectivity"; at least I could not find anything. So I looked for the Swedish meanings of the words, and I found these:[3]

Object (from ob-iaceo) lay in front of, just opposite to, throw against, throw towards, expose to, give up, abandon.

Subject (from sub-iaceo) lay under, throw upwards, practice underhand tricks, cheat, hide, spirit away. . . .

From our shared student perspective I could nod with recognition: "Yes, objectivity, science means exposing people, treating them in a careless way". To put people in different experimental settings with the aim of the study being unknown to them, that is to "surrender" and "abandon" them as we students looked upon it at that time. Several years later, I found our student apprehensions excellently expressed by Abraham Maslow in his *Psychology of Science* (1968). Maslow classes the extreme objectivistic position as a "mechanomorphic Weltanschauung"; I will return to this concept later on.

But then, the old etymological meanings of the word "Subject" made me frown a little, and suspect Heidegger for leaving out any elaborations of that concept because of the concept's inherent meanings.

To be subjective would mean to "hide the spirit away", or to "cheat". These were not the meanings we students attributed to the concept! On the contrary, we meant that to be subjective was to be honest, and absolutely not to hide a personal meaning away!

I was disappointed, but on the other hand I could now really grasp the meaning in the Hegel text – about the artist who should express all her subjectivity in order to become objective – for everyone to see. That's really being honest.

Anyhow, the great advantage in looking at the meanings of the concept of movement – of throwing, of "Geworfenheit" – was the intuitive kick I got from that perspective. I don't know if I was thinking, or if the thoughts were given to me; but my *phanta zein* showed me new possibilities in this imaginative variation of meanings. So I cannot distinctly say if I made use of a method, or if I suddenly dis-covered something that was hidden the moment before.

Because this was to me a moment of "Wesens-schau" – I suddenly saw the subject – object split, unfold, and blossom into something quite different – I will take a wordly walk around here for a while, the way

Heidegger's texts taught me. My topic is the fact that I cannot say if I used a method, or if something happened to me. . . .

Another meaning to be found in the concept of subjectivity struck me when I read the etymological expressions. "Throw upwards": Obviously the concept had something to do with "coming up", perhaps "growing up", breaking a surface of some kind? Or coming into a situation? In Swedish we have an expression "Jag kastades in i situationen" – "I was thrown into the situation". The meaning is that you enter the situation without having any control over that. We also have another expression, "att ge sej i kast med någonting", which very directly translated means "to give oneself a throw into something" with the intent to start a project. Aha! Listen to that! The same expression, but in Latin! To "start a project" is to pro-ject, to throw oneself forwards!

And if you cannot throw yourself from that point of departure, perhaps you are submerged under that "surface" of possibilities? You have imploded in your own subjectivity? You have no energy left, you are fading away in that hole of yours.

It also struck me, that a problem we had in our student discussions was that "subjectivity" could be so different, often depending on where you had grown up.

So I discovered the ever-there so-called con-textuality of "subjectivity". Then there is the meaning of the concept as we know it in the English word, "being a subject of the Kingdom of Sweden". I can make this subjectivity more concrete, by telling you that I am also a subject of Halmstad town, the West Side. Well, with my being a pedagogue, as you know, my thoughts went back to my very early years, and my first subjectivity, in the sense of "being subject to" my mother and father. I ended my geographical and historical journey there, sensing the meaning of power in the concept of subjectivity. We students experienced the same dimension as subjects to our teachers.

I continued my phantasy of "being a subject of". As I had dis-covered the "growing up" sense of the concept of subjectivity, I proceeded – which means that now I am deliberately thinking my way. As you understand – whoops! – "under-stand"? In what way do you stand "under" me? Does the word imply that you have been where I have been? On or in the same spot taking in impressions and experiences? Is that why you "stand under" these words of mine, these objectifications of my experiences? so that we share and duplicate a part of life?

My subjectivity is a flow of impressions from my context, my surroundings. And I noted that I thought – sensed from my experiencing body "from where I sit", as anthropologists say. In my above-mentioned essay, "Om verkligheten", I gave examples from different cultures. Different cultures have so many different words – objectifications[4] for phenomena. I am most impressed, and at the same time I was surprised at my Swedishness. The Swedish vocabulary was obviously not complete!

The Eskimos have twenty-five different words for "snow". We Swedes have five. The Arandas in West-Australia have twenty-five periods of the day. We Europeans have eight in common, I think.

Different so-called primitive cultures have time-categories closely connected to their lived world. "Now-is-the-time-to-milk-cows"; "they made love during the-time-to-get-water". There has been a very close connection between subject and their surroundings. At the same time I have come to understand the little mechanical device we use for producing time. (That is the difference between the cultures I spoke of above and ours. They do not have any machines for producing time. Time is produced by the way they live.) I quite saw how the clock so effectively separates our subjectivities from our surroundings![5]

Time . . . time and temporality is also a frequent topic in discussions of the subject-object relation, and in phenomenological thinking at large, I presume.

An example of this linear-regressive thinking is given by Dunne. An artist is painting a picture. He notices that something is missing in the painting, himself. So he "moves back" and paints himself into the picture – moving backwards in eternity to get himself into the picture.[6]

This problem is also dealt with in old Eastern thinking. In Hatha yoga[7] we find the concept of "Purusha", which is a kind of consciousness wherein the difference between object and subject has vanished. Instead of "moving back from the picture" the wise women of Hatha yoga "stepped into the picture", or paradoxically, they also knew that Purusha was not a part of the picture, that what you have in mind is not the mind.[8] Noema is not noesis.

Toni Packer[9] writes about Zen:

> Zen is consciousness that understands itself clearly and completely from moment to moment. It is the complete seeing of clarity – to see and act freely, with no inhibitions, without the bonds of the self. It is the emptiness of no-dichotomies; it is wisdom, love and empathy in unconstrained activity.

Pirzig, another person influenced by Eastern thinking, writes:[10]

That was a useful metaphor for whoever studied scientific objectivity. Whenever the marine chart was not in accordance with his observations, he rejected the observation and followed the chart. In accord with that which the consciousness thought it knew, it had build a static filter, an immunity defence that rejected all information that did not fit. To see is not the same as to believe. Believing, is on the other hand the same as seeing.

Talk about a noetic mind that has been locked onto the noemata of printed matter.

Even Tina Turner sings about the impact of "seeing": "What you see is what you get!" I think that the problem with this conceptualization of Self and consciousness and temporality has to do with using metaphors of "seeing". We can use other metaphors, as well as other senses, and thereby create other conceptualizations.

Sound is always immediate and here and now, so also is feeling. They are im-mediate. Merleau-Ponty uses as a metaphor of the Self-as-Subject-and-Object somebody's shaking hands with herself. The left hand takes the right hand, and so the body-mind can simultaneously feel that it is subject and object for it-self.[11]

Or as Tse-Tse says in one of his Thirty-three Happy Moments:[12] – "Happiness is to have some spot on your body that itches, and to scratch yourself on that spot!" – an im-mediate conceptualization of the object-subject relation. The itching is the selfsame point of an object-subject as when the two hands recognize (or as we say in Swedish, directly translated: "feel again") themselves; if there were no body there, no I-tching.

But, Merleau-Ponty talks about the positive side, instances wherein you know that you are subject and object. We can use the concepts in another, much more problematic "sense".

I got this example from a psychiatrist; a person so split in her emotional experiences, that her logic of sense followed this path: My face is wet with tears, I am crying. Because I have a wet face and am crying, I must be sad and unhappy.

Can you, dear colleagues, think of a more frightful split between Self-as-Subject and Self-as-Object? "Inside"? one and the same body?

For me the "sense" or moral of this anecdote is that we have to be careful in our application of the "subject-object" split. We must really be conscientious about where in the worlds of phenomena we apply this conceptual split . . . because when we do, we at the same time create a contradiction.

My students' reactions and reflective writings on our video-interviews made me aware of the truth that it takes some time to learn about the possibilities we have to really get to know ourselves. Husserl writes a lot about ever growing intersubjectivity, but I would like to add "intra-subjectivity", just to have an academic term that points out the road to self-knowledge.

Later on, I met the perhaps most well-known example of this linear, regressive metaphor and thinking in George Herbert Mead's theory of the Self as "the I" and "the Me". Let me put you into his way of reasoning:[13]

> The simplest way to treat the problem would be in terms of memory. I talk to myself, and I remember what I have said and perhaps the emotional content that belonged to it.
>
> "I" in this moment exists in "Me" in the next moment. Where I cannot turn around fast enough to catch myself, I become a "Me" because I can remember what I said. . . .
>
> We can go back a few moments in our experience, and then we are dependent on pictures from our memory. So that "I" in the memory was there as a spokesman for that self which existed a second, a minute or a day ago. As it is there, available, it is a "me", but it is a "me" which was an "I" at an earlier point in time.

"Where I cannot turn around fast enough to catch myself. . . ." Well, this is a puzzling metaphor. Why turn around? You are already there, aren't you? A fascinating "turn-around-of-sight" (a not very good translation of the Swedish concept "synvända") was given to me by a student who practised a special kind of meditation that he had learnt in India. These students of meditation were taught to create *memories* of the future – a very elegant phenomenological turn – an evasion and overthrow of the experience of mechanical linear time. The students were literally concentrating and shaping a consciousness that "the future is here and now".

PRACTICAL USE OF THE SIGHT-SEEING-LOOKING-METAPHOR: HOW I USE VIDEO-RECORDINGS TO PROBLEMATIZE THE SUBJECT-OBJECT DICOTOMY

The few lines written above have enormous practical significance. The meanings of these metaphors have a deep personal significance; they have to do with how we know ourselves. Because of this, I will make a little stop in my theoretical story, and tell you something about the deep personal meanings that might blossom from the abstract dichotomy of "subject" and "object".

In a combined course on the theory of science and research methodology that my colleague Christer Bjurwill and I give at the university, my part is interview training. I use video recordings, and the students and I look at the varied video-interviews together, and talk about what we see. Parallel to this, we have a literature-seminar series around Pirzig's book *Lila*, dealing with morality and the theory of science.

Now, even though videos are part of everyday life today, many students are not used to "seeing themselves". Their reactions sometimes make me think of stories from anthropologists who took Polaroid pictures of so-called primitive people, and got startling reactions. People reacted as if their souls had been taken away from them, and put into that little piece of paper.

Some of our students react to the videotapes as if they were exposed or unmasked, or even as if their souls had been taken away. . . . God knows from what. I mean, we around them cannot see any difference from what we see with or without video. But this personal objectification, i.e., becoming a picture, obviously has an enormous importance for some students. As some students put it in their concluding paper after the course: "We were no longer afraid of our video Selves". And in this situation, wonderful metaphors of self-knowledge blossom. The students are fascinated by discussion of the object-subject split as they experience it. One student exclaimed in a high voice: "Do you see me as I see myself on that video!"

Well, dear student, I don't know if we do – but don't worry, it is O.K. You see, this was the problem Husserl had, so it is an old question, which is not yet solved. We are a whole bunch of fellows with your question: "Do the people around us look upon ourselves the way we look upon ourselves?"

Now, Husserl had no videos, so his example of this alien feeling you experience was the encounter with another person. He called it a meeting with the alien, and he thought a lot about how we could recognize ourselves in meeting with somebody alike, but alien. Here is another body, yes. But is it another soul?

With videotape, you can even meet your own self as if it were a person alien to you! Remarkable, is it not? I mean, you are quite often together with yourself, aren't you?

The students also experience the Husserlian distinction between presentation and appresentation, as concerns their Self. Here again, I can make the students aware of these two concepts thanks to the videotapes. Several students take up the fact that in everyday life they cannot see themselves so "objectively", since they cannot see the back of themselves, or that which is behind them. But on the video – they can! What is appresentation in everyday life becomes presentation in videotime! As a student said: "The videotape is a more objective picture of me, because I can see my own back!" Three other students wrote: "It gives you a lot to look at yourself in an objective manner when you study the videotapes. Then you have a possibility of looking at yourself that you do not have when you are in the middle of the situations. We think that we can learn a lot from this".

THE SPLIT OF THE OBJECT-SUBJECT SPLIT

Well, here I will start to end my dealings with the subject-object dichotomy. The world was divided in two: the subjective, living human inside and the objective outside, mostly consisting of not-living, not-human phenomena.

And at the gateway was the concept of the individual, the link between the two worlds, the inside and the outside.

The "individual Self" was also unique; "in-dividare" is Latin for "not possible to divide". If it was a Self that was divided, it was not a healthy Self, in this old conceptualization of "individualism". The term "dividual" would in Greek be "schizo". The most well-known author who has described and studied this link is in my opinion the psychiatrist Ronald Laing.[14] In Swedish the term "divide" also has the sense of "share". Translating the Greek word schizo in this meaning would give us "shareable", denoting that a *dividual* is a person who has no clear borders with his surroundings; the *dividual* or the schizoid looses herself in the surroundings. Very directly translated, the concept "individual" is someone/something standing out from its surroundings and environment and even from other species of the same kind.

There are no connections, the individual is not "share ware", to use an electron metaphor. The patter of "not-sharing" is crucial to the concept of individualism. Now, the Latin word "in-dividare" is in Greek "a-tomos". The atom: the smallest part in the Mechanistic physical universe; the individual, the smallest part in the Mechanistic human universe.

Laing writes about psychiatry's individualistic language:[15]

> The terminology nowadays refers to, on one hand, the human being isolated from others and from the world, that is, an entity that is not fundamentally (= essentially) in relation to others and to the world, or it refers to misleading concrete aspects of this isolated entity. (My translation from Swedish.)

As he continues, Laing takes literally the egological theme in the Husserlian phenomenological method,[16] when he talks about theory:

> The loss of personality in a theory that aims at being a theory about personality is just as misleading as the schizoid's depersonalisation of others and is in the end just as much an intended act. Such an objectification gives us false "knowledge", even in the name of science.[17]

As scientists and philosophers, we must write in such a way that it is evident that we are living, experiencing and feeling humans. Not machines or computers.

Thanks to Heidegger, I found in the history of the words subject and object a hidden conceptual meaning of positions and movements, important for the creation of metaphors of knowledge. I then coined two other terms:

ad-ject = lay close to, throw to, add, enlarge, wish, intend (the term opposite the term ob-ject);

super-ject = throw over, lay over, transcend, exaggerate, wash over (the term opposite the term sub-ject).

What happened now? In my mind, I split the individualistic atom into something else. My thoughts then took up another form of metaphors. Instead of thinking in mechanomorphic images, I seem to use electromorphic images. This is, of course, a change in my subjective thinking, but it is also a change in my superjectivity, in my sharing in our collective way of apprehending and conceptualizing. Changes in conceptions of our World-Images, changes of my logic of images, in my ideos – logos. Superjective knowledge I propose to be analogous to what Michel Foucault called "the archive";[18] plus images, concepts, representations. That in turn means that the concept of the superject also is analogous to Popper's World 3:[19]

> World 3 then, is the world of ideas, art, science, language, ethics, institutions – the whole cultural heritage, in short – in so far as this is encoded and preserved in such World 1 objects as brains, books, machines, films, computers, pictures, and records of every kind. Although all World 3 entities are products of human minds, they may exist independently of any knowing subject – the linear B scripts of the Minoan civilization have only recently been deciphered – provided they are encoded and preserved in some potentially accessible World 1 form. (Hence the crucial difference between the knowledge in people's heads and the knowledge in libraries, the latter being far and away the more important.)

First I will give another connection to Foucault's concept of the archive: "It is not a returning to the secret of origin; it is a systematic description of a production as an object".[20] To illustrate the link between the two gentlemen's concepts, second, I want to point to the starting-point, or the secret of origin, which in my conceptualization is a question of adjectivization. As with the Minoan scripts – their world, their type of adjectivization to the world has vanished – so it takes some hermeneutic detective work to "find out" and make their meanings ours.

Third, Popper's notion about the importance of libraries provokes me to try to imagine the consciousness of a cyborg! Instead of the slow communication we are engaged in now, I could have informed you of my paper in parts of a second and we could have had an amazing "chat" that I cannot even dream of. . . . Husserl's dream of worldwide inter-subjectivity is coming closer to reality.

There is however a noticeable difference between the concepts of "the archive" and "the superject", as I see them. The former is bureaucratic and Cartesian in the sense that it deals with "talk and written language", with what can be spoken and said. The latter is more computercratic and Husserlian because it deals with images and representations; with "what can be seen and made visible". (Concerning the difference between Descartes and Husserl, Descartes did not know himself in the Husserlian way, for he stated: "I think, therefore I am" and we get the impression that the I is purely intellectual and thinkable, don't we? Husserl ends his *Cartesian Meditations* with the oracular utterance from Delphi: "Know thyself" and as the Greeks at that time were thinking in terms of bios-logos,[21] I assume the oracle included the body-mind. Husserl's superjective message was that of intersubjectivity, a phenomenon hard to achieve when you proceed down the path of Cartesian Doubt, or what do you think?)

But I want to point to the Foucault concept again, for he takes up the theme of power or will: "What can be spoken and said". This dimension always lies behind images of the world – but is also intertwined in the reason for choosing a world-image.

I can distinguish four distinct "World-Images": anthropomorphic, organomorphic, mechanomorphic and electromorphic. They grow up from the different worlds we live in due to changing technologies; this connection also gives the connection between the concepts of superjectivity and adjectivity – how we adject to the world we live in as we see it. The dimension of power and will is apart from these superjective world-images. The power and will dimension belongs solely to the concept of adjectivity.

When Leonardo da Vinci drew a picture of human muscles as mechanical devices, he saw two world-images transcending and mixing, just as William Harvey did when he conceptualized organic blood circulation using the mechanomorphic image of the heart as a "water"-pump.[22]

Nowadays we can see all three world-images mix in the concept of the "cyborg", creations like the Terminator or Robocop – have you seen the movies? One of our first superjective combinations of organomorphic and electromorphic images was Superman. Examples of the anthropomorphic worldview are so familiar I believe, that it is unnecessary for me to give any.

I also want to stress, that the different superjective world-images often are used in a metaphorical way, but some "images" are indeed made real, made into objects.[23] The substance of the superject-dimension is collective information. But I want to stress the fact, that this superjective information set out on its journey to superjectivity from single subjects. Here it is really true that the world is man-made. From a single individual's phantasy, to thousands of people's everyday life.

As with the instances of Leonardo da Vinci and William Harvey, a new world-image has an enormous impact on our reality, as Eskola nicely shows us in the following quotation. Eskola also makes a point of how important a role the phantasy has in research:[24]

> In an empiricist milieu the researcher can be trained to make observations, arrange them, explain for such things that can be collected from tables and develop "theoretical" thoughts against the background of what already is observed or as a kind of ornamentation of the results. If such a milieu is permitted to dominate for a long period of time, it can make psychologists and sociologists incapable of doing genuine theoretical work. The phenomenon might be illustrated by a metaphor: in past times, when you still travelled by foot, under sail or by horse, the above-mentioned, chained to his facts researcher would probably have been able to construct new and better skiis, sledges or sailing ships but never ever such totally new inventions as the steam locomotive or the automobile.

Here I will return to Maslow's conceptualization of extreme objectivism as belonging to a mechanomorphic Weltanschauung.[25] The Marxist concept of reification as well signifies this kind of superjectivity. Here I want again to point to the fact that power or will is intimately connected to the world-imagery you "choose". I deliberately put the word "choose" within quotation marks, because it takes some self-knowledge to be able to choose world-imagery. Initially you are given world-imagery by your families and later on by the system of education, your institutionalized ways of adjectivization. It might be important here to make it clear, that when I think of the criteria of the theory of science, I always look at the criteria in a very egological way, i.e., I use the criteria to describe persons acting as scientists. This is the only way I can think of that will make science humanistic. I do suppose that this was the meaning of a text from Husserl:[26]

If I create a scientific theory, then this complicated activity of rationality along with the forms of being that according to rationality belong to it, is of a "Wesen"-kind that is not possible in every possible ego, but only for an ego which is "rational" in that specific meaning that presents itself when the ego becomes worldly in the "Wesen"-form of human being (*animal "rationale"*).

I can also return to Dunne's serialism. Using electromorphic imagery, we can stop the linear, mechanomorphic time as it is conceptualized in our mechanical analogue wristwatches where time is the space "around the clock" – and "walk" into the ever flowing Now of digital wristwatches.

The concept of the Individual Self is expanded into the concept of the Omnividual Self, which is a pairing,[27] taking in part of Other Selves and the World around.

Something else has come to me, prompted by the metaphor of "splitting the atom" – "Splitting the individual". The split I make creates, makes stand out, the adjectiveness of the subject. Now, when you split the atom, a lot of energy is created. What is the case with "the individual"? By changing my superjective conceptualization of "the individual", will other metaphors, with other inherent meanings, suggest themselves, that might explain phenomena in new ways?

Immediately I saw (in a Wesenschau?) the Self letting go of destructive introjections, of destructive adjectivizations to the world. And when that happens, a lot of mental energy is freed! (The concept of introjection is used in psychodynamic theories, and means that the Self takes in events from the outside world, the classic example being to take in the image of a father or mother – the first Husserlian pairings a human makes, I presume.) I will come back to this metaphor and its implications for the theory of personality. The metaphor also indicates that we are differently adjected into the world. It is time to really differentiate our existences-in-the-world. We have to consider the Lebenswelt's different connections to subjectivities, through different forms of adjectification.

For me it is an obvious truth that humans can become either destructively or constructively adjectivized. As a pedagogue, I think we learn how to be-in-the-world in different ways, that we are not just "thrown into it". Perhaps some of us are, though. I will return to this under the heading "The Concept of Projective Identification".

On leaning back and reading what I have just written, I get a peculiar feeling of recognition. I can sense that my electronic metaphors were just modern imagery for something old. I think it is something I have read. Back home, polishing my mirror, I suddenly see the book behind me: *Zen Buddhism and Psychoanalysis.*[28] Ha!

Language of knowledge before electronic world images. And of course, there they were – my figures – but from a totally different culture, expressed by D.T. Suzuki in Japanese. He translates the Japanese concepts to English, points to the fact that some of the Japanese terms express movements and positions. His first figure in that book is a line going in one direction; his second figure is a line going in the opposite direction; his third figure is the two lines together, one above the other.

Linear and dualistic thinking. . . .

In Suzuki's fourth figure, he has bent the two lines, with the argument that it symbolizes something living, not something mechanical; and in his fifth figure, the two lines have become a rotating circle; rotating because it is really a symbol of life.

So my thought pattern followed the McLuhan route; the electronic world gives us back the organic patterns that disappeared in the era of the Mechanism.

I ended up with a figure very much like a very old symbol, the yin-yang symbol of the Japanese concept "sho chu rai". In Suzuki's words: "The logical abstraction, the Logos, is now coming out of its cage and becoming incarnated, personalized, and entering directly into a complex world, as 'the goldhaired lion' ".[29]

I don't know about that lion, but Zen seems very much like Husserl's ego-logic. What an adventure to study these possible connections between two conceptual worlds!

In reflecting on my own reasoning above, I would like to present you with a quotation from Moustaka: "Noesis refers to the act of perceiving, feeling, thinking, remembering, or judging – all of which are embedded with meanings that are concealed and hidden from consciousness. The meanings must be recognized and drawn out".[30]

Noesis, in Moustaka's words "the subject-correlate", and "ideal" side, i.e., my noesis, could clearly be split into two parts – the superjective world idealistic/image side, which made me present the subject-object phenomena in the way in which I did, going from mechanistic metaphors to electronic metaphors to organic(?) metaphors a la Suzuki. And so there is the adjective part of this mind play, my strivings with the languages, trying to understand what the different authors really mean and intend behind their different languages, given their connections to their time and place and values in the world. Is this meaning in Husserlian philosophy the same as this meaning in Zen? Husserl repeatedly talks about the close connection between the noema and the noesis, the universal epoche; is he really conceptualizing anything other than Zen? Is this not another way of saying it, another language for the message to let go of all your sens-ational ex-periences? After this, what is there? Chaos? Emptiness? No, Husserl argues that real life becomes accessible to us through the epoche.[31] And Zen?

To complete the picture, the conceptualization of the human as being individual has as its starting point a single person, a lonely person. But we, the humans, are not lonely. We are not lonely in the aspect of here-and-now, we always have other people around us. And we are not lonely in the sense that we are aliens in the world; we all have a history of others. I do not meet a "single" subject when I meet an-other person. Of course, I see the other's objective body, but I also meet parts of a superjectivity, parts of that person's cultural heritage, of which that person is a message. I also meet that person as an adject, the way that person relates to me, a message of that person's "subjective" history. Husserl also starts with the solitary self, but very soon he talks about "the other". I would like to start with, as a minimum, two persons, two persons who at the same time look at each other and ask themselves: "Who is that?" and at the same time start to react and act towards each other.

This is the starting point of the conceptualization of the omnividual, a strong emphasis that we are "the same", that we are more alike than different, and that we are not alone. Husserl has a very humanistic paragraph, the fifty-fifth of his *Cartesian Meditations*. (The first time I read this, I experienced the same thrill as I once experienced at a "Philosophy of Care" seminar at the University of Linköping, with Lennart Nordenfeldt as chairman. For about two hours we discussed the question: "Can a fox be looked upon as a person?" And yes, our conclusion was, a fox can be conceptualized as a person. After the seminar, a colleague said: "My promenades with my dog will never be the same. . . .")

Well, what I think is funny with the §55 of Husserl's meditations, so astonishingly "childish", i.e., in a very positive way, an absolutely pure-looking epoche and curious observing look when he describes "the first form of objectivity: the intersubjective nature". Perhaps the following four quotations give a moving picture of a man who suddenly realizes that he is aware of the presence of another human being in an "animal", full-sensed way – not just as a picture to look at.[32] Here are my translations from that section:

If we hold on to the factual and always actual experience of the alien, we find that the sensorily perceived body without any hesitation is perceived as the body of the other, and not only as an indication of the other; is not this a mystery?

It's about one and the same nature, with the difference that it presents itself [as if I should be on the same spot][33] as this alien corporeal body. The body is one and the same central body, given for me as "over there" and for the other as "here", and the whole of "my" nature is the same as that of the others.

That what I see is not a sign, not just *an analogon* or in any natural sense a representation; it is the other. And that which was primordially experienced, that corporeality "over there" (and even just one aspect of that corporeality) is the body of the other, but looked upon from my direction.

In relation to the animal the human being is constitutively the normal case, just as I am the original norm for all human beings. Animals are constituted to me essentially as abnormal "modifications" of my humanness, even though we must distinguish between normality and abnormality also among the animals. Overall, it is a question of intentional modifications, that make themselves known in the structure of meaning. All this calls for a deeper phenomenological explanation, but this common remark is sufficient for our purposes here.

That deeper explanation I think we came close to at the above-mentioned seminar. I do think that even a fox can experience that holy moment of recognition of a psycho-physical entity's sameness when it recognizes an-other as a fox.

But Husserl wrote about us – and at this point in his text his words shiver with recognition of the full feeling of a living other, not just a picture to look at. When I read these pages, I think of Confutse, who once asked himself how he could know that he was not a butterfly dreaming he was Confutse. I don't know how many of you have experienced that accompanying feeling of loosing your sense of really being here, when you start to doubt whether what you experience is real?

And again, we can step into the superjectedness of electronic world-experiences. I am sure Husserl would have had great fun with a couple of cyberspace-spectacles on. To really see into a three-dimensional picture world, Husserl would have loved it!

And when I translate Husserl's following paragraphs into my modern Swedish, I understand Husserl is talking about the concept of information in this text. I think that is what he means by the expression "logical-ideal objects", that can be created again and again, at different points of time. To take your cyberspace-spectacles on and off, on and off.

For me, in my world, it makes a lot of sense to read the concept of information into Husserl's text when he describes the higher levels of intersubjective community and solidarity. Husserl did not grasp all of the information that we have; that is why he uses a mixture of organomorphic and mechanomorphic concepts to catch the meaning: "spiritual objectivities". "Spiritual" is a common organomorphic term, since it conceptualizes something living that cannot be seen, but felt. "Objectivities", the most common mechanomorphic term, signifies dead and calculable entities. "Information": an electromorphic term for dead(?) messages.

Well, these higher forms of intersubjectivity lead to different forms of superjectivities; patterns for arranging information.

The concept of superjectivity gives us the possibility of seeing that "truths" have a local meaning and "area". For example, I have two students, Martin Göran and Fredrik Olsson, who have asked pupils and their teachers how they look upon their sports education in school. They have asked them with the help of questionnaires, of course. This kind of research is often looked upon as "objective", since you turn verbalizations – the answers – into numbers. So it seems like measuring an object, but it is not; it is to present a lot of people's understanding from a special viewpoint – that is, you present an issue, a noema, that is intersubjective, and when the group can be looked upon as large (several thousands of people), the noema has become superjective. By the way, from a phenomenological time-perspective Göran's and Olsson's research can also illustrate how the concept of the Future varies among intersubjective fields, at the same chronological Time.

The everyday life of pupils contains halls for aerobics; you find a couple of them in Halmstad. A lot of pupils, of course, the girls, want to have aerobics during the schoolday. They exercise outside of school naturally. But during their schoolday, there is no aerobics. Now, a group of visionairy school developers around Ann-Margret Corsander have created "a school for the future – the Frösakull-school," which will start next year, in spite the fact that many regional politicians do not share that future. Can you guess what kind of hall is built into that school? Quite right, a hall for aerobics! A nice illustration of the fact that there can be a close connection between time and place! And we find a remarkable intersubjectivity between the pupils and a few visionary school developers. On the other hand, there is that intersubjectivity shared by teachers and politicians. Also, the example illustrates the power dimension of the concept of superject. If the pupils had had political power in Halmstad, presumably aerobics halls would have been built some years ago in our schools. The noesis of the youth is the noesis of school visionaries. They live in the same fields of information, their "spiritual objectives" are the same!

Some words about the concept of adjectivity. Its meaning is the opposite of objectivity, i.e., here all the statements above about objectivity are turned around. I used some lines to talk about the objectification of my thoughts into words, enabling my thoughts to be seen and read by everyone. Objects go together with expression. So adjectivity goes together with im-pressions, and of course empathy. Adjectivity is something that exists in relations. When Zen devotees use paradoxes, they create adjectivity. I almost wrote "they create adjects", but that would have been to objectify the concept. Adjectivity is a process phenomenon. The concept deals with acts, with action. Therefore it does not exactly belong to a bureaucratic world. If numbers belong to the world of objectivity, the world of adjectivity contains acts, performances and, if objectivized in symbols, the phenomena of adjectivity are caught by stories and poems. To catch a meaning is to describe it in a way that creates associations, feelings, meanings. It is to use phenomenological variation in a poetic or paradoxical way, so as to catch meaning.

The adjective way of making science and sense can change the rules of a society. In Sweden we had a remarkable author, Ivar-Lo Johansson. He wrote a novel about the agricultural labourers in old times, living in the dreadful system of wages that was called "statar-systemet". In his novel, he really grasped the soul of that destructive system, and the novel created a climate of popular opinion that made that old "statar-system" cease to exist. It is a novel, so by definition it is fiction; but since Ivar-Lo Johansson himself had been living that life, it is also a story told by a participating observer – or better – an observing participator. So it is also methodologically anthropology. Ivar-Lo also wrote about the prostitution in Stockholm – and that novel changed the laws about prostitution in Sweden at that time. Adjective knowledge.

The knowledge of empathy, of "tuning in", of imitating are adjective forms. From the field of acting I have learnt that actors first must "act in" a role; then they can "act out". The first thing actors do is adjectivize, they must put themselves into a situation, into a context; then they express and make the role objectified for us to see. . . .

The birch tree
in greenish shimmer
welcomes the sunwarm steps
of the ant.

THE DIALOGUE – EDUCATION AND THE BIRTH OF AN OMNIVIDUALISTIC SCHOOL

During 1978–1982 I walked around in the Hageby school as a participant observer. The Hageby school district was a mixture of blocks of flats and private houses, of asphalt and woods. About one thousand pupils had their daily occupation there, in and outside the school buildings. These years I experienced a lot of situations that made me think about the teachers' perspective on a theory of science. My impression was that they shared our students' views, that they were also trapped in confusion or dogmatism concerning the concepts of subjectivity and objectivity. The traditional opinion was that the education should be objective, which to many teachers meant that they were not allowed to have a personal opinion. If they had, their pedagogy became subjective and suspect in the eyes of their colleagues, became untrustworthy. Well, that was what they said in discussions and meetings. But, when I had the pleasure of visiting them in their classrooms, most of the teachers were very subjective in their views and statements . . . so subjective that some of them did not care especially much about the viewpoints of their pupils.

The concepts of subject/object were much discussed in Sweden in the middle of the seventies in the school world.[34] At that time "education by dialogue" swept over our country. A much used figure for this education was that the teacher in her relation to the pupil should behave like a subject in front of another subject. This was in contrast to the more traditional monologue teacher-pupil relation, which pictured the teacher as subject in front of the pupil as object. The pupil was, as we say in Sweden, an "object for education". I don't know if in other languages you have that expression?

To start with the concept "teacher as subject – pupil as object", what does that mean? The conceptualization here is on a very high level of abstraction, and as I am almost a fundamentalist empiricist, I would like to unfold and open up this hard-package of information (that is what an abstraction always is), and try to go "zu den Sachen selbst", to the meeting-point of sensuality and sense.[35]

First, the abstraction has the character of a statement about Swedish educational culture. Everyone taking part in discussions assuming this concept understands it to be a cultural statement, that it characterizes the most common relationship between teacher and pupil. During the four years 1978–1982, we even measured the relationship. We handed out a questionnaire to over 1,500 pupils of grades 4 to 9 and asked them about the characteristics of their working-forms. We also distributed a similar inquiry to their teachers. And we found that yes, the most active participants in the educational relationship were the teachers. And, yes, the monologue form of relating was the most frequent.

For some funny reason the pupils experienced more of the object-relation than the teachers, if we look at the answers on the questionnaires. The teachers' opinion was that they related more in a dialogue, in a subject-subject relationship. The subjectivities of the teachers and the pupils were different. A confusing situation, perhaps. I talked to some boys in the higher grades about how decisions were made at school. They believed that the school was a dictatorship, and they wanted anarchy. Myself a democrat, I could not understand their viewpoint. But then it occurred to me, that given the way they were adjected to the school, their dichotomy made sense. They experience one teacher deciding what to do – and that is of course a great liberty; you do not have to share decisions with anyone else. Hence, freedom is loneliness, solitary persons without any control – is that an OK description of anarchy? Freedom is lone individuals with all the power. Why on examination, the whole organization of the school was built on the concept of individualism, persons in isolation; the teachers work alone in the classrooms, and the pupil's work was guided by individual examinations.[36,37]

We get at another aspect of the relationship if we look at the cornerstone of traditional education, the textbook. In Sweden there are special textbooks selected for each grade. If you as a teacher venture into investigative education and leave the selected textbooks, you are moving into giving a more "subjective" education. And that is not so good, as education should be "objective". All pupils should know the same things. They should copy their teacher's knowledge and use one and the same textbook. The curriculum of the seventies had recommended working forms based on investigation and problem solving, activities that use the subjectivity of the pupils, but the teachers being hooked into the conceptual dichotomy of subjectivity and objectivity felt confused. Angry voices were heard, and the terms "objectivity" and "subjectivity" came to be social markers of desired order.

"Objectivity" was the marker of truth and trustworthiness, the concrete and heavy facts. "Subjectivity" was the marker of very uncertain knowledge, knowledge that changed from time to time. Public opinion was that knowledge is not a social phenomenon, that knowledge is products not processes.

And today we have a new curriculum in Sweden, published in 1994 and available on the Internet. For the first time we have one curriculum for the whole school system, including education for adults. And in this new curriculum there are some statements about the forms of education, statements that forbid the old education "ex cathedra". This is very remarkable for a curriculum plan.

All curricula we have had before were *recommendations* about educational forms. And, very very important for addressing the conceptual individualism, the teachers are ordered to work in teams!

And there are also lines about forms of knowledge. It is not just the old intellectual kinds of knowledge that the school system shall impart. Intersubjectivity in the form of social competencies are today a form of knowledge spoken of in the Swedish curriculum. I imagine that Husserl would have been satisfied. Together with this also goes an emphasis on the pupils' making science and sense on their own, and an emphasis on the contextualization of knowledge, all of which fits with the concept of adjectivization. And it also means that individualistic thinking is supplemented with omnividualistic thinking!

The new way of thinking is objectified in the way the new schoolhouses are built, for example, the above-mentioned Frösakulls school. No old classrooms, no cath; instead libraries, group-rooms, laboratories, different workrooms, "a nature room" where teachers and pupils can go right in from the outside with dirt from the forest on their shoes.

And, of course, a completely new teacher role. The teacher is now a skilled supervisor, a tutor, she knows about group dynamics and works preferably with groups of 6–8 pupils. All over Sweden we can now see the effect of this new omnividualistic curriculum. The old teachers, used to working in the individualistic dimension, are stressed and put themselves on the sick list to an extent that different regional health insurance offices have reacted!

A pupil in this omnividualistic school will not gravitate toward dichotomous thinking about decision-making in terms of dictatorship versus anarchism, believe me!

Ending my example of how an educational system creates a superjective concept of individualism, I must give you two quotations cited in the Sweden Academy's Wordlist (1933) that give me the impression that the concept of the omnividual has been there lurking, all the time:

A . . . necessary law for nature is its Individualism; because every real object is an Individuum, and just because of that belonging to nature (Höijer 1799, my translation).

The Atom, a plant, the Earth, all are Individuals, the Universe is an Individuum; it thereby follows that everything that exists is part of the great Individuum (Wikner, 1826 my translation).

I could say the same about the omnividuum.

A SPECIAL ASPECT OF OMNIVIDUALISM, THE PHENOMENON OF PROJECTIVE IDENTIFICATION

Husserl talks about "the pairing" in his fifth Cartesian Meditation (§51).[38] It is an interesting text, about the ego and the alter ego, on how we recognize each other. There are two sentences that made me stop, and think: "To be precise, it is about a living, reciprocal awakening, a reciprocal excising of objective meaning. This overlaying can be total or partial, it has always its degree, and its limiting case is that of 'sameness'".

I have found the concept of projective identification very interesting. I have discussed this phenomenon with colleagues and friends, psychiatrists and therapists. As I am very much an empiricist, I have never been satisfied with the final explanation that feelings are transported to another person. (Here I use the term projective identification, but there are other terms for the phenomenon I mean; subconscious identification, induced countertransference, parallel processes.)[39] Feelings transported to another person? How? In any case, the phenomenon really shows that we are omnividuals. If we can feel the same.

This much I know. Reading Melzer's text, "Anal Masturbation and Its Relation to Projective Identification"[40] made me frown. I absolutely could not understand using that metaphor to explain the phenomenon of the transportation of feelings. My adjectivization to the phenomenon of the transport of feelings was the theater, with actors awakening feelings in an audience.

I find it interesting, that Husserl's phenomenology has this as a cornerstone – the sameness and the mutual sharing of meaning and feeling. If I use electromorphic metaphors instead of anthropomorphic metaphors, such as those that Meltzer uses, I can understand this process of sharing simply in terms of signal systems. Amongst psychotic patients the phenomenon is extreme. And it is exciting to read how one single person can make several people around her start to act and feel in ways that reflects her history.[41] This is a true piece of adjective phenomenon. It is performed and acted out, it is even non-verbal, and it is relational. And it exists in between two or more persons.

If I conceptualize the phenomenon in terms of signal systems, I can understand the transportation of feelings in a far better way. The first thing is that we as humans react with our body to other bodies, as Husserl saw. This means that I do not just see the facial expression of another, I immediately imitate it for a snapshot of a second, and thereby the signal I see becomes a feeling that I can sense with my body. This is the immediate identification Husserl speaks of.

Now, this was a destructive example of omnividualism; to stay in your own subjectivity you must use your individualism.

University of Halmstad
Sweden

NOTES

[1] E. Husserl, *Cartesianische Meditationen. Eine Einleitung in die Philosophie*, p. 12 (Göteborg: Daidalos, 1992) (Swedish translation).

[2] M. Heidegger, *Teknikens väsen* (Halmstad: Raben & Sjögren, 1974), "Varat och Tiden".

[3] B. Tjellander, "Om verkligheten" D-essay, Sociologiska institutionen, Linköpings Universitet.

[4] B. Magee, *Popper* (London: Fontana, 1973), ". . . and since any criticism he may put forward, or proposed change, or solution to a problem, has to be formulated in language before it can be tested or even discussed; any such proposal becomes an objective proposal . . ." and further down: "This underlines the enormous importance of objectifying our ideas in language or behaviour or works of art" (p. 71).

[5] I have had great pleasure in reading Lewis Mumford and Marshall McLuhan on the topic of the production of time.

[6] J. B. Priestly, *Människan och tiden* (Aldus, 1967).
[7] L. Zeilich-Jensen, *Hatha Yoga* (Capricornus, 1966).
[8] A. J. Deikman, *The Observing Self*, p. 21, Swedish translation (NoK, 1986).
[9] T. Packer, *Seeing without Knowing. Writings on Zen Work*, p. 8, Swedish translation (Via förlag, 1989).
[10] R. Pirsig, *Lila – en Studie i Moral*, p. 365, my translation back to English (Alba, 1992).
[11] J. Bengtsson, *Sammanflätningar*, p. 73 (Daidalos, 1988).
[12] Lin Yutang (ed.), *The Wisdom of India* (New English Library, 1964).
[13] G. H. Mead, *Medvetandet, jaget och samhället* (Argos, 1976).
[14] R. Laing, *Det kluvna jaget* (Aldus, 1968).
[15] *Ibid.*, p. 15.
[16] Husserl, *Cartesianische Meditationen, op. cit.*, p. 12 (Swedish translation).
[17] *Ibid.*, p. 20.
[18] M. Foucault, *Vetandets arkeologi* (B.O.C., 1972).
[19] Magee, *Popper, op. cit.*, p. 61.
[20] Foucault, *Vetandets arkeologi, op. cit.*, p. 158.
[21] Heidegger, *Teknikens väsen, op. cit.*, p. 67.
[22] S. E. Stybe, *Idehistoria* (Halmstad: Raben & Sjögren, 1961).
[23] An exciting article on the concept "cyborg" is Charles Laughlin's "The Evolution of Cyborg Consciousness", www.carleton.ca./claughli/cyborg.htm 1996. Laughlin argues for "the real thing", and is irritated over the metaphorical use of the concept. He believes that the cyborgs will have another adjectivization to the world than a pure human being – and that seems plausible to me. One result of the communication between cyborgs is of course the almost collective consciousness they will create. They can just plug in on Internet, you know. . . . A guy who plugged in on Internet was the artist Stelarc. In an article in *Svenska Dagbladet* 1996–11–11 it is described how the artist connected his body to Internet in such a way that his muscles reacted to the signals that were sent to him from all over the world! The leap to cyborg consciousness was hereby excellently performed, I think. Stelarc felt the communication, but he did not think it.
[24] A. Eskola, *Personalighetsbegreppets förvandlingar* (Halmstad: Raben & Sjögren, 1987), p. 130 (my translation).
[25] A. Maslow, *The Psychology of Science* (South Bend: Gateway, 1966).
[26] Husserl, *Cartesianische Meditationen, op. cit.*, p. 85 (Swedish translation).
[27] *Ibid.*, p. 85.
[28] E. Fromm, *Zenbuddism och psykoanalys* (Halmstad: Raben & Sjögren, 1965).
[29] *Ibid.*, p. 79.
[30] C. Moustakas, *Phenomenological Research Methods* (Sage, 1994), p. 69.
[31] Husserl, *Cartesianische Meditationen, op. cit.*, p. 38 (Swedish translation): ". . . this phenomenological epoche, or this bracketing of the objective world – it does not place us in front of a nothingness. On the contrary, something is made accessible to us, or more clearly; that which becomes accessible to me, the meditating one, is my pure life, with all its pure experiences and pure meanings, the universe of phenomena, in the wide and specific meaning of phenomenology" (My translation).

Well, Husserl is a bit excited here, isn't he? And why does he choose to use the term "pure"? And why does he use opposites at the same time – "wide and specific meaning"? And why does he use the term "the meditating one"?

[32] Concerning my use of Husserl's text – please observe that I live in another time and place and culture than Husserl did. In this adjective sense, I really do not "under-stand" Husserl. . . . The four quotations are respectively from pp. 126, 128, 129 and 130.
[33] (Or perhaps under-stand that spot?)
[34] H. Eklund, T. Erasmie, K. Forslund and G. Hultman, "Dialogpedagogik och samhällsutveckling", *Lärahögskolan I Linköping*, Institutionen för Pedawgogik, June 1976.
[35] By the way, as this story deals with meanings, living in words, – the meeting-point as I expressed it above is more closely expressed in English than in Swedish. We do not have any word like "sense" in the meaning above. If I would have translated it in a very Swedish way, I should have used the world "rationality" instead. The statement would then have been like this: "and go 'zu den Sachen selbst', the meeting-point of sensuality and rationality". The meaning is perhaps the same, but in the latter way of stating matters I leave the phenomenological credo of a sense-bound rationality.

Perhaps the lack of a term in our Swedish language for this close connection between body-senses and rationality illustrates why public opinion on science is as it is in Sweden? Very positivistic, that is. It is interesting to look at concepts as they are expressed in different languages. I can ask you in English: "Is anybody here?" and I can take a "body-count" to see how many "any-bodies" are in front of me. This is the concept of the body rendered in German as "Körper". I could, of course, also ask: "Is any mind here?" but then I would have difficulties in counting how many minds I have in front of me, as I enter the primordial world of information-of-sameness. And for me it is a question, when I change to German, whether I have "Leibs" or "Geists" in front of me. I can however postulate that most of the minds/Geists/Leibs are "phenomenological" in a sense. . . .
[36] B. Tjellander, "Arbetsformer", Inst.f.pedagogik och psykologi, Linköpings Universitet, LiU-PEK-AR-12, 1986.
[37] B. Tjellander, "Stämningsbilder-en intuitiv värdering av elevers arbetsmiljö" Inst.f.pedagogik och psykologi, Linköpings Universitet, LiU-PEK-AR-14, 1986.
[38] Husserl, *Cartesianische Meditationen, op. cit.* (Swedish translation).
[39] See Springmann 1986, Ogden 1987, Rosenfeld 1987, Feinsilver 1987.
[40] D. Meltzer, *The Psycho-analytical Process* (in Swedish) (NoK, 1995).
[41] S. Belin, "Projective identification and a little more . . . – an attempt to describe a specific and common phenomenon in teams working with early disturbed and psychotic patients." *Psykisk Hälsa* no. 2, 1992.

GRZEGORZ GRAFF AND KAROL DZIEDZIUL

THE MATHEMATICAL HORIZON OF THE FUTURE

INTRODUCTION

In his famous masterpiece *Geist der Utopie*, Ernst Bloch broke with old philosophical tradition that found the fundamental truths of human existence in primordial principles and placed the future at the centre of his philosophy.[1] He preferred to base his thought not on what had passed, but on what was coming, what was being awaited and realized on the universal horizon of hope.

This work is an attempt to view modern mathematics from the perspective proposed by Ernst Bloch. Catching the essence of its present and possible future participation in coming changes, anticipating the direction in which reality may turn, predicting the behaviour of individuals and societies, these are our concerns.

We would also like to discuss the problem of prognosis in the social sciences, which have been raised by Alasdair MacIntyre in his book *After Virtue*.[2] MacIntyre questions the ability of these sciences to provide the right prognosis of social phenomena. He also remains very sceptical about mathematics as a device for foreseeing, arguing that even mathematical predictions that concern the discipline's own development almost always fail.

On the opposite side is the literary vision of Isaac Asimov presented in the series called *Foundation*. Unlike MacIntyre, Asimov much overestimates the power of mathematics, describing so-called psychohistory, the science which, on the basis of mathematical equations, as able to predict the future behaviour of mankind within the compass of thousands of years.

Our text is a collection of reflections on modern mathematics treated as an instrument for making predictions and is an attempt to examine its abilities to uncover the paths that lead us to the future.

A-T. Tymieniecka (ed.), Analecta Husserliana LXVI, 157–163.

THINKING ABOUT THE FUTURE

Scientific predictions are based on two foundations: empirical knowledge and general laws. MacIntyre's critique of the efficiency of the social sciences in constructing correct prognoses concentrates on the inability of the social sciences to formulate generalizations. Their general rules function apart from evident counterexamples, they are expressed without quantifiers and the range of the used variables is often indefinite. It seems that most of these objections result from the fact that generalizations in the social sciences are made with words which come from everyday language. The ambiguity, the lack of precision and clarity, typical of verbal generalizations, substantially reduce their predictive usefulness. Without a doubt, it would be more effective to think about the future in a formal language. It is natural to ask whether mathematics is able to face this challenge. First of all, it is important to note that applying the mathematical apparatus requires redefining the examined objects in terms of mathematical relations. This brings us face to face with a serious problem: although we are able to operate using the language of formulas, we cannot actually think in that language. As a result mathematical "thinking" about the future must involve a model. The construction of a mathematical model is connected with simplifying: we are forced to choose only some facts from many-sided, abundant events and phenomena in the sphere of life. The next necessary step is the formalization of these selected facts; only then may the mathematical machinery begin to act. The enterprise of prediction depends not only on the efficiency of the mathematical apparatus, but also on the ability of the investigator to appreciate and isolate those influences that shape the future most powerfully. The best translation of physical time into mathematical terms is an extremely important part of this transfer. The connections between "real" time and its mathematical counterpart reveal the essence of mathematical prognosis, indicating the way in which time processes are understood in the framework of a formal system.

PROGNOSIS AND THE STRUCTURING OF TIME IN MATHEMATICAL MODELS

Bertrand Russell, the eminent representative of logicism, claimed that mathematics can be reduced to logic and consists only of an obvious collection of tautologies. There is no room for time in this approach.

"Pure mathematics says nothing about time", concluded Russell,[3] all connections with reality being situated in the realm of other exact sciences. There are however, opposing conceptions of time which find the sources of mathematical thinking precisely in the phenomenon of time. Brower's intuitionistic logic abandons the axiomatic approach and is based on an aprioristic experience of time. The fundamental construction of mathematics, that is the set of natural numbers, appears to be a result of the human experience of time. Two numbers: one and two, which found the structure of this set are taken, in Brower's opinion, from the intuition of "two-ity": the breaking up of life into two segments of different quality divided by time's passage. In this way time, or more precisely, its perception, is according to Brower the main cause of mathematical inference. The evolution of mathematics, which has followed from a clear and undoubted intuition connected with time, completely contradicts Russell's conception then. The present authors would add this opinion of theirs, that the development of mathematics has been a consequence of human hope in anticipation of the future.

The theory of dynamic systems illustrates very well the special role of time in modern mathematics. Temporal interpretation is so natural there that it is used not only in various models, but also in the terminology of the theory itself. This means that one of variables weighed in strictly theoretical considerations, on the purely abstract level, is called "time". Although there is no need to use this term, it has such rich and stimulating connotations that it helps with the realization of some important relations within the theory, giving researchers additional heuristic help. While talking about dynamic systems it is worth listing two fundamental forms of mathematical modeling which are commonly used in all the sciences, and whose sources are in this dynamic systems theory. Linear models, very popular in many applications due to their simplicity, have the following shape:

The dynamic of changes in the system at the moment t = the present state of the system at the moment t + parameters.

For thousands of years scientists were sure that this simple and elegant model represented the true structure of the universe. Then it was realized that a more complicated nonlinear model was much more valuable:

The dynamic of changes in the system at the moment t = the state of the system from the initial moment till the moment t + parameters.

It is interesting that Husserl's understanding of the continuum of time perception takes the form of a two-dimensional diagram of impression and retention, which resembles the scheme of a nonlinear model.

The theory of a dynamic system has often been an area of evolution leading from mathematical theorem through time interpretation of the entered variable to physical ways of understanding it, all ending up in metaphysical explanation.

The best example of this process, one which strongly supports the apprehension that the lineage of mathematical thinking lies in forecasting, is the history of Poincare's theorem of return. The first, mathematical level, describes the behaviour of the iterations of some kinds of maps. The introduction of the interpretation of one of the variables as time results in the formalization of an intuition about isolated mixed gas which states that after a "long period" the system would return to its previous stage. Finally, the widest metaphysical interpretation of this theorem expresses faith in an immemorial and endless cyclical structure of the universe.

There are the branches of mathematics in which time is not directly present, but in which, however, its influence is distinct. The theory of fuzzy sets examines objects at the moment of their transformation from one structure to another. The apparatus of this theory allows discussion of the state of the objects at key instants, when the type of their structure is just changing. The whole mechanism recalls Zeno's famous observation on the flying arrow, which captured on a fuzzy photo, stays in many places simultaneously. This helps us to simulate fuzzy reality, and consequently we are able to create, in the framework of a model, a situation which is very similar to its real counterpart.

The theory of games made its appearance as a scientific reflection on the conflicts between individuals and groups in societies. The language of this theory does not mention time, nevertheless the main task accomplished in this mathematical field is that of finding the optimal strategy that allows a player to reach the privileged position. This, of course, cannot be done without careful and drawn out prognosis. What is more, the prediction should be constructed on the level of metastrategy: the choice of the appropriate plan depends on the strategy of opponents. The models of game theory involve time in a special discrete form. The strategy takes the form of the ramifying tree of choice. Every move is a reaction and an answer to the move of the opposite side. The main advantage of this approach is the opportunity for global analysis of the system. Sometimes it is even possible to check out all variants of the changing situation, which is a great help in the process of making the right decision.

The future may sometimes be perceived as a single entity. Jan Łukasiewicz was the first scientist who proposed to use three-value logic as a device for describing the future. The third logical value refers to the class of sentences that are neither true nor false, but may become one or the other. The interpretation of the third value in terms of indeterminacy may, however, bring about paradoxes. If a given sentence is non-determined then its negation is non-determined also. The conjunction of these two sentences is also non-determined, which contradicts intuition. Every sentence with its negation yields a false statement. This example, revealing the complicated structure of forseeing, constitutes an argument that thinking about the future as a modality transcends the barriers of common sense, and that we should not impoverish the metalanguage by reverting to classical, two-value logic.

A different approach is required when we want to predict the future of large populations. In that case, mathematical statistics is the most successful instrument. Two major points of difference distinguish statistics from other branches of mathematics. Firstly, the results of statistical methods concern a population as a whole. That means that we have no information about the future behaviour of an individual. Secondly, the time structure of the equations used in statistics is characterised by a special kind of dichotomy. Equations describing a society never really function "at the same time". We always observe a drift, measurable by the duration of the interaction between the variables of an equation. But statistical relations look as if a change in one quantity immediately causes alteration in the others. In fact, there is a delay connected with the propagation of an informational wave in a society. Unfortunately, it is extremely difficult to obtain models that take into account this subtle, but important factor.

THE PREDICTIVE POWER OF MATHEMATICAL MODELS

Nature is mathematical. There is a strong connection between the formal world of mathematical structure and the reality we experience. The ontological priority of formal constructions, their precedence in the sphere of phenomena guarantee the correctness of mathematical prognosis, indicating that its power comes from two sources: empirical data, yes, but also from the heart of reality, from the world of pure mathematical ideas. It is interesting to ask about the foundation that sanctions this fact. The pillar on which the modern mathematics was built is the theory of sets. This base suffers, however, from the lack of an important feature: consistency. The famous result obtained by Kurt Gödel has made us sure that it is impossible to prove the consistency of set theory in its own framework.

The theory of sets may be perceived as a house built on sand, but the apparent weakness of the theory establishes its non-triviality. The essence of mathematical prognosis is hidden in the shadows of uncertainty, which makes Russell's reduction of mathematics to logic impossible, and which assures us that mathematics is able to deliver non-evident knowledge about the future.

It is almost a rule that if an uncovered relation that refers to the future is time dependent, the certainty of prognosis is limited. Absolute conviction may be achieved only in the field of phenomena that are either necessary or impossible. The factor of time always introduces an element of risk that cannot be eliminated.

The epistemological value of time dependent theorems may be very different. The classical understanding of "knowledge" did not allow using that term to describe a collection of information that included a dose of error. Although there is a kind of uncertainty in mathematical prediction, we are able to calculate its measure and evaluate the precision of our forecast.

Statistical methods represent that type of knowledge: probabilistic knowledge, where not only a prognosis, but also additional information about possible dispersion are needed. Statistics resembles the stairs of certainty which we climb, realizing at every step the distance from perfect prediction.

A new horizon of interpretation of mathematical exactness has been created by chaos theory. As we mentioned before, classical, deterministic models were usually built by simplifying reality, throwing away those parts of a system which looked unimportant. The common tendency was to replace nonlinear systems with linear ones. The latter were equipped with a very convenient stability: small changes in initial conditions led to small differences in the final prognosis. The prevalent view that this reduction is the best solution was shaken by the American meteorologist E. Lorenz. While modeling the periodicity of climatic changes, he noticed the so-called butterfly effect. A slight change, even so inconsiderable as that made by butterfly wings may bring about an alternation in the direction of the evolution of a whole system.

Ecology then became the field of competition between two theories. One claimed that populations are stable and must always behave in a periodical and predictable manner. The other argued that thousands of biotopical impulses effect the completely accidental development of a population.

Chaos theory advanced a new way of understanding these and similar processes. Deterministic models, fruitfully applied to these kinds of problems, reveal seemingly completely chaotic structures which in fact contain some "islands of order". We are able to find and describe them. The irregular behaviour involves only a part of a model, one that can be fixed. We are now in a better position than Socrates, because we know exactly the limitation of our ignorance.

The problems of prediction and of attempting to present time in a formal way have been, as we have seen, a very lively stream in modern mathematical thought. The realm of the exact sciences has always been one of looking ahead. The declaration of Ernst Bloch[4] that the linear representation of history is a mistake because history needs a space wider than a line sounds so very true in this context. We deeply believe that precisely mathematics is able to deliver or create that space.

Technical University of Gdansk
Poland

NOTES

[1] E. Bloch, *Geist der Utopie* (München-Leipzig, 1918).
[2] A. MacIntyre, *After Virtue. A Study in Moral Theory* (Notre Dame, 1984).
[3] B. Russell, *My Philosophical Development* (London, 1959).
[4] E. Bloch, *Ernst Bloch. Politische Messungen* (Tübingen, 1965).

HENRYK SZABAŁA

THE INDIVIDUALISM OF TWENTIETH-CENTURY PHENOMENOLOGY AND EXISTENTIALISM

Twentieth-century existentialism has its roots in the philosophy of phenomenology. This is impossible to understand unless we first present the essential features of phenomenological thinking (reasoning). Phenomenology was always oriented towards the systematization of its problems, and despite differences in its formulations it is possible to grasp the *eidos* of this school. It is impossible to do so with existentialism, however. Existentialism was antisystemic; therefore one might speak of different, often difficult to compare, existentialisms. But these different existentialisms eagerly sought their origin in phenomenological reasoning. This was an unreciprocated love because phenomenologists definitely cut themselves off from any connections with existentialism whatsoever. But let us pass by the emotional relations between these philosophical currents and take up the essential features of these positions.

Phenomenology came into being in the twentieth century, precisely in the years 1900–1901 with the simultaneous appearance of three works by authors then unknown to each other: *Logische Untersuchung* by Edmund Husserl, Alexander Pfänder's *Phänomenologie des Wollens* and Max Scheler's *Die transzendentale und die psychologische methode.*[1] But Edmund Husserl provided the decisive brandmark for this trend in its development with his distinct method of approaching philosophical problems and the special language with which he presented the results of his research. His writings influenced in may ways Heidegger, Sartre, Maurice Merleau-Ponty, Jean Wahl and Gabriel Marcel, and in Poland Roman Ingarden and Anna-Teresa Tymieniecka.[2]

Edmund Husserl is often called the Descartes of the twentieth century because much like Descartes himself he declares his spurning of all science and wishes to begin everything all over again. His starting point too is the cognitive subject and more specifically its consciousness. To reach full, unfalsified consciousness he has first to throw away the shells of apparent knowledge. Husserl starts "with the decision to first expel all beliefs binding us, among others, all our sciences".[3] This method which suspends all binding beliefs, places them, so to speak,

A-T. Tymieniecka (ed.), Analecta Husserliana LXVI, 165–175.

in "brackets", a strategy he calls *epoché*. Epoché is refraining from judgment on the subject of the existence or nonexistence of the world in its "conscious reality".

Although I know nothing about the nature of the things appearing in my consciousness, nor about the nature of the perceiving subject itself, the substance of my real cognitions are given me from the source, imminently and undeniably. The task of a human being is to look into these clear phenomena as they appear in our consciousness, because only they are imminently present, only they are here. The task of phenomenology is describing those phenomena as they appear in our consciousness without the exertion of investigating the matter of what they really are. Ontological certainty, in Husserl, is consciously suspended; one should pass by all existential superstitions to gain this certainty through epistemological insight.

This insight, by which we gain cognitive certainty, we obtain by eidetic reduction, because the main aim of describing a given phenomenon is finally discovering in it some universal rule. This universal rule exists in this single phenomenon as its essence, what Husserl calls its eidos. This ability to discern the essence of phenomena we obtain by eidetic intuition. It is the direct experience of the essence, the idea of the phenomenon appearing directly, evidently.

A question arises. Is there a way leading from the world of consciousness to the world of transcendental reality in Husserl's philosophy? Is it possible to experience the object itself and not only the phenomenon of the object? In the opinion of the author of *Ideas* there is no such way, and there is no possibility of such an experience. This is what he writes:

> experience is not some hole in the infinity of consciousness, a hole through which the world already existing before any experience looks; experience is not about accepting into consciousness anything from the outside either. How would I be able to articulate intelligent statements, not noticing in such a way the acceptance itself, to see then consciousness as well as everything else from the outside – so not experiencing it? [. . .] Experience is something in which, for me, the experiencing, the experienced being "is this here" and is this here as this, which it is – with all its contents and the way of being given to it by experience itself through creating achievements by its intention . . . all of this lies intentionally contained in the consciousness itself as this, present and potential intentionlessness, about which structure I can always *ask*.[4]

This position Husserl defined as transcendental idealism, the meaning of which reflects the position that the intentional movement of consciousness not only recognizes objects but also constitutes them. Conscious constitution is the act of creating meanings. In this way the subject, the human being is given the highest prerogatives, because the very existence of the world then has validity only when it gains meaning in transcendental consciousness. Any "objectivity" is reached solely within transcendental consciousness, because only it is the source of absolute certainty. Husserl, who throughout his life looked for complete certainty in his philosophy, finds it only in himself, in his consciousness, because only it is a necessary being and the source of certainty of other beings. Only our own consciousness is given in an absolute way.

Husserl's philosophy saturated with the gnoseological aim of finding certainty is in fact an anthropology in which a human being finds himself again and again. Husserl, starting from subjectivity's attempt to reach a common world, cannot step beyond the subject. The world of consciousness becomes the only world, an autotelic one in which the quest for a way to the real world is given up. However, man, by giving up a subjectivity involved with natural commonplaces, an alienated abridged subjectivity, now reaches his base. He is no longer defined by anything outside him and arrives at an irreducible absolute, *Ego cogito cogitata mea* – pure thinking, the self thinking. So, according to Husserl only man has ontological legitimacy, only he is the truth.

* * *

Max Scheler was the first philosopher who applied the phenomenological method to various philosophical problems, especially to ethics. Scheler's thoughts about universal man are especially interesting, and they became a source of permanent inspiration for different shades of existentialism. The anthropological problem – who and what is man, what is his place in existence as a whole – is the main problem in Max Scheler's philosophy.

Scheler uses the phenomenological method to grasp the meaning of man, or to use an expression from phenomenological language, he seeks to find the eidos of humanity. To do that, he looks for constitutive features, traits characteristic only to man and not transferable to any other being. Bringing out these features will make it possible for him to constitute the idea of man.

Scheler, using phenomenological analysis shows, man in two alternate orders, i.e., the natural order and the axiological order. In the natural order man belongs to the animal world; in the axiological order man exists as a person, stepping beyond the animal world and irreducible to it.

Man, in Scheler's opinion, is the least successful of nature's creatures, is as a being basically a *faux pas* of evolution. Man in nature becomes a living fugitive from life.[5] This is so because of all beings, man has the worst capability for adapting to nature itself. In this line of thought, man is a chronic invalid as nature's creature. But Scheler sees a positive element as constituting the foundations of humanity. This is because man, who is a "complete dilettante" in nature, has in contrast to animals, no specialization, and has compensated for this lack by making tools and complete systems of signs making it possible for him to survive in different conditions and situations. Not having instinct, man compensates with intellect which is a substratum of this instinct. Thus, an organic dilettantism has saved him from becoming a "specialist" like an animal. Thanks to a lack of specific organs and functions adapted to concrete surroundings, man was given a certain freedom of adjustment.

By developing intellectual activity and making all sorts of tools, man gained some free time which he does not have to fill meeting the demands of conditions or situations. Disengaged from natural surroundings, man became an "open to the world" being. All this constitutes preparedness for something, the creation of the possibility "for something which can happen only to man".[6] Contrary to an animal, man is transcendency, is a becoming.

"The becoming of man", says Scheler, "is rising by force of spirit towards opening to the world".[7] He goes on to say,

> This becoming conscious of himself, this new insight and the centering of human existence enabled by spirit, points to the second creative feature of man. Thanks to the spirit the creature which we call "man" may not only widen his surroundings into the dimension of the world's being but may also – and this is most original – objectify his own physiological features and every single psychological experience, each of his vital functions. Only because of this can this being freely dispose of its own life.[8]

Man steps beyond and above life due to the fact that in him the spiritual factor has defeated the biological factor. Man is a being who is constantly on the road, he is constantly moving towards something. Thus, says Scheler in *Zur Idee des Menschen*, the question about the meaning of man is undecidable. Man is an unknown being. This undefinability of man belongs to his meaning.

A human being, says Scheler, cannot be treated as a thing or substance, but reveals himself in actions, is a directly co-experienced unity of experiencing. The human being is a being who constantly becomes and creates.

Each person is an individual whose "specific character" penetrates different actions. He does not owe it to the substance of the experience or the character of the object field at which his actions are directed. There is no connection with the body. The "individuality" of a person comes from the person as such. As an organic-psychological creature man is not yet an Individual. He becomes one only thanks to a personified form of being that prevails.[9]

Because of his nature each person has his own unique axiological character. A person is always placed on the ground of values. There exist two spheres in a person: intimate and social, but they are two sides of an indivisible unity. The intimate sphere is hidden deep in the self. It is, as Scheler said, the sphere of loneliness. Only the chosen sometimes have access to it. The social sphere is that of the actions which find their realization in the commonwealth. These are actions which demand the existence of other people. Neither of these spheres of the human person has priority with Scheler. Of course between these spheres there may be harmony or disharmony, which reflects on the values of a given person. Each person is entitled to hold moral values.

Man as a personal being is unique, the only being in the universe who is individual, who has his own freedom in the name of which he can declare disobedience to the whole world. Man can, according to Scheler, be a creature who says NO to the outside world as well as to his own blind instincts for the good of spiritual and absolute values. Scheler writes:

Man is a living creature, who in relation to his life which excites him, may by force of his spirit behave *ascetically*, stopping and silencing instinctive impulses, i.e., denying them their nourishment in the form of perceptions and performances. In comparison with an animal, which always confirms real being with (its) "yes", even when it is repulsed by it and runs away from it, man is "*the one, who can say no*" (*Neinsangenkonner*), the "*life ascetic*", a constant protestant against any naked reality.[10]

* * *

Between phenomenologists and existentialists there are close relations. It is hard not to notice the inspirational influence of Husserl's and Scheler's philosophy on Heidegger's, Sartre's or Jaspers' philosophy. But despite existentialists' liking for phenomenology, phenomenologists decidedly cut themselves off from existentialists for there existed basic differences between them in the understanding of the world and of man as such. The cognitive attitude itself of each movement stood out in contrast. No wonder then that Husserl, in one of his letters to Ingarden described Heidegger as his "antipode". Phenomenologists believed in the objective experiencing of the truth, believing that the world which we are reaching in the form of phenomena has its constant visage and values that really exist independently.

We, according to phenomenologists, are specifically "Columbuses" discovering new lands of cognition and the horizon of human values exists no matter what the quality of eyesight. In phenomenology persons have a mission to fulfill, to discover their own axiological essence.

Existentialists are basically subjective, discarding all formulas of objective truths. Saying that only man himself is the truth, they break off all links with their teachers. Man in existentialism ceases to be "the great discoverer" of truths laying beyond him, and begins to be their creator. Any acknowledgement of the existence of "objectively acknowledged values" is held in existentialism to be just one more form of human captivity, a factor of man's false reification. In existentialism there is no essence of man, no mission to fulfill. Man creates his own essence, imposes on himself his own mission in the world.

In phenomenology man realizes his freedom through the prism of the values which constitute him, but existentialists do not know the boundaries of their freedom and even if they see some boundaries they are not consistent enough to then question their professed rule that man is freedom and the horizon of his freedom depends solely on him. Phenomenology is a philosophy of optimism, of faith in the cognitive capabilities of man. It holds a beautiful hope that man has a sense of value and goodness. Scheler said that "only the happy do good". Existentialism is a philosophy of anxiety, worry and sickness over existence, of storms of despair; it is a never-ending labour of seeking authentic existence. In phenomenology others are the aim of our personal acts, but in existentialism Sartre's declaration, "Hell is other people", comes up again and again.

I think that despite these basic differences there exist between existentialism and phenomenology many similarities. Existentialism and phenomenology are linked together by a common philosophical belief that the starting point of cognition, evaluation and even the whole of being is always man and his thinking. Human activity, human freedom, human responsibility, human actions, and finally the human world, all of these are created through the prism of a real man. Philosophy becomes anthropology here in all its dimensions. It is impossible to discuss all the trends of that philosophical movement we call existentialism.

In the simplest way, one might say that existentialism is the philosophy of human existence. The basic question is: what is a man like and what is the sense of his existence. Those and other problems are fundamental in Martin Heidegger's philosophy. Because of the brevity of this paper I shall confine my deliberations to him, making here essential shortcuts.

Heidegger postulates, on the basis of Husserl's phenomenological method, the creation of a "fundamental ontology". The basic question of this fundamental ontology is "what is a being when it is?"[11] Therefore the task of ontology would lie in considering different possible meanings of being. The question, *What does it mean "to be"?*, is basic for our knowledge. It is the starting point for all knowledge; it is a cardinal dilemma of our existence. For a human being may have different degrees of self-identity, higher or lower, authentic or hypocritical, but he always has some possibilities of being, which he deliberately chooses or involuntarily faces.

Man whether he wants to or not must have some attitude towards this being because here is the matter of his life. Heidegger says "understanding of being is in itself a description of being".[12] By the term "being" Heidegger means the way of behavior of a human being.

The essence of being lies in existence. Existence is a way of being relying on reflection, on turning towards oneself; it is reference to one's own "is". And what is important about this reflection on self is that it is not entered into of our own choice, because our existence itself was not our own choice, it has been given to us. The man here called "being" decides about his existence.

"The question of existence is the ontological matter of being".[13] To speak more clearly, man in choosing this or that form of life gives the answer to the question, "What does it mean to be?" not with his mind but with life itself. In doing so, he more often than not does not even realize it. This is because man very often forgets himself. His attention is drawn from himself to other things, other people. He forgets that he is the basis of the sense of his whole world.

Being is here as it were the background or horizon against which what is appears. Thus the understanding itself of being happens through that which appears in the background, through other beings, the self included. Being or *Dasein* may be comprehensive only through the prism of relations towards the self, towards other people and towards things, at the same time understanding the being of self, others and things.

The special character of existing being is that it constantly steps beyond its current factual being towards potential being, the forthcoming. Heidegger calls this state of stepping beyond being transcendency. *Transcendere* literally means to step beyond, to step towards. Thus, transcendency means going over, stepping over and transcendental describes that into which the stepping over is made, that which is to become accessible and graspable and therefore requires his stepping over. It is man's nature to transcend. Otherwise he would not be man. He does not appear as somehow existing primarily in this world, only occasionally performing the act of stepping over. That is not so, because his existence is for man the source of stepping over. Transcendency is the basic shape of being, on the basis of which one can behave towards being.

Man living in the world is also a builder of this world. The constitution of man's being creates his actions. Man, in action discovers the world and himself. This is why Heidegger often repeats that man is governed by discovery, given to him, although it is never forced. "Man", writes Heidegger, "becomes free in as much as he belongs to the area of participation, in this way becoming one who hears and not one subjugated who listens".[14]

The constitutive condition of the action which discovers, in other words truth, is freedom. Man does not choose freedom as he does not choose existence, he is thrown into it. In the nature of being there exists a specific tendency to escape from the self, to throw oneself into the whirlpool of the world, to melt into apparent action. It is an attempt to identify oneself with the world, with other people. "The term 'others'", says Heidegger, "does not mean all the rest except me, from among which 'I' comes up; others are rather those from whom I am not different and among whom I am".[15] The world of others is in fact a world of impersonal forces, one in which our own "I's" become alike. Those others in common experience are not some particular persons, individuals, but only representatives of a nameless power of everyday world, an anonymous force, which tells us to act in a certain way. All obey this force and those that oppose this force do not gain by it.

If our being cares for our welfare, that is for our food and clothing or care when ill, we must obey those institutions on which the "constitution of the individual being as a co-being" is based.[16] Therefore, if we want to live well, we must obey some nameless norms of conduct hidden in the impersonal reflexive pronoun *self*. At that moment our real being, assenting to common being falls under the rule of others, specifically that impersonal *self*.

A man who submits to this global rule ceases to belong to himself, his individuality ceases to be real, because it exists as a *Man-selbst*, as an ordinary mediocrity. Total mediocrity is the self's way of being, and what we know under the term "public opinion". Public opinion rules the world, it decides the level of being. It is insensitive to all individuality and the uniqueness of the human being. People readily obey common rule over the Self, because they then rid themselves of a sense of responsibility. This means that this Self eases the commonness of being. Because, although everyone is different, none is himself. Man stops belonging to himself, becomes a passive object, lets himself off from organizing a life for himself, and loses his own axiological ground. In this way unreal existence is born.

Unreal existence is man's escape from the truth about himself. Only an understanding of the nothingness of the world, an understanding of the individual's passing in time, makes it possible for the individual to break with the structure of the global Self and to direct his efforts towards himself. This state, which is a result of the awareness of unavoidable death, Heidegger calls the state of fear. Fear makes us aware of the fact that everyday actions in fact mask our real being. When this fear appears, we begin to understand that we have been thrown into a strange world of things and other people, we begin to understand our own loneliness. Everything then becomes fragile and uncertain in its being; only our death and our loneliness are certain. Our being becomes being unto death. The omnipresent Self wants to hide this truth. Because even death is entangled in the web of dependencies of the Self. It is said that "one dies", meaning that it is always someone else that dies, never I. Really death was never present in my being. Telling people that they should live as if they were to die tomorrow, prompts them to really live, in their own name, with real courage.

Heidegger recalls the words of the post Friedrich Hölderlin: "Man lives poetically on this earth" ("Dichterich wohnet der Mensch auf dieser Erde"). Because only man, mortal man knows that he lives. "The ability to understand life is paradoxically the fruit of man's mortality. To understand must be as an entity which I shape in my freely projected acts, I must know where I am heading".[17] And I know I am heading for death.

Only man knows that he is nothingness and that everything ends at the time of his death. Therefore he knows that he is food for himself, that he is the one that gives sense to things. He himself and only by himself begins to understand the miracle of existence, only he by his creative word lights up the being of things on the horizon of his being. Only in the face of fear does real existence awaken, existence which does not submit to the pressure of the Self. The person then knows that there is nothing to be afraid of. Real existence has a sense of freedom and does not think in imposed categories, but in categories flowing from within. The person then knows that he is expressing himself and only himself.

But Heidegger notes that because of fear the freedom given to man is fragile and not durable, it has to be defended every day. It is not given definitively, but is set before man. Man must fight a battle for his freedom every day and always in his own name because his person is marked by all the choices he makes and those which he wants to make.

Summing up Heidegger's thought, one might risk the statement that his philosophy treats with the possibilities of man. It presses for the realisation of our own potential values, which lie within us. The world is potentially dangerous and alienating. A real human being is lonely, lives in his own name, and in his own name gives sense to other beings. The human individual lights up the horizon of the world. Only he passes through being, giving sense to the future as he turns towards death.

University of Gdansk

NOTES

1 See R. Ingarden, *Wstęp do fenomenologii Husserla* (Warsaw, 1974), p. 7.

2 W. Mackiewicz, *Filozofia współczesna w zarysie* (Warsaw, 1991), p. 203.

3 E. Husserl, *Medytacje Kartezjańskie* (Warsaw, 1982), p. 9.

4 Husserliana, v. XVII (The Hague, 1974); quote from: "From Husserl to Levinas", *Wybór tekstów z ontologii fenomenologicznej*, ed. W. Stróżewski (Kraków, 1987), pp. 182–183.

5 M. Scheler, *Mensch und Geschichte* (Zurich, 1929), p. 40.

6 A. Wegrzecki and A. Czerniak, "Wstęp" in M. Scheler, *Pisma z antropologii . . .*, p. xi.

7 M. Scheler, "Stanowisko człowieka w kosmosie", in *Pisma z antropologii . . . op. cit.*, p. 85.

8 *Ibidem*, pp. 86–97.

9 A. Wegrzecki, *Scheler* (Warsaw, 1975), p. 84.

[10] M. Scheler, "Stanowisko człowieka . . .", *op. cit.*, p. 103.
[11] M. Heidegger, *Bycie i czas* (Warsaw, 1994), p. 18.
[12] *Ibidem.*
[13] *Ibidem.*
[14] M. Heidegger, "Pytanie o technikę" in *Budować, mieszkać, myśleć* (Warsaw: Czytelnik, 1977), p. 243.
[15] M. Heidegger, *Bycie i czas*, *op. cit.*, p. 169.
[16] *Ibidem*, p. 173.
[17] T. Langan, "Filozofia niemiecka", in *Historia filozofii wspóczesnej*, p. 153.

WŁODZIMIERZ PAWLISZYN

IS PHENOMENOLOGY AS A SCIENCE POSSIBLE?

Reading Heidegger's Viewpoint

" 'Phenomenology' means: ἀποφαίνεσθαι τὰ φαινόμενα – to let that which shows itself be seen from itself in the very way in which it shows itself from itself. [. . .] 'Phenomenology' neither designates the object of its researches, nor characterizes the subject-matter thus comprised. The word merely informs us of the 'how' with which what is to be treated in this science gets exhibited and handled. To have a science 'of' phenomena means to grasp its objects in such a way that everything about them which is up for discussion must be treated by exhibiting it directly and demonstrating it directly."[1]

What is phenomenology? Heidegger declares that it is the proper way of performing the act of looking at objects, phenomena. Proper means to be directly faced with them. Phenomenology is the proper method of philosophy. Philosophy is ontology, universal ontology, "real" ontology – fundamental ontology which is the foundation of every theoretical consideration, of every investigation. It is phenomenological ontology because it is based on phenomena.[2] To philosophise is nothing other than just taking the trouble to understand the meaning of everything which *is*.

Therefore this philosophy is the phenomenological science of Being.

"Being, as the basic theme of philosophy, is no class or genus of entities; yet it pertains to every entity. [. . .] Being and the structure of Being lie beyond every entity and every possible character which an entity may possess. Being is the transcendent pure and simple. [. . .] Every disclosure of Being as the transcendent is transcendental knowledge. Phenomenological truth (the disclosedness of Being) is veritas transcendentalis."[3]

But as such – as phenomenological, essential inquiry into the meaning of Being – philosophy grows from a Dasein hermeneutic. In this hermeneutic is set forth human existence or being in spite of the question of the meaning of Being. Inquiry into human essence, becoming, is, according to Heidegger, the necessary introduction to the elucidation of the question of Being. Therefore authentic human being recognized in the proper way is the essential meaning of philosophy.

A-T. Tymieniecka (ed.), Analecta Husserliana LXVI, 177–185.

"In the phenomenological conception of 'phenomenon' what one has in mind as that which shows itself is the Being of entities, its meaning, its modifications and derivatives. [. . .] Least of all can the Being of entities ever be anything such that 'behind it' stands something else 'which does not appear'."[4] However every single entity as well as the sum of entities are manifestations of Being – Being announces itself in every single entity and in the sum of entities. Heidegger distinguishes as human being Dasein, the entity that is ontically sensitive to Being while ontologically enabled to understand his ontic situation.[5] "The phenomenology of Dasein is a hermeneutic in the primordial signification of this word where it designates this business of interpreting."[6] It turns out that the understanding of Dasein's Being, in other words, the notional description of the existential situation of Dasein, leads to the understanding of the meaning of Being in general.

As we know, *Being and Time*[7] realizes in fact an analytics of Dasein, is an inquiry of the dimensions of human being in consideration of the meaning of Being in general; the analytics of Dasein ought to be an interpretation of the question of Being in general.

But we ask: are these investigations carried out in a phenomenological way? Are they in general possible within phenomenology? Is the way of proper philosophy (fundamental ontology) possible only in the phenomenological approach? Heidegger suggested the meanings of these terms.

* * *

Let us imagine; we are walking by the shiny shop window of a florist. We have just noticed, suppose, roses, beautiful and fresh, at the very moment that they are opening. "Let us buy," we decide. We go in and order; we take them into our hands, and . . . "My dear, they look lifelike" – to our amazement.

Here something which is a piece of workmanship, which was produced by a man, appears to be what it is *not*: namely, it presents itself *as* a work of nature, it looks *as* if it were a work of nature. We have noticed fresh, real flowers – they are revealed at present. But in a while we already know that these are not fresh flowers – here is a semblance. An artificial flower has imitated a natural, living flower – which of course it is *not*. This artificial flower *looks like* a natural flower – here is *a semblance*! When an artificial flower as an object in human hands has been recognized *as such* – here is *a phenomenon*! When we unmask the semblance – when we strip off a mask – the semblance is just phenomenally the phenomenon. We have imagined the phenomenal situation and the seeming one.

Now, let us imagine that something is manifesting: we are surrounded by the dark, complete black except for two lights. Let us suppose we can see it exactly! Here we describe the phenomenon: two lights close by against a background of black space. But when we come to know that the lights are on a running express train, this phenomenon described by us becomes *a manifestation* of the running train – two lights are recognized as a train, the lights are indicative of the express. The lights in the darkness are the phenomenon; they appear as such in themselves. But our train is manifest as the phenomenon of two lights in the darkness. The train itself is not manifesting itself; thus it is not the phenomenon (that-which-shows-itself-from-itself). Two lights in the darkness are, on the one hand, the phenomenon under consideration of themselves; on the other, considering the train, they are a signal of something which does not show itself from itself, but only in a manifestation. While the train, according to our imaginary situation, does not show itself, but – Heidegger declares – here is manifesting. . . .

But here it is possible to imagine a situation occurring further on when after the phenomena of lights in the darkness, and the otherwise fundamental supposition that this train is an express, it turns out that our guess is pointless – that it was not an express train, but, let us say, a freight train. . . . The phenomenon of lights in the dark it is then agreed is a manifestation of something that does not show itself, albeit that phenomenon of lights in the dark is manifesting an express train. But after a while, let us suppose that just phenomenally, it turns out that it has not been an express, but the freight train. Now, what is the phenomenon of two lights in the darkness when considered as manifesting something hidden behind them?

A thing hidden behind a phenomenon is not a semblance, because the phenomenon of semblance occurs when an entity appears to be that which it is not. Our phenomenon, two lights in the darkness, does not show in general either an entity hidden behind it (the freight train), or "a thing" supposed in the first moment (an express train). Our phenomenon was only a signal that announced something, either an entity manifested in fact, or something supposed at first by mistake.

But our phenomenon (lights in the darkness at the first moment seen, and the consideration of something in fact hidden behind them) is also not a manifestation. The phenomenon of manifestation occurs when something does not show itself, but announces, augurs itself. However, at first our phenomenon is taken as a signal of something, which is not there in fact; so, it does not announce and manifest itself – because it is not there. After all the manifesting of a phenomenon consists in its being hidden, when something does not show itself, but in fact is there. . . . On the other hand, something's being there at the first moment is not inherent to our phenomenon. And if from this our phenomenon from the first moment something inherent *in it* is absolutely missing, it cannot be inherent to announcing itself (it does not show itself). Thus, on the one hand, our phenomenon has been taken as a signal of something at first, but this "something" in that situation is not there (our supposed express). On the other hand, there is something in fact (the freight train) which does not signal itself at first in the phenomenon that has been inquired into by us.

So, the phenomenon presented here is neither semblance, nor manifestation, according to the Heideggerian meanings of these terms.

It is obvious that it is also not a phenomenon in the primary meaning of the world φαινόμενου (that-which-shows-itself-from-itself).

Therefore, what is the phenomenon we have imagined: the phenomenon of two lights close by each other against a background of black space considering that what is hidden behind them at that moment is not what we supposed by mistake? Is this a false manifestation? Or perhaps an apparent manifestation? Let us call it – we assume that Heidegger is in agreement with our decision – an illusion. And just at that moment (phenomenally) when it becomes clear that our first supposition is false (that in fact here is a freight train), we acquire the knowledge that our previous supposition (about the express) is an illusion. Of course we entertained an illusions at the time we were mistaken. "The thing" of our illusions could become an illusion, a phantom. . . .

* * *

Heidegger speaks of his ontology. Here we keep in mind the programmatic declarations of the author of *Being and Time*, about a sphere which would only be shaped. It would not be a new standpoint, nor an ontological trend; it would not be in general any ontology to be set *beside* the previously existing ontology, but an ontology which has to stand *before* any others – first ontology, fundamental ontology.[8]

"The object," which is the subject of fundamental ontology, is (ought to be) the Being of entities, or the meaning of Being in general. The question about the meaning of Being is the fundamental question of the whole of philosophy.[9] We must undertake this question in a phenomenological way.[10]

* * *

The word "phenomenology" consists of two parts: "phenomenon" and "logos."[11]

What does "logos" mean? According to Heidegger, the meaning of the word "logos" has been shaped over the progress of history, which obscures its proper meaning.[12] The fundamental meaning of "logos" is "speech," and it means: to make a vision possible, to reveal what is spoken of.[13] As such, "logos" could be true or false. At the same time, truthfulness is described by Heidegger using the notion of *aletheia* – from ancient Greek – which means unconcealment.

In its purest and most original meaning the "true" is something which for the most part does show itself and could never obscure itself. Simply, sensory perception is always true – every time it is not exposing.[14] In this way we disclose by the simplest perception an entity as an entity which *is*. This perception is also a vision and here speech is always true – the purest possibility of the vision (entity as Being). But "logos" as "speech" could also appear in a plurality of references to entities – entities are doomed to the plurality of equal aspects of a vision.[15] In this meaning of the Greek *aletheia*, unconcealment is at the same time the condition of concealment; the truth is the foundation of falsehood.[16]

According to Heidegger, the ontological falsehood, concealment is every concealment connected with an actual un-concealment; therefore it is the potential exploration of unknown aspects of entities. Thus, in every falsehood is potentially a truth: something which is concealed could be unconcealed. And also: every unconcealment was or will be concealment: the truth *grows itself.*

If we understand the truth as *aletheia* as opposed to the classical comprehension of truth and falsehood (from Aristotle), every "thing" is truth and falsehood as well: it shows itself (as a truth) and at the same time makes impossible the truth of other aspects of itself. These aspects are hidden and in a sense false. In the ancient attitude of mind truth and falsehood (un-concealment and concealment) are of equal worth.

So, according to Heidegger, speech can be considered as making possible a vision of one aspect of a particular entity, as well as hiding other aspects of that entity. Therefore speech in this form involves at once truth and falsehood.

However, as Heidegger says, it is only because of these ancient understandings of the truth (such as aletheia) that we can understand in general the possibility of a science established on the possibility of a pure look – "real" philosophy.[17]

Let us again thoroughly study, as suggested by Heidegger, the meaning of "phenomenon." We could see if in an excess of possible meanings (for "logos" and "phenomenon") has been hidden "phenomenology's" constitutive impossibility. . . .

* * *

So, what is – could be – a "phenomenon"? According to Heidegger, the word "phenomenon" is from the Greek: that-which-shows-itself-from-itself. That which shows itself could do so in two ways: firstly as that which it is – "a thing" as an entire and unique "thing" – primary and positively,[18] secondly, as that which shows itself as what it is *not*: a semblance of something, an ontological untruth about something.[19] A semblance is not a phenomenon. Therefore every "thing" could have its unconcealment, truth, but also its semblance, falsehood, concealment.

But the "thing" can not only show itself – truthfully or untruthfully – but can also merely announce itself, which means to become manifest, but not to manifest itself – only to be manifested by something that could appear. Here the "thing" itself is not given, but only its announcement, its signal.[20]

All of these cases assume the "thing" in itself.[21]

We can notice that at any particular moment it is not given whether we have encountered the very thing, or its semblance, or the announcement of this thing. And that means that in any of these cases it is not given in an act whether we have encountered the very thing, its semblance, or its announcement.

If our reasoning is correct, we have to say further that in effect the "phenomenality" of phenomenon is not given by the phenomenon! In other words, it is not given (phenomenality of attainment) whether we have just encountered the phenomenon or not. As soon as the phenomenality ("a piece of information" that this is a phenomenon) itself is approachable, only in a manner other than through the phenomenon, could phenomenology as a philosophical programme or a philosophical method (the "how" of thinking) have any success?

If we are right that the phenomenality of a given phenomenon (distinguishing whether it is semblance or a manifestation) is not given in an act, and that a phenomenon thus does not notify us of (its) phenomenality, then phenomenology's task is never done. Therefore phenomenology is impossible!

* * *

It was stated by Heidegger that there are many alternative ways of seeing an entity or the kinds of manifestation manifesting things. As long as there is the main difficulty that not every kind of manifestation (not every type of "phenomenon") is phenomenon proper, we cannot know that. That is because it seems that this is a matter connected with the world of meanings, not with the world of things, one that is not situated on the side of a reality, but on our side. This means that even should things somehow appear, they manifest themselves beyond the distinction between phenomenon and non-phenomenon; if they simply manifest, that means they always manifest properly. The attempt to differentiate between phenomenon and non-phenomenon (between what Heidegger called semblance, manifestation, announcement, manifestation only) is not an ontological, but our epistemological concern.

In this *excess of meaning possibilities* for a "phenomenon" (there are a lot of kinds of manifestation, and only one of them is a phenomenon proper, and this phenomenon does not inform us about its property, whether it be a phenomenon), there is contained the conclusion that "phenomenology" is impossible.

And in the excess of possible meanings for "logos" as well (speech logos) as the medium of everything has the power to not only uncover an entity, but also to cover it, a situation revealed also in the truth-falsehood dimension, truth being *aletheia* (and we cannot arbitrate whether a particular act of speech covers or uncovers the truth), there is contained the conclusion that "phenomenology" is impossible.

In excess of possible meanings for "phenomenon" and "logos" both, there is contained the constitutive impossibility of "phenomenology."

As soon as phenomenology has to be the methology and guarantee for fundamental ontology, then fundamental ontology becomes impossible!

Also when Heidegger accepts a "concrete thing" as a phenomenon (given the Being of entities in general), that does not annul this impossibility. According to Heidegger, Being can "only manifest itself." From where then would the phenomenality of Being arise?! When Heidegger says that it arises from anxiety, that description, if anything, increases and does not smooth away the difficulties.

Therefore, is it not the case that here phenomenality is *reasoned out*? The phenomenon: Being itself is actually reasoned out! Where is the carnality, or substance, of this phenomenology? But phenomenology cannot just be reasoned out.

Is phenomenology possible? And how? It seems that it has to be either that everything is given as phenomenon, or that phenomenality is not given. . . .[22]

* * *

We have been reading and entering into the spirit of Heidegger's project of phenomenology, into the "how" of that "real" philosophy asking for the meaning of Being – fundamental ontology. We have been considering Heidegger's suggested meanings for the terms employed by him in *Being and Time*.

Here is the result of our investigations: doing phenomenology – the "real," proper way of "real," proper philosophy – using the appropriate phenomenological method, is impossible! The phenomenality of phenomena is not arrived at in the phenomenological way. Here phenomenality is worked out by a reckoning, constructed by a reason. . . .

So, what is, or could be, "phenomenology"?!

Institute of Philosophy and Sociology
University of Gdansk

NOTES

[1] M. Heidegger, *Sein und Zeit* (Tübingen: Max Niemeyer Verlag, 1993), pp. 34–35. (M. Heidegger, *Being and Time* [Oxford: Blackwell, 1967] §7.)
[2] See *ibid.*, p. 38, and p. 35.
[3] *Ibid.*, p. 38.
[4] *Ibid.*, pp. 35–36.
[5] See *ibid.*, pp. 15–19.
[6] *Ibid.*, p. 37.
[7] See *ibid.*, pp. 39–40.
[8] See *ibid.*, p. 27.
[9] *Ibidem.*
[10] *Ibidem.*
[11] *Ibid.*, p. 28.
[12] *Ibid.*, p. 32.
[13] *Ibidem.*
[14] *Ibid.*, p. 33.
[15] See *ibid.*, pp. 33–34.
[16] About the question of truth, see *Sein und Zeit*, *ibid.*, pp. 212–230.
[17] *Ibid.*, p. 34.
[18] See *ibid.*, pp. 28–31.
[19] See *ibidem.*
[20] *Ibid.*, pp. 29–31.
[21] Heidegger declares that the question of the world's existence is beyond all dispute.
[22] Perhaps these phenomenological impossibilities are the best medium in which to interpret the turn in Heidegger's thought: to think about Being, but now trusting inspiration – if we cannot track down Being (by any method) – we should give in order to be gifted by Being, we should trust the mercy of Being. . . .

JAROMIR BREJDAK

SELF-INTERPRETATION OF TIME AS A RULE OF INDIVIDUATION IN SCHELER'S, DILTHEY'S AND HEIDEGGER'S CONCEPTS OF MAN

Das Historische ist immer nur ein Rohmaterial, das erst der, welcher es sich zueignet, aufzulösen weiß in ein posse um es sich als esse zu assimilieren (Kierkegaard, *Gesammelte Werke* IV 405).

INTRODUCTION

At the beginning of this century phenomenology underwent a change under the influence of the philosophy of life. The reception on Bergson's or Dilthey's works, whether by Husserl, Scheler, Heidegger or Ingarden, led to a change in the understanding of apriority, which became more and more generative of the grammar of history. The consequence of this change was a new way of understanding the essence of man. The paradigm of substance, still present in Husserl's concept of consciousness, was superseded by the paradigm of time,[1] changing views on the process of individuation. Whereas both Dilthey and Heidegger would search for the invariableness of substance rather in the actuality of man and his relationship with the world, Dilthey would find it in the existing permanence of the semantic context (the category of power and meaning), and Heidegger, in the permanence of executing the same ways of being, in particular being towards death. Scheler, on the other hand, – in accordance with the phenomenological principle that *intentio* and *intentum* have an identical source (*Gleichursprünglichkeit*) – would highlight the relationship between the world-blind drive to live and the powerless spirit, i.e., the complex multifariousness of the existence of a human being as a person.

This work deals with three ways of individuation which derive from the struggle to solve the problem of life in its historical dimension. Each of these three approaches – Dilthey's, Heidegger's, and Scheler's – represents a different view on the same problem. The purpose of this essay is to point out a certain convergence between them and to synthesise them in the manner the title of this paper indicates.

A-T. Tymieniecka (ed.), Analecta Husserliana LXVI, 187–200.

Let us begin with a short outline of the assumptions of these three philosophical constructs. In each of them the cosmological principle is always *history* in the form of life in Dilthey, in the form of being in Heidegger, and in the form of suprasubjective spirit in Scheler. This cosmological principle is expressed in the shape of a person in Dilthey, being in Heidegger, and a single and collective person in Scheler. In each case man, whether as a person or as a being, is the potentiality of history, living in a world of possibilities already created and being created by history. Since the main character of history is man, the understanding of history has to mean understanding past existences by living them through reconstructively, simultaneously experiencing oneself as time. This understanding is not an intellectual process; it is an existential one that enables individuation and existential growth.

1. A MULTITUDE OF PRINCIPLES: TIME AND THE MEANING OF A PERSON

1.1. *Scheler's Individuation through the Person*

The unique position of man in the cosmos follows from his presence in concrete (material) axiological spaces, and from his relationship with different values and their mutual relations. This relationship is a secondary category; primary is the constitution of man by given axiological spaces. Here phenomenology goes beyond the transcendental approach, revealing that superior acts – starting with the feeling of vitality – are not man-made productions but cosmic tendencies with which man, like a Stoic, can agree or not. This unique dynamic of the universe in the human being is called a person by Scheler. Persons are executors of the history which happens because of them. A person is at the same time a fundamental exemplification of human individuality. Individuation happens in a general way through religion, culture, nationality (the *Gesamtperson*), and through a specification coexisting with those influences (the *Einzelnperson*).

Let us have a look at the consecutive aspects of human reality with its stages of individuation, slight at the beginning and later becoming stronger.

In Max Scheler's phenomenological perspective the reality of man presents four aspects: *a physical body* (*Leibkörper*), *a body* (*Leibseele*), and *I* (*Ich-Leib*) and *a person* (*die Person*).

The feelings of the sensory sphere, which can be localised, correspond to the physical body; they are punctual (through their organs) and momentary; they are only indirectly related to the *I*, but there is no relationship between them and the person. They cannot be separated from the content attributed to them; they are a present factual state, not a function or an act. That is why recalling cannot restore the same sensory feeling again, for example, a sensual pleasure such as itching or pain. Related to sensory feelings are such values as pleasure and usefulness.

Feelings of vitality (*Lebensgefühle*) correspond to the body. They are not attributed to particular organs, but to the whole organism. They are homogeneous, not complex, in character (as in the case of sensory feelings) and demonstrate a certain degree of intentionality. Through them one can directly get to know the feelings of vitality of other creatures (animals and birds), which constitutes thus a basis for a community (*Gemeinschaft*). These feelings have an axiological meaning while revealing a vital sense inaccessible to discursive thinking, for example, fear or uneasiness preceding a danger (or good/bad frame of mind). These feelings correspond to correlative values such as nobleness and good luck.

The *I* includes psychic feelings, not localised within the body (examples: cheerfulness, sadness). They may be intentional in character, thanks to which they can constitute a basis for more radical community. They are directly linked to the ego, the human psyche. These feelings disappear from the stream of attention. Beauty/ugliness or truth/falsehood are values of psychic feeling.

What is characteristic of a man described at the level of his body is his relative self-seclusion, which is typical of living creatures in general, a certain egocentrism of life (*Selbstbezogenheit*) expressed through the individual's "imprisonment" in vital functions. That is why the surroundings of man as a psycho-physical individual are called an environment (*die Umwelt*) which he cannot leave. This state has been expressed well by the anthropological idea of sin in Augustine, who calls this concentration of life in itself as *curvatio in seipsum*. Yet man can overcome this isolation, which keeps him alive and allows him to cultivate his environment, so that he may go out from it into the world. Then man becomes a person, a spiritual being, defined as a bundle of intentional acts which settle him in a world (*die Welt*) that is correlative to a person that is an aprioristic yet historical cosmos of values and meanings. This process was described by Robert Spaemann as the "awakening of life to reason" (*Erwachen des Leben zur Vernunft*). *The overcoming of self-imprisonment is at the same time the individuation of man.* The value that corresponds to the spiritual feeling is holiness.

Before analysing Scheler's idea or ideas on the person, it is worth mentioning that spiritual feelings (love, hate, preference), in a way similar to that found in the case of psychic feelings, are intercepted by attention, and that they are the ones most independent of our will.

Scheler's man is fundamentally communal (*Gesamtperson*), embodying community's three basic forms: religious, national and cultural. Primary acts of sympathy or secondary experiences make each individual life live with the life of the community, which is co-present and co-operating within man. (Research on the development of the child's personality has confirmed Scheler's view that the activation of individual acts is genetically subsequent.)

According to Scheler, individuation starts at the level of intentional acts, namely in:

1. *the body* (*Seelenleib*) through
a. *the environment* (*Umwelt*):
"Und uns selbst unbewußt formt sich – im weitgehenden Maße – unser Erleben gemäß der Gliederung der Beachtungsrichtungen unserer Umwelt. Auch die Sprache und ihre psychologischen Bedeutungs einheiten werfen ihr gliederndes, artikulierendes Netz zwischen unsere Beschauung und unser Erleben" (*Sympathie* 246); and
b. *tradition* (*Tradition*):
"In seiner zeitlichen Ausdehnung heißt das, was sonst Milieu heißt, *Tradition*, d.h. die in uns noch lebende und wirksame Geschichte, die gerade bewußte Erinnerungen an die wirksame Erlebnisse *ausschließt*" (*Formalismus* 166, Bemerkung 1).

The *Ordo amoris* is an effect of individuation at the vital level. "Wer den Ordo amoris eines Menschen hat, hat den Menschen". The Ordo amoris in the dimension of space is called an environment (*Umwelt*), in the dimension of time – a lot (*Schicksal*): "*Einheit eines* – durch die Jahre hindurch – *durchgehenden Sinnes*, die sich uns als eine *individuelle Wesenszusammengehörigkeit* menschlichen Charakters und des Geschehens um ihn herum und in ihn hinein".

2. *persons: the individual person, the collective persons of church, culture and nationality* through:

a. *the world* (*Welt*):

For Scheler full individuation begins with the overcoming of the fragmentariness and particularity of the perspective determined by our body. It is worth mentioning here that – in Scheler's theory – man experiences the reality of the world through the resistance which his will and vitality encounter.[2]

This primary "experiencing the reality, or experiencing the resistance of the world, *precedes* any realisation (*Be-wußtsein*), any re-presentation (*Vor-stellung*), any perception (*Wahrnehmung*)".

This *preceding* is essential, because in a way it pre-cedes the sense of what will emerge to human cognition. (Heidegger's question – not about being but about its sense – allows us to grab the primary rhythm of the emergence of sense; the proper [i.e., full] sense or an improper sense, i.e., one reduced to the level of the surroundings dominated by the time perspective of the present.)

Scheler's phenomenological reduction is designed to liquidate that particular perspective determined not by the hasty judgements of the cognitive subject but by his psycho-physical condition, or, to be more exact, by the resistance which his will and vitality encounter in the world. This reduction is not only abstention from *judgement* on the reality, it stalls the very moment of *experiencing* the reality, by reducing the subject in the spheres of his psycho-physical drives and his will. This transcendence of the psycho-physical drive of life occurs thanks to: 1) *ascetic* withdrawal from the sensual sphere; 2) *humility*, i.e., switching off the volitional sphere; and 3) *love*, which expels man from his private little nook into the world.[3]

The experience described by Scheler's phenomenological reduction is the *sine qua non* condition of individuation in its most profound sense; its equivalent in Heidegger is *being unto death*.

"Alles als gegenwärtig Erlebte ist wesensnotwendig gegeben auf dem *Hintergrund* jener *Total*gegebenheit, in der wieder das *Ganze* des indyviduellen Ich-zeitlich ungeteilt intendiert ist" (*Formalismus* 435).

The psychic sphere of *I* is not the only one to be inferred from an experience, it is also the person as a unique constellation of values that is expressed in each act. This unique constellation of values is called by Scheler *individual destiny* (*individuelle Bestimmung*).

Schicksal wie Milieustruktur wachsen . . . aus zielmäßig wirksamen – aber nicht aktiven, frei bewußten Akten des Urteilens, Wählens, Vorziehens, sondern automatischen, aber unter fremder Mithilfe noch ablenkbaren – Vorgängen des psychovitalen Subjektes im Menschen hervor. Dahingegen ist die *individuelle Bestimmung* eine an sich *zeitlose* Wertwesenheit in der Form der *Personalität*.

The person in this definition is above time, but is not ahistorical, being an organ of history with greater or lesser content of historical values. The soul of history is systems of values, i.e., rules that build an ethos; this soul is expressed through historical patterns (which will be discussed later). The idea of *Ressentiment* is very important for our considerations; it represents an inadequate motivation reflecting values not belonging to the sphere of *individual destiny*. The purpose of philosophical therapy in Scheler's style is authentic individuation consisting in identifying *individual destiny*, i.e., the true types of personality proper for concrete individuals, types which are carriers of the values that allow each to develop his real individuality. A change allowing the realisation of individuation is possible thanks to *repentance* (*Reue*), which is a potentiality of a person as a historical being; a person's acts are now different, because they are articulated by a different constellation of values. Scheler proposes the idea of "repentance" which, even though it cannot undo an act that has been done, can change its sense, at the same time changing torment into "rebirth" (*die Wiedergeburt*) and opening – with the key of a hermeneutically "expiated" past – a perspective on the future.

It is essential for our thesis to point out that a person – as *das principium individuationis* – is a category from the sphere of history, is determined by history, and is hardly dependent on our will. Yet it does depend on our will to make the effort to interpret ourselves as history, thus making our authentic development possible. This conclusion has led us to another aspect of individuation, i.e., to meaning.

1.2. *Dilthey's Individuation through Meaning*

The main principle of Dilthey's concept of the world is that of life articulated in history in the form of personality types based on outlooks on life (*weltanschauliche Typen*), a prototype of Scheler's ethos. These types constitute the physiognomy of a given epoch; they shape the spirit of a given epoch – its ways of being religious, its regimes and its culture. Man in historical perspective seems to Dilthey to be a person, a being who breaks through the limits of his finiteness and becomes a potentiality for history, which gives the person a certain uniqueness. This transcendence of our finiteness takes place through coming to understand another man, which opens a perspective of freedom (*Gesammelte Schriften* VII, 216).

Das Geheimnis der Person reizt um seiner selbst willen zu immer neuen und tieferen Versuchen des Verstehens. Und in solchem Verstehen öffnet sich das Reich der Individuen (*Gesammelte Schriften* VII, 212). Ein Reich 'der unermeßlichen Fülle singulären Gestaltungen' in dem 'sich die Menscheit auslebt' (*Gesammelte Schriften* V, 241).[4]

These personality patterns play an essential role in forming an individual life. Yet because of their intersubjectivity, it is not possible to find in them the sufficient condition for the principle of individuation, only a necessary one. Besides, Dilthey in looking for a justification for the historical humanities, introduced the assumption of homogeneity in quality between persons, whose differences then result only from their quantitative reference to a particular personality pattern.

In no heterogeneous individual manifestation of life can there occur anything that would not also be included in the original endowment of personality. The same functions and components appear in all personalities; people's dispositions differ only in the degree of their intensity (Zusätze aus den Handschriften).[5]

In this situation Dilthey will ascribe the direct individuating power to the constituting semantic context, which is for Dilthey primarily a fundamental existential category and secondarily a semantic category. Because Dilthey's semantic category (*Bedeutung*) was the result of an unprecedented phenomenological description of the phenomenon of life, we shall concentrate on it for a while.

In *Der Aufbau der geschichtlichen Welt in den Geisteswissenschaften* (1910), Wilhelm Dilthey presented an impressive piece of phenomenological analysis, continuing what he had initiated in *Idee über eine beschreibende und zergliedende Psychologie* (1894): a new kind of philosophical anthropology, or rather philosophy in general, *defined as an action* in opposition to the static distance of the *theoretical attitude.* Dilthey accused Husserl of practising "psychological scholastics" in Brentano's style, which does not describe life adequately and fully, but tries to force life into the fossilized categories of "object", "content", or "attitude". Dilthey's ideas are *categories of life* (*Lebenskategorien*) *in statu nascendi*, where life is both *genetivus obiectivus* and *subiectivus.* These categories would be the prototype of Heidegger's *existentials.*

Dilthey decidedly rejected the classical idea of substance in the description of human reality, because its "being in itself" – *subsistere* – does not reveal the human reality whose essence is *temporality* (*Zeitlichkeit*), attended by the possibilities of the already experienced individual past and by the historical possibilities encircling the individual as the horizon.

Temporality is the first category which Dilthey extracts from the description of life. This temporality will appear discursively in three perspectives: as the past in the act of *objective formulation* together with the category of *meaning*, as the present in the act of *feeling* together with the category of *value*, and as the future in the act of *will* together with the category of *purpose*. The unity of these three acts and three perspectives on time is founded in the category of *meaningfulness* and *meaning*, and not that of substance.

Meaningfulness (*Bedeutsamkeit*) is a category which describes the constitution of experienced moments in a coherent whole, in *a unity* (*Zusammenhang*); it describes the emergence of a certain semantic continuity out of the amorphic mass of experienced events and moments. At this stage meaningfulness has nothing to do with the linguistic-semantic idea of meaning; it deals with the primary fragments of the experienced life, which still do not belong to anybody, because the process of the constitution of the semantic unity is for Dilthey at the same time a process of individuation. "Was meine Anschaung (das Wort in weitestem Sinne genommen) umfaßt, von dem wird ein Teil durch die Bedeutsamkeit in den Blick erhoben, apperzipiert, dieses wird dann unterschieden von den nicht apperzipierten geistigen Vorgängen. Dies ist, was wir Ich nennen, und es besteht das doppelte Verhältnis: ich bin und ich habe". (*Aufbau* 285)

This pre-reflexive unity is something totally dynamic and is described as *an immanent teleology*. If this unity involves a certain intention of materialising something, that will be defined by Dilthey as *a power*. This power makes a semantic unity crystallise around its centre of gravity. At this level the power of life starts creating the meaning of experienced moments. *Meaning* (*Bedeutung*) is not a semiological union of the symbol and symbolised, nor is it situated on the platform of linguistic communication, but is a structural unity of a part and the whole of life.

Meaning is a relation between a part and its complementing whole. At the existential level this whole is the context of life; at the historical level it is the prevailing spirit of the times.

Because these contexts are usually open, their defninteness will be modified by the future: “Our view on the meaning of life keeps changing. Each plan of life is an expression of the meaning of life. What we consider to be future purpose stipulates the semantic qualification of the past” (*Gesammelte Schriften* VII, 233).

For us it is essential to understand that the meaning of a single experience already implies a certain wholeness, a certain teleological *unity of life*, i.e., a fundamental way of existing for a human being that makes its further *development* (*Gestaltung/Entwicklung*) possible. This semantic unity of life is Dilthey’s alternative to the metaphysical idea of substance. Existence is here a horizontal movement of temporality in its three perspectives, and a vertical oscillation between a comprehensive formulation of the temporal horizon and fragmentary experience of particular events. “Immer schwebt das Verstehen zwischen beiden Betrachtungsweisen” (*Aufbau*, 288). In this permanently reconstructing semantic context that keeps changing a certain direction starts to crystallise whose continuation is an expression of the human person’s identity. In this way the second aspect of individuation can be discussed; it takes place in correlation with “the spirit of the epoch”. Nevertheless individuation primarily occurs through individual history, or rather its interpretation, which we are.

A significant shortcoming of Dilthey’s analyses is a certain automatism in the constitution of meaning therein. The formulation of this new meaning – as Rollo May writes in *The Discovery of Being* – always reveals the possibility and necessity of some personal decisions and so requires a new orientation for the person’s movements within his Gestalt in relation to the world and the future. This shortcoming is remedied by Heidegger’s analyses.

1.3. *Heidegger’s Individuation through Time*

Dilthey’s category of meaning became the basic category of the existential analysis of the Apostle Paul’s letters that Heidegger made in his lecture *Einleitung in die Phänomenologie der Religion* (1920/1921). Heidegger noticed two ways of human being in Paul’s antithesis of flesh (*sarx*) and spirit (*pneuma*), which make it possible to constitute two separate semantic contexts. The noetic correlative of these contexts in temporal characteristics will be time for Heidegger, as a way motivated by the temporality of the present or of the future. The pneumatic way of human being – *fides qua creditur* – has made the process of existential becoming (*genesthai/genethenai*) or the subsequent past (*Gewesenheit*) possible. The phenomenon of becoming oneself through faith, which is an opening of man to the future, led Heidegger to formulate the principle of the individuation of being three years later in the lecture *Begriff der Zeit:*

die Zeit ist das rechte principium individuationis (26).

Heidegger's time is a way of being, initially deprived of its correlative, which subsists in Scheler's idea of the person. This noetic time is a basis for noematic time's coming into being; in Heidegger's analyses the latter recedes into the background.

For Heidegger in his *Being and Time*, one of the most fundamental existentials of being, now recognized as *the* way to constitute being, is that of existential meaning or, to be more precise, of semantic context (*Bedeutungszusammenhang*). This context is a coherent constellation of the possibilities of the world, which possibilities start individuating themselves in the form of existence. In Heidegger's philosophy the so-called turn is to be found in the presentation of being in two semantic contexts (*Bedeutungszusammenhang*) correlating to their fundamental ways of being: the improper context blinded by the present, and the proper context revealing the future in its individuating dimension. Heidegger, in opposition to Dilthey, pays more attention to the conditions, or ways, of being which make it possible to constitute the proper semantic context, and consequently the authentic individuation of being, which is motivated not by the anonymity of "*oneself*" (*das man*) but by *the historicity of being itself.*

1.3.1. *Death – the* Sine Qua Non *Condition of the Authenticity of Being*

"*Moribundus* gives *sum* its sense". It is death that will reveal the proper way of being, its *determination* (*Entschlossenheit*), showing the simultaneous powerfulness and powerlessness of being that can choose and realise different potentialities of life, but cannot choose the way of life itself. Our finiteness is of such a nature that we cannot overcome the nothingness through our decisions and will. Heidegger's proper making of oneself is carried out "not because of human decision or will"; theologically speaking it has the character of grace.

The transition, effected in the act of approaching death (*Vorlaufen in den Tod*), causes, simultaneously with the disappearance of the meaningfulness of the world, the extinction of the everyday "in consideration of what?" (*worumwillen*) which is the improper motivation of being. *The death of improper being emerges from the nothingness of proper being, thus making it possible to adequately interpret existence and being in general.* This is the key point in Heidegger's analyses of being and of our considerations on the question of individuation. Death individuates – says Heidegger in *Begriff der Zeit* – making everybody equal before the tremendous impossibility of being (27). The experience of setting the perspective which concentrates the meaningfulness of the world around man is the fundamental experience of mankind. In his *Glück und Wohlwollen,* Robert Spaemann shows it to be a kind of metanoia:

> . . . die Entdeckung der radikalen Unbezüglichkeit des endlichen subjektes läßt diese Subjekt in einem Glanz erscheinen, der nicht sein eigener ist. Voraussetzung hierfür ist, daß ein Wesen herausgetreten ist aus der Zentralität des bloßen Lebens, für das alles, was ihm begegnet, nur eine Bewandnis hat als Funktion für ein Subjekt, das als bewandnisloses Selbst verborgen bleibet. In diesem Heraustreten, in dieser metanoia erst, wird das Selbst als die fundamentale allen "Wert" begründete Wirklichkeit sichtbar (127).

Heidegger's alternative to the idea of substance is *making*/processing (*Geschehnis*), defined as a continuity of making the right ways of being – conscience, fear, being unto death – which then smash the fossilised scheme of existence and bring us to stand guilty in front of the motivating potentialities of history. In this way being becomes *destiny* (*Schicksal*) allowing us to participate in *historicity* (*Geschichtlichkeit*) and allowing others to be themselves (*Sein-lassen*) through *fate* (*Geschick*).

2. THE UNITY OF PRINCIPLES: ANOTHER'S TIME AND ONE'S OWN TIME

Dilthey was the first to notice how it is possible to experience oneself in a more profound way through understanding another person: "to understand is to find I in you" (*Gesammelte Schriften* VII, 191).

Dilthey declares, ". . . man gets to know himself only in history" (*Werke* VII, 316). In this way he points out the individuating potential in history. This theme was later explored by Scheler and Heidegger. Scheler's ethics is an ethics of the personality patterns that constitute man by co-making and subsequently fulfilling (*Nachvollzug*) the intentional motivation of a historical personality pattern. It is an ethics of "*self-iden-*

tification with a person", a total transference of oneself to and in that person. The transference of personal substance results first of all in a unity in the sphere of thinking, wanting, feeling, and consequently shapes (*Einsbildung*) and transforms one's ego into the master's; this creates a continuous dynamic chain of newer and newer reproductions of his spiritual form in the material of one's own psychic date.[6]

To adopt a personality pattern is to adopt its time horizon. Heidegger pointed out that self-being in a time horizon means being *guilty* (*schuldig*) in front of the historical potentialities creating the horizon, that is to say, that historical self-being is based on the fundamental experience of guilt over previous existences, which thanks to the destruction of their perspectives can in a repetition (*Wiederholung*) initiate a dialogue with us as contemporarily living creatures.

For Heidegger this existential *birth* initiated by an existential *death* involves the return of self-being back to the possibilities of previous existences and is motivated by the power of the possibilities given to man in *repetition* (*Wiederholung*). "Nur faktische eigentliche Geschichtlichkeit vermag als erschloßenes Schicksal die dagewesene Geschichte so zu erschließen, das in der Wiederholung die 'Kraft' des Möglichen in die faktische Existenz hineinschlägt, das heißt in deren zukünftigkeit auf sie zukommt" (*Sein und Zeit*, 395). Heidegger's destruction of the idea of substance uncovered the foundation of self-being in time as a simultaneous making (*Geschehen*) of three ways of being: 1) *approaching death*, which smashes the fossilised contexts of former meaning; 2) *repetition* of previous historical possibilities, which are the beginning of a revival of existence; and 3) *in a split second*, which focuses on the initiative of acting.

This destruction must result in the *rebirth* (*Wiedergeburt*) of being, revived from *eternity*, which is defined by Heidegger not as a co-presence but as a motivating cooperation between previous existences and us today and the activation of our previous time horizons in relation to our contemporary ones. The correlative of this operation is *our being guilty*.

Man pflegt daher dem Seyn Gottes die Ewigkeit zuzusprechen. Aber was heißt "Ewigkeit", und wie ist sie in den Begriff zu fassen? Das Werden Gottes läßt sich nicht nach einzelnen Abschnitten am Nacheinander der gewöhnlichen "Zeit" aufreihen, sondern in diesem Werden "ist" alles "gleichzeitig", gleichzeitig aber bedeutet hier nicht, daß die Vergangenheit und die Zukunft ihr Wesen aufgeben und zur reinen Gegenwart "über"-gehen, im Gegenteil: die ursprüngliche Gleich-Zeitigkeit besteht darin, daß Gewesensein und Künftigsein sich behaupten und gleichursprünglich mit dem gegenwärtigsein als Wesensein der Zeit selbst ineinander schlagen. Und dieser Schlag der eigentlichen Zeitlichkeit, dieser Augenblick, "ist" das Wesen der Ewigkeit, nicht aber die bloß stehengebliebene und stehenbleibende Gegenwart, das nunc stans (Schellings Abhandlung, 136).

SUMMARY

After analysing the above approaches it is apparent that time is the power that sorts out unique human individuals in the human species. Time is a way of human being in front of the transcendence articulated in the living organism as a person, or is a way of being, or is a meaning in relation to the transcendence. Man, as a being with revived time, is inscribed in the history of a generative grammar consisting of the previous existences inscribed in the previous time horizons, which co-create the time of people living now.

The condition which allows previous existences to enter into the time of people living now is the abolition of their egocentricisms together with their meaningfulness, the deepest layer of Dilthey's or Heidegger's existence. Both in Scheler and Heidegger, the fulfillment of this condition is not possible for man. This means that both time in its passive individuating role and the conditions of active individuation are beyond man's power; man as revived time is an interpretation of his own time, which suggests the title of this work.

University of Szczecin

NOTES

[1] The forerunner and executor of this change was Henri Bergson who pointed out that invariableness, constancy, motionlessness – the main attributes of substantiality – are a fiction resulting from the pragmatism of man who out of the need to accomplish useful work starts thinking in the category of space and not that of time, of eternity, understood as a continuous creative act present in the acting memory.

[2] In a way this recalls the attempt to existentially justify the fundamental principles of Fichte's logic (*Wissenschaftstheorie 1794*): the identity A = A is founded through the intentionality of will disturbed by no resistance; the contradiction A = B is founded through the resistance which the intentionality of man encounters in contact with something else, different from this intentionality.

[3] Scheler's reduction is reminiscent of some Buddhist techniques and of Protestantism, which clearly expound on nothingness in experiencing the whole.

[4] See Heinrich Anz, "Hermeneutik der Individualität. Wilhelms Diltheys hermeneutische

Position und ihre Aporien", in Hendrik Birus (ed.), *Hermeneutische Positionen* (Göttingen, 1982), pp. 58–89.

[5] Both Max Scheler and Martin Heidegger would expound here on the finiteness of being instead, which for Scheler results in the moral "tragicalness" of the person, who is only a single perspective – one of many "good" ones – of the undivided divinity (*Seitenansicht der Gottheit*), condemning the person to ethical imperfection. Heidegger expressed this through the nothingness of the outline (*Nichtigkeit des Entwurfs*), which consists in the fact that each choice will be shown to be a limitation (*Wahl-Auswahl*).

[6] Max Scheler, *Wesen und Form der Sympathie*, pp. 96.

REFERENCES

Dilthey, Wilhelm: *Der Aufbau der geschichtlichen Welt in den Geisteswissenschaften.* Einleitung von Manfred Riedel. 3. Auflage. Frankfurt/Main 1990. (quote as *Aufbau*)

Dilthey, Wilhelm: *Gesammelte Schriften V* Die geistige Welt. Einleitung in die Philosophie des Lebens. Este Hälfte: Abhandlungen zur Grundlegung der Geisteswissenschaften, hrsg. v. G. Misch.Stuttgart/Göttingen. 6. Auflage 1974.

Dilthey, Wilhem: *Gesammelte Schriften VII* Der Aufbau der geschichtelichen Welt in den Geisteswissenschaften, hrsg. v. B. Groethuysen. Stuttgart/Göttingen. 5 Auflage 1977.

Heidegger, Martin: *Der Begriff der Zeit.* Vortrag vor der Marburger Theologenschaft, Juli 1924, hrsg. v. H. Tietjen. Tübingen 1989.

Heidegger, Martin: *Sein und Zeit.* 16. Auflage. Tübingen 1986.

Heidegger, Martin: *Schellings Abhandlung "Über das Wesen der menschlichen Freiheit" (1809)*, hrsg. v. H. Feick. Tübingen 1971.

Kierkegaard, Sören: *Gesammelte Werke* in 12 Bänden, hrsg. v. H. Gottsched und Ch. Schrempft. Jena 1912.

Scheler, Max: *Der Formalismus in der Ethik und die materiale Wertethik.* Neuer Versuch der Grundlegung eines ethischen Personalismus. 4. durchgesehene Auflage, hrsg. v. M. Scheler. Bern. (quote as *Formalismus*)

Scheler, Max: *Wesen und Formen der Sympathie.* 6. durchgesehene Auflage. Bern. (quote as *Sympathie*)

Spaemann, Robert: *Glück und Wohlwollen.* Versuch einer Ethik. 2. Auflage. Stuttgart 1990.

SECTION III

THE TRANSITIONS OF SENSE: BODY, ORGANISM, CONSCIOUS LIFE

CARLA CANULLO

THE BODY AND THE SELF-IDENTIFICATION OF CONSCIOUS LIFE

The Science of Man between Physiology and Psychology in Maine de Biran

Interest in the "science of man"[1] is one of the themes around which the entire speculation of Maine de Biran (1766–1824) revolves, constantly reaffirming the centrality and importance of the study of man and his complex constitution. We will offer here an outline of this theme, starting with the definition of Hermann Boerhaave,[2] "*homo simplex in vitalitate, duplex in humanitate*", which Maine de Biran makes his own and in the wake of which he clarifies the various levels of anthropological constitution as they come up.

1. *HOMO SIMPLEX IN VITALITATE*: PHYSIOLOGY AND THE SCIENCE OF MAN

Biranism, for many reasons, is marked by a great interest in the problems debated in physiological science insofar as physiology studies "physicality". This interest is motivated by the role physiology plays in anthropology. Maine de Biran notes, in fact, that "la nature humaine, sous quelque face qu'on la considère, ne peut se manifester pleinement qu'à celui qui possède le système intier des connaissances physiologiques et médicales".[3] Moved by this knowledge, our author co-founded with respected doctors the *Société médicale de Bergerac*, and presented before this group important communications and addresses now gathered in volumes entitled *Discours à la Société médicale de Bergerac*.[4] The originality of the Biranian position compared with that of his contemporary physiologists lies in the fact that the philosopher first occupied himself with founding a physiological science concerned with physicality, a science totally autonomous – as much in its method as in its object[5] – from any residue of metaphysical analyses. This article proposes two tacks: first we will examine the basis for Maine de Biran's criticism of the physiologists of his time, then we will analyze the original Biranian formulation of physiology.

A-T. Tymieniecka (ed.), Analecta Husserliana LXVI, 203–223.

1.1. *"Vitalitate" between Metaphysics and Physiology: Maine de Biran's Polemic with the Physiologists of His Time*

It is possible to identify two great themes in Biran's critique of his contemporaries in physiology. The first concerns philosophical themes and marks again the difference, proceeding from an essentially methodological question, between our author's thought and those systems of thought; the second critique concerns more specifically the physiological theories of the time.

Regarding the first question, it should be noted that Maine de Biran never abandons his original goal, in this context, by sifting out the influence exerted by some philosophical systems on the concept of physiology. For example, his critique of the Baconian system should be read in this light; here the method of the natural sciences is applied in the same manner in the science of man. A second critique concerns the system constructed upon the hypothetical-nominal classification of the science of the human intellect. This is the error of Condillac[6] and, on this same track, *l'idéologie*, which thus betrayed its own original vocation of being a science of facts and not of abstract hypotheses. The third system which Biran examines and distances himself from is that which is premised on an a prioristic definition of the subject, the heritage of most classical metaphysical positions.[7] Regardless of the correctness of Maine de Biran's judgement, for which question it would be interesting to examine the Biranian approach to the history of philosophy,[8] this line of reasoning is of particular interest because its conclusions draw us to the heart of the anthropological and physiological prospectives. The point of arrival – in the first instance – is an abstract conception of man, a conception to which these systems lead, and this because of a kind of initial vice of the hypothesis – this as well is abstract – from which these systems depart. The Condillacian system takes into consideration above all the receptivity of the subject and does not adequately understand the activity of the same, precisely at the juncture at which it examines the birth of consciousness.[9] In these pages, Biran also criticizes Locke's thinking, for while he did well to point out the reflectivity of the human intellective process, he did not adequately include the reality of the intellective acts, either exemplifying them abstractly of reducing them to *titres nominaux*.[10] Finally, in so-called "metaphysical analysis", every distinction can be related only to an abstract reality of possibilities, so that, in the end, science comes before and is more important than existence.[11] For Maine de Biran, the problem is not the legitimacy of the systems under consideration but the premises they use regarding the science of man. When one goes on to examine the final results of these systems, their influence on some of the most important physiological theories of the time emerges. These took two directions: the first involved a *reductio ad unum* of the human faculties.[12] After proceeding from the analyses of the ideologues, it led to a kind of alliance between physiology and metaphysics,[13] an alliance which first of all was based on an incorrect conception of sensitivity, assumed to be the only principle of consciousness. The second approach, poles apart from the first, involved the repeated fragmentation, falsely deduced, of the activities of the intellect.

The alliance between physiology and metaphysics drew its rationale from the Cartesian distinction of substances,[14] a distinction on which was premised one of the most widespread physiological theories of the time, namely, the theory of Stahl; the second motive favoring this combination was its reduction of the human faculties to a *sensorium commune* which in some way should be a sign leading back to the unity of the acts of the I. Physiologists wed the thought of Stahl and Descartes in their assumption of a unified principle heading all of man's organic and psychic functions, that is, the soul. Taking as a point of departure the Cartesian distinguishing of substances into *res cogitans* and *res extensa*, and given the position that bodies are only capable of a receptive passivity before that which is active, in this case, the soul, it necessarily followed for these physiologists that the soul totally determines any movement, thought or sentiment, and must be the only center for propulsion and motor activity.[15] To Maine de Biran this unification of the laws of the body and of the spirit so that a single principle is at the origin of all states of unconsciousness must have seemed absurd, transgressing the most elementary rules of human observation.[16] The same is true for the claimed existence of a *sensorium commune*, that is, for the search for a single, indivisible and identifiable site of the soul or the I. This way of picturing things arose from the need to explain the unity of the I. However, there arises the problem of justifying the perception of resistance to oneself, that is, that which "dans la langue plus commune et dogmatique"[17] is understood as the body-soul relationship.

This claimed and presumed reduction of the human faculties, through a kind of internal contradiction that does not distance the two hypotheses very much, leads to a kind of decomposition of the faculties which is not really demonstrable in man. Once the brain is identified as one of the possible sites for the *sensorium commune*, not only the bodily but above all the emotive and intellectual activities of the subject are progressively parcelled out into nerves and terminals. The theory of the *sensorium*, which began from the need for the unity of the I, ends up justifying one of the most common and widespread assumptions, that is, that everything has a location, that this location is in space and that it is thus identifiable through a knowable *point de vue extérieur*.[18] This hypothesis gave rise to some important physiological theories, such as those of David Hartley and Charles Bonnet, who, even though they used different methods, both traced all the subject's acts and sensations back to the cerebral nerve fibers.[19] Maine de Biran them devoted particular attention to analyzing the craniology of Doctor Gall, a German doctor whose theories enjoyed particular favor. Our philospher discussed these theories in an important work, *Observations sur les divisions organiques du cerveau considérées comme sièges des différentes facultés intellectuelles et morales: des rapports qu'on peut établir entre cette division et l'analyse des facultés de l'entendement: examen du système du docteur Gall à ce sujet*,[20] in which he explained his aversion to theses which appear so incontestably false. Gall accounted for the human passions, the appetites, the subject's moods and intellectual activities on the basis of the structure of the cranium and its protuberances; these divisions presupposed the divisibility of human *entendement*, allowing the attribution of each thought function to a seat in the brain. Thus, on the one hand, he saved the need for the unity of the sensorial and intellectual organ, and, on the other hand, he justified physiologically the multiplicity and variety of its effects. A few years later Maine de Biran will raise the same criticism in his *Rapports du physique et du moral de l'homme* (*Mémoire de Copenhague*)[21] grouping Gall's theses with those of another contemporary physician, Dr. Willis, and placing them in the category of those who try to justify the faculties of the spirit by attributing to them an organic seat.[22] While Willis did not treat exclusively craniology, he did posit the location of the various intellectual faculties in different seats of the brain,[23] in this way using an analogous subdivision based once again on abstract principles.[24]

This is the panorama of the physiological doctrines Maine de Biran compares his own thinking with and from which he distances himself, not sharing their claim to reduce to a physical "cause" the origin and richness of conscious life.[25]

1.2. *The Simplicity of Vitality and the Immediacy of Affections. Physiology As a "Mixte" Science of Man*

Biranian attention to physiology is prompted, as already explained, by recognition of its particular value as the basis for an authentic anthropology. Maine de Biran admits, however, that this task will be achieved only from the starting point of a correct premise which above all permits clarification, both in the preliminaries and in the outcome, of what his real constitution is. Maine de Biran's reflections are accompanied by the thought-provoking formula of Boerhaave: *homo simplex in vitalitate, duplex in humanitate*.[26] The starting point of the Biranian theory of physiology is observation of the state of vital simplicity, and it is constantly concerned with evaluating and analyzing the modality of existence of this state; in other words, this physiology takes as its starting point a level of physicality which is not reflective and is pre-conscious, and at the same time proposes to ascertain the authenticity of this state: to do so means observing phenomena which emerge solely due to man's observation, excluding any abstract formulation or one derived from presuppositions which are extrinsic to man. So already from these first sketches we can understand the terms of the debate with the physiologists so far cited. Our intention, however, will be to follow the development of this polemic with attention to the anthropological theory which it yields.

For Maine de Biran, the possibility of the existence of a vital state without consciousness is an irrefutable fact: "La lumière intérieure de la conscience n'éclaire pas subitement tout l'*homme venant en ce monde*: la condition nécessaire sur laquelle se fonde le moi, première dans l'ordre des faits de l'existence *pensée* ou *aperçue*, ne l'est point dans l'ordre des phénomènes d'une vie *animale* ou purement sensitive".[27] The method the author follows in arriving at this affirmation is scientific and medical observation, based on examples which prove the possibility of a state of absolute physical passivity. Man's first vital instincts are extremely close to the order of animal instincts and lack consciousness: "L'homme commence à vivre et à sentir sans connaître sa vie (vivit et est vitae nescius ipse suae)";[28] or in other words, there are states completely lacking self-awareness: "Ces sensations dénuées de personnalité, que j'appellerai *affectiones pures*, peuvent bien, il semble, être considérées comme les résultats les plus immédiats de fonctions, qui concourent et *conspirent* dans une vie générale organique".[29] There are states in which man lives in a kind of suspension of consciousness, for example, the oneiric state, delirium, under the influence of temperament – which escapes all conscious control – and mental alienation.[30] In addition, in this context Maine de Biran dedicates particular attention to *obscure perceptions* (*perceptions obscures*), composed of two elements, the material action of objects and the concurrence of a superior force which the I perceives inasmuch as the force is different from itself, that is, these are "sensations et perceptions complètes qui exigent, même de l'origine, le concours de l'activité d'une force supérieure à l'organisation, informent l'âme ou le *moi* des existences étrangères et de la sienne propre".[31] These "obscure perceptions" are divided into two kinds. The first are immediate influences on the senses, evidenced in the senses by a sort of instinctive liking which conditions the subject. The second are internal affective impressions, which are derived from the interplay of the individual's organic vitality and are related to nervous system centers, some of which direct the brain and manage the active part of man, while others manage the passive part and direct the individual nervous system centers (*ganglions nerveux*). These obscure perceptions represent a latent and often confused activity, the founding premises of which are also in this case related to the state of the suspension of self-awareness and to the consequent absence of a state of self consciousness. The location and cause of the internal affective impressions are particularly difficult to identify, as seen clearly in the example of the hemiplegic treated by Dr. Rey Régis; struck in any part of the body, he felt only a general sensation of discomfort, or felt the sensation diffused all over the body without being able to identify the sensation's site or cause. Our author associated this state with common drowsiness. Now, the end to which these Biranian analyses lead is of no little importance: to the recognition that alongside conscious life is the life of a totally passive physicality, autonomous as much from consciousness as it is from the will, a positive state absolutely free of reflectiveness. The proposal of such an organic state of innate simplicity and its differentiation from the state of consciousness contradicts as much the *reductio ad unum* of Stahl's vital principle as it does the identification of a single principle for organic and inorganic phenomena; it also contradicts Gall and Willis' fragmentation of the faculties and their vision of the passions and intellectual activities as originating in distinct organic centers. To this recovery of a pure and unconscious vitality Maine de

Biran links the possibility of the authentic reform of physiology[32] and its right to debut in a 'mixed' and complete science of the human faculties,[33] a science which includes physiology next to psychology, that is, the science of the conscious man, with which physiology must interact and not interfere. In this way the Biranian operation, by restoring to physicality the role paradoxically usurped by its own indiscriminate exaltation, leads us to the threshold of conscious life, even while not confusing itself with conscious life. The state of vital simplicity is an original life which asserts itself, being positive in itself,[34] autonomous from consciousness and from the will,[35] and which finally can be defined as absolute passivity next to conciouness.[36] With this affirmation we arrive at the last of the points to which the itinerary proposed for the recovery of physicality leads us.

Given the premise of the absolute irreducible passivity of this physicality, self-awareness will only emerge in its originality when it reveals the contemporaneous presence of this element, when it does not attribute to it any interference in the birth of the subject; this stands in contrast, we once again emphasize, with the theories of the physiologists whom Maine de Biran opposed. In this way the philosopher clearly distinguished the physiologist's field from that of the psychologist: this conception, intending to establish and safeguard the limits of physicality, in its turn, allows light to be cast on a subject that can say "I" without censuring any level of existential expression, thanks to a survey which does not misrepresent the original human peculiarity.

We wish to conclude this first part dedicated to the Biranian recovery of physicality and his formulation of physiology, using the words of our author: "N'est-ce pas d'un semblable état de vitalité simple qu'il faut partir, si nous voulons assister pour ainsi dire à la naissance du sujet pensant, et chercher les limites qui le séparent de ce qui n'est pas *lui*, ou *de lui*?".[37] And "en explorant dans son étendue le domaine de la sensibilité physique, et s'attachant surtout à cette classe d'impressions vitales obscures dont l'analyse si difficile reste pourtant si imparfaite, la physiologie parviendrait à circonscrire elle-même plus exactement ses propres limites: en nous faisant mieux connaître l'homme come être sensitif, simple dans sa *vitalité*, elle conduirait, mais s'arrêterait nécessairement, à la sphère de l'être pensant et voulant, double dans *l'humanité*".[38] Thus we find ourselves at the threshold of *humanité*, which we will now find in the signs which characterize conscious life.

2. *HOMO DUPLEX IN HUMANITATE*: THE LIFE OF "CONSCIOUS" MAN

"Le mode essentiel fondamental de la personnalité ou du *moi* individuel, vraiment premier dans l'ordre des faits de conscience, peut bien ne pas l'être dans l'ordre de phénomènes d'une vie purement sensitive que nous pouvons observer de dedans en déhors".[39] Having established and confirmed the simultaneous presence in man of two orders of fact, and having examined the purely sensory state, let us now see in what consists "le mode essentiel fondamental de la personnalité".

"Qu'est-ce que le moi?" Abbot Morellet asks Maine de Biran,[40] and our philosopher replies "je n'ai pas pu repondre".[41] And yet, all his meditations revolve around the clarification of this theme. Thus in what sense can Maine de Biran not respond? There is no response to this question, which presumes precisely the attempt to objectify the I, saying what "thing" it is. The only viable road is in "se placer dans *le point de vue* intime de la conscience, et, ayant présenté cette unité, qui juge de tous les phénomènes en restant invariable, on aperçoit ce *moi*: on ne demande plus ce qu'il est".[42] Now, clarifying what the I is does not differ from clarifying what it means for man to be *duplex in humanitate*, thus completing the formulation of the anthropological definition with which we began. Regarding innate simplicity, it is said that its principal distinguishing feature is the fact that man, even while living, is not conscious of his own life, that is an impediment to a kind of return to himself, a reflectivity by which he not only lives but "feels" at the same time that he lives. This reflectivity defines the state of consciousness, thus permitting us to reach the heart of this philosophy[43] and of its definition of the subject. Reflectivity implies a halving by which it is possible to return upon oneself, this first halving (*dédoublement*) above all regards the "fact" itself of the I. "Fact" is that which exists for us: "tous ce qui existe pour nous . . . ne nous est donné qu'à titre de fait".[44] Precisely this meaning of "for us" implies a datum of fundamental importance, that is, the feeling that we are already in some way distinct from that which is given us, because we can consciously perceive it. This distinction is at the origin of the feeling of the existence of oneself, in psychology called *consciousness*, defined by Maine de Biran as a state of ownership of oneself (*compos sui*). This ownership implies the activity or at least the possibility of activity with respect to what of oneself is gathered as given passivity. Already the distinguishing of consciousness from reality and from itself as a purely sensible being specifies it as active ownership and differentiates it insofar as it gathers a reality perceived as other than itself, in a sort of dialectic of differentiation which makes this same passivity perceived. At this point, we have arrived in the wake of the analyses of "fact". "Fact" is divided into facts of external sensation (that is, all the phenomena of reality) and facts of internal sensation. This second order of facts is what Maine de Biran defines as *fait primitif du sens intime: fait primitif*, that is, the first evident to ourselves – *du sens intime*, from sensation, which by definition excludes any external interference and is totally defined in itself. Such a fact is the I, which does not know itself and recognize, except in an immediate relationship with that which modifies it, and which perceives insofar as something is different from itself. Maine de Biran defines the manner in which this self-recognition happens as *aperception immédiate interne*, that is, as that act of consciousness contemporaneously to which the I acquires consciousness of itself. Immediate internal apperception is contemporaneously first and simple in regard to the I, which, differentiating itself from reality, gathers itself; but the manner in which this happens is essentially reflection, *dédoublement*; it is rooted, in regards to its possibility, in the fact that man, *double* in his own humanity, is gifted with a life of relation and of consciousness (*vie de relation et de conscience*). But in what way does the I propose itself and identify itself? What is the modality of immediate apperception? Maine de Biran responds, thus proceeding to a new starting point, that the I gathers itself only as *action force, effort*: "En descendant en moi-même, je cherche à caractériser plus expressément quelle est cette pensée primitive, substantielle, qui est censée constituer toute mon existence individuelle, et je la trouve identifiée dans sa source avec le sentiment d'une action ou d'un effort voulu". [45] In the same way the I "ne se connoît qu'autant qu'il agit".[46] Thus individuality is recognized insofar as it is the beginning of an acting force. The identification of the I with this active force, at the root of that which is able to apperceive itself, is the *primum* constantly sought by Maine de Biran, the primitive fact which is the origin of the I, the *mode actif* of the constitution of the subject. This active mode is precisely *effort*, with which we can clearly identify the I.

But the proposition of *effort*, and thus of the I, would remain incomplete if only its active nature were underlined. It implies, being distinct and yet united in a way that cannot be disregarded, an organic or material resistance as the reason for the exercise of the force itself. In fact, referring to the I, Maine de Biran affirms that "ce mode actif est celui de l'effort, lequel, s'effectuant librement et suivant certaines conditions que nous verrons déterminer par la suite, comprend dans l'apperception (*sic*) immédiate de la force vivante qui le produit sans cesse le sentiment corrélatif d'une résistance organique ou matérielle, nécessairement et primitivement perçue hors de la force comme son terme d'application, hors du *moi* comme son object".[47] Therefore the I, insofar as it is active, is not absolute force but *effort*, an exertion which in its birth implies a resistance coessential to its very identification. Without this resistance, one could not speak of the I, since it would be impossible to speak of the one way in which the personality is born and identifies itself. However, before proceeding to a deeper understanding of the nature of this resistance, it is important to stress one of the most interesting implications which derive from this theory of *effort*.

The first meaning which follows from the very definition of the latter *is that the I is will*. Let us see how Maine de Biran arrives at this important conclusion. The I as *effort* represents the original nucleus of the subject inasmuch as the subject is self-determined and inasmuch as it is not derived from organic causes, which were analyzed by our author in other works, that is, in the physiological sphere. With this *foyer* of the personality, overcoming the substantialism of the Cartesian *res cogitans*, Biran at the same time overcomes Condillac and his concept of consciousness and sensitivity, introducing an interesting motive to the heart of the question of consciousness itself, that is, that of motility and, as a kind of consequence of the latter, introducing the free will with which *effort*, and thus the I, is identified.

Effort is essentially *effort moteur* which allows consciousness to differentiate itself from that which is moved.[48] "Toutes nos facultés actives sont originaires de la motilité".[49] True, there exist instinctive muscular contractions involving absolutely no intervention of consciousness, as already seen regarding innate simplicity. But man is *simplex* and *duplex*. Thanks to the possibility of duplicity, the instinctive action of some organs, by an equally instinctive habitual repetition of movements, "comes out" of the pure sphere of impersonal sensations, acquiring the capacity to enter immediately into action, thanks to a sort of spontaneity. Precisely voluntary action is that in which movement, beginning with the spontaneous, begins to be perceived as freely produced and totally self-determined: the will is born in the first moment in which the power to effect action in the body is perceived. Reciprocally, if the body were not perceived as resistance, movement would not be voluntary: "Le mouvement n'est senti ou accompagné de résistance qu'autant qu'il est volontaire"[50] and "la lumière de la conscience naît surtout et peut-être uniquement d'un contraste".[51] From the will, the passage to the birth of the I and of the personality is brief, or rather, simultaneous: the force which distinguishes itself from blind sensory perception acts on the organism, distinguishing itself from the organism, becoming *moi* by itself and not by any extrinsically caused action. It begins to know itself insofar as it begins to act freely, so that the origin of the will and of self-awareness comes to reside in the possibility of making a movement that, precisely because it is voluntary, is poles apart from sensory movement. Two important consequences for the will follow from this identification of the will and of personality. Since the will is in the first instance moving (*motrice*) action, motor exertion, it will come to identify itself with *effort*, which Maine de Brian often defined as *voulu*. Secondly, following the equation which identifies *effort* with the I, the will and the I will be identified reciprocally insofar as they are the original plexus of the personality. But, having arrived at this identification, we must proceed down the second road to which such self-identification leads. The first path, as we have just seen, brings us to equate the self-identification of the conscious life with free will. We must follow the second path in order to complete the very definition of *effort* and therefore of the I, specifying how resistance enters into this same act which originates the personality. And we will see shortly how this resistance enters as the body itself.

3. THE BODY AND THE SELF-IDENTIFICATION OF THE CONSCIOUS LIFE. THE HEART OF THE FULLNESS OF THE I

In June, 1815 Maine de Biran wrote in his *Journal intime*: "Le développement de cette proposition toute nouvelle: que le moi ne peut être conçu *sans la notion de substance passive*, m'a coûté quelques efforts, mais c'est l'idée mère qui doit changer la face de la philosophie".[52] The definition of *effort* implies, as we have just seen, the double convergence of resistance and of the acting force as the origin of the beginning of personality. Destutt de Tracy summarized the Biranian definition of the I in a letter to Maine de Biran: "Vous pensez bien qu'il y a un sentiment du *moi* ou conscience *d'être* distinct de tout sentiment . . . ; vous le faites consister dans le sentiment de l'action de cette force vive sur ce qu'elle anime et meut; et vous croyez que ce fait (ou cette supposition) est nécessaire pour que nous acquérions la connaissance distincte de notre corps et de ses parties".[53] Let us see the basis for Destutt de Tracy's affirmations. Maine de Biran's first proposition says that *effort* extends itself to all parts of the muscular system subject to the will's action. Everything that is part of this sphere of activity is part of the fact of the consciousness and is the very object, mediated or immediate, of the apperception of consciousness itself. In contrast, all the kinds of vital simplicity mentioned above, are not part of and will never be part of the sphere of consciousness. Why cannot the vital kinds of this simplicity be part of this shpere? Or, in other words, what characterizes the organs that *are* part of it? Are we not always speaking of the same physicality? Immediately, the body is felt as an indistinct extension, an internal mass. It is important to add the adjective "internal" because it signals a first distinction which will also provide the basis for distinguishing the pure physicality of the body itself. This bodily extension is a continuous resistance. Now, the concept of extension invites us to hypothesize a sort of localization of impressions, the only alternative to the obscurity of the same: "Cette étendue intérieure du corps . . . est le lieu de toutes les impressions internes, qui ne peuvent être perçues ou senties par le *moi*, que sous cette forme locale, sans laquelle elles demeureraient toujours confondues dans la combinaison vivante, comme affections simples ou perceptions obscures",[54] that is, they would still remain on the level of pure unconscious passivity. Let us see how the progressive localization of the parts of the body happens. Their identification requires the exercise of their own immediate sensation (*sens propre et immédiat*). The will, which is identified precisely after movement passes from instinctualness to free determination, acts upon the muscular system. The muscular system is already in and of itself distinguished into parts upon which the will acts. Through this sort of exercise aimed at various terminal centers, internal immediate apperception distinguishes places and localizes impressions themselves, so that the I begins to consider the parts of the body as if they were each apart from the others, an immediate consequence of point by point knowledge of the body understood in itself, then, as a group of organs subject to the power of the I, we can say, paraphrasing Bernard Baertschi's definition. The difference from vital innate obscurity – the object of physiology – is at this point evident: speaking of the state of total bodily passivity and citing the above-mentioned example of the hemiplegic treated by Dr. Rey Régis, Maine de Biran noted that the paralytic could not distinguish the various parts of the body and consequently could not localize sensations. But having established the difference between pure passive physicality (indistinct mass) and the body itself (localized and localizable physicality), we still must clarify the problem of the collocation and role of passivity regarding the state of self identification of the conscious life (*homo duplex in humanitate*) and thus regarding the I. Clearly the body cannot be held to be an element *a latere* of consciousness: in fact, we know that *effort* is will *and* resistance. Having set out from consideration of movement as the origin of consciousness prior to the I and the body, Maine de Biran roots the two terms in a unity which cannot be disregarded, in this manner "embodying" the I and "subjectivizing" the body,[55] when he introduces into the heart of *effort* the resistance which in its turn is resistance *in* a facual internal sensation. We will clarify this delicate point returning once again to the analyses of movement. The latter is action by which the I imposes itself; movement is the possibility of action upon a resistance which, to draw from an evocative definition by Xavier Tilliette, is exercised on a body which the I has comprehended. The same author speaks of the Biranian body as "forclos, reployé sur sa lumière interne".[56] The term *forclos* perhaps indicates the most distinctive character that the body has in Biranian psychology: to be *forclos*, *reployé* involves the *effort itself* which we have always constantly identified as a force which "acts upon". Now we see that the reality which the force acts upon is not extrinsic, but an intrinsic defining one from the very birth of consciousness, the I: yes, consciousness is force which imposes itself, but, by the very definition of *effort*, it would not play this role without the original presence of the resistance upon which this force – by its very condition of origin – acts, thus constituting an original duality which is the I, and establishing the definitive centrality of the body in the self-identification of conscious life. The body which enters into this life – a presence which lies before the voluntary and free motility that constitutes *effort* – is the *very body itself*, that is, the whole of the organs under the power of the I, that body of which Maine de Biran states "jusque dans ce fait de conscience (admis comme primitif) où le moi est censé n'exister pour lui même que dans la relation à une force étrangère [. . .] nous pensons qu'il y a un rapport plus simple et antérieur à celui-là".[57] François Azouvi notes on this passage that such anteriority, understood as the origin of a relationship which is the I, causes us to continue considering the I on a level in which the duality of the first fact is recognized. The I and the body itself are distinct. However, we know that the I is *effort* which is a "relation immédiate du moi phénoménal au corps propre".[58] The often called upon resistance implicit in the very definition of *effort* is thus the body itself understood as that one body which appertains to *effort*; that is, it can be subject to the power of the I and distinct from unconscious corporeity.

The I is *effort voulu*, the will which acts upon a resistance; the I is consciousness, and to become *conscium*, *compos sui* means to recognize and admit into the heart of the act itself for which the I "ap-propriates" itself, that *body itself* which makes possible for one the very fact of ap-propria-(tion). But this movement is possible only for the original duality of the I as *effort*. Following Maine de Biran, we thus find ourselves at a threshold which leads to a hypothesis of subjectivity, at the origin of which is an original duality of activity and passivity, a binomial which characterizes the reflective self-identification of conscious life.

4. THE BODY IN PHYSIOLOGY AND IN PSYCHOLOGY. PASSIVITY AND SELF-IDENTIFICATION OF CONSCIOUS LIFE

Having arrived at the conclusion, let us review the steps proposed here in order to clarify our author's anthropological hypothesis, a theme which we indicated at the outset as being central. We began with scientific and medical observation, which Maine de Biran proposes as the method of physiology, specifying the existence in man of a state of absolute unconscious passivity. To characterize it Biran borrowed an evocative definition from physician Hermann Boerhaave: *Homo simplex in vitalitate*. This study becomes necessary at the moment in which the anthropological problem poses itself in terms of the clarification of the levels which constitute man himself. However, next to unconscious vitality there exists the level of psychology; here method shifts from the scientific to the reflective, and the problem is to seek the origin of the I's imposition of itself and gathering of itself independently of external interferences. Thus we are at the level of consciousness which is identified with *effort*. Now, precisely here we meet resistance, or better, the body. Or perhaps more precisely, we *re-find* the body which is also the object of medical and physiological analysis. Should we perhaps posit the existence of two types of corporeality? This certainly is not the Biranian hypothesis. Rather, the author invites us to look at the difference between the concept of the body without the state of consciousness, and the concept of the body in the presence of consciousness, defining the body in the second case as "the body itself". But this is not simply a terminological distinction or contextualization within different methods. In fact, the problem shifts to the very origin of the self-identification of the conscious life, a self-identification which finds its very origin in *effort*, as we have clearly seen. Finding ourselves then in the definition of *effort* as acting force and resistant passivity, we find ourselves at the same time at a level in which conscious life is comprehended at its origin as a duality, in which the *body itself* is recognized as a resistance co-essential to the definition of and to the birth of consciousness because no free action force exists as such unless it is from the outset exercised on a recognized passivity, at the same time giving the body and passivity a new centrality which announces interesting points of departure for the *science de l'homme* which so occupied Maine de Biran in the course of his meditation. We can define this *science de l'homme* as *mixte*, mixed both because it embraces different levels and because it invites us to rethink a subjectivity in which the first act of the imposition of the I is originally a duality of free will and the body itself, a duality which represents the complex *foyer* of anthropology.

University of Macerata

NOTES

[1] Cfr. François Azouvi, *Maine de Biran. La science de l'homme* (Paris: Vrin, 1995).

[2] Hermann Boerhaave (1668–1738); a doctor and scientist whose naturalistic doctrines influenced La Mettrie, who attended his school. This philosopher spread his writings among members of the *idéologues* movement, and thus they came to the attention of Maine de Biran. The citation is drawn from *Praelectiones academicae de morbis nervorum*, Vol. 2 (Leyden: Petrus vander Eyk and Cornelius de Pecker, 1761), Part II, pp. 496–497.

[3] *Mémoire sur les perceptions obscures ou sur les impressions générales et les sympathies en particulier lu à la société médicale de Bergerac* (15 novembre 1807), in *Discours à la Société Médicale de Bergerac*, Vol. V (Paris: Vrin, 1984), p. 10. The writings gathered under this title bear witness to the great interest Maine de Biran had in this theme, and to its importance in the context of his thought.

[4] *Discours à la Société Médicale de Bergerac*, cit. This article will use the edition of Biranian works published under the direction of François Azouvi (henceforth cited as

E.A.), except for some works from the edition edited by Pierre Tisserand (henceforth cited as E.T.).

[5] "Dans l'état actuel de nos connaissances, nous sommes bien fondés à distinguer les trois sciences qui ont respectivement pour l'object les propriétés ou mouvement des corps bruts, les fonctions ou phénomènes des corps organisés vivants, enfin les phénomènes qu'offrent les êtres doués d'intelligence et de volonté, c'est-à-dire, la physique, la physiologie et la psychologie" (*Rapports du physique et du moral de l'homme* (Mémoire de Copenhague), Vol. VI, E.A. [Paris: Vrin, 1984], p. 91).

[6] *Mémoire sur la décomposition de la pensée*, Vol. III, E.A. (Paris: Vrin, 1988), pp. 339–348.

[7] This triple distinction is clearly enunciated in *Mémoire sur la décomposition de la pensée,*, cit E.A., pp. 336–361. The title of the section in question is *De l'analyse des facultés humaines dans divers systèmes de philosophie. Project d'une méthode particulière de décomposition.* In this context, in addition to the thought of Condillac, Maine de Biran cites that of J. Locke (*ibidem*, E.A. pp. 348–354), Descartes, Leibniz and Kant (*ibidem*, E.A., pp. 354–361). Maine de Biran's grouping of these three systems may seem strange, particularly the classification of Kantian thought with metaphysical systems. Actually, the author wants to underline the analogies in these philosophers' thoughts in their conceptions of man. In fact, "malgré les divergences des ces systèmes de métaphysique dont nous venons de parler, on les voit néanmois placés dans une sorte de point de vue commun" (*ibidem*, E.A., p. 359). This common point of view is the supposition that "des facultés qui préexistent dans l'âme, avant leur exercice complet ou la réflexion possible de leurs actes; mais ils les réalisent plus expressément comme *formes* permanentes, sous lesquelles le sujet sentant et pensant aperçoit ensuite sa propre existence, ou se représente les existences étrangères" (*ibidem*, E.A., p. 357).

[8] Maine de Biran never wrote a history of philosophy. We only have brief writings and comments on works and philosophers who preceded him or were his contemporaries, gathered in the Azouvi three-volume edition: *Commentaires et marginalia: XVIIe siècle*, Vol. XI-1, *Commentaires et marginalia: XVIIe siècle*, Vol. XI-2, *Commentaires et marginalia: XVIIe siècle*, Vol. XI-3 (Paris: Vrin, 1990). Of interest is the judgement expressed by Joël Ganault, curator of the last cited volume; in the introduction he affirms that comparing Maine de Biran with his predecessors and contemporaries "deux modes de confirmation interne et externe, sont à distinguer. La confirmation externe d'une pensée, anhistorique et analytique, extrait un énoncé de son système d'origine, et essaie d'en déterminer le degré de vérité par l'appel à une instance de référence extra-systématique, l'expérience, le sens commun, le sens intime" (*ibidem*, p. viii).

[9] The relationship between Biranian thought and that of Condillac deserves a separate chapter; it is certainly not our intention to cover it in just a few lines. We know from our author's constant references to the Condillacian opus how important he was to the formation of his philosophy, in particular for the initial and original position on sensitivity as the *faculté de sentir*. Thus we affirm Henri Gouhier's observation that Maine de Biran broke from Condillac's philosophy on this point: "Si dans ses premiers écrits, Maine de Biran 'critique assez catégoriquement' centaines parties de la philosophie de Condillac, n'oublions pas que cette critique prend forme à l'intérieur de la philosophie de Condillac. Avant d'être aristotélicien, Aristote est un platonicien qui corrige Platon: de même Biran, avant d'être biranien, est un condillacien qui corrige Condillac" (Henri Gouhier, *Les*

conversions de Maine de Biran [Paris: Vrin, 1948], p. 65. See as well, in *ibidem*, pp. 65–88, "L'école de Condillac" and "La critique de Condillac". Finally, the article by Bernard Baertschi should be seen: "La statue de Condillac, image du réel ou fiction logique?", *Revue philosophique de Louvain* LV (1984), pp. 335–364. In this article the author clarifies how Maine de Biran faces the theme of consciousness, in contrast with Condillacian "sensitivity".

[10] *Mémoire sur la décomposition de la pensée*, cit., E.A., p. 353.

[11] "Toutes les distinctions ou précisions métaphysiques qu'on pourrait faire de ces points de vue, ne se rapportent donc qu'à un monde abstrait de possibles, où la science est placée avant l'existence même. Mais dès qu'on cherche à les joindre au monde des réalités, elles y demeurent sans application . . ." (*ibidem*, E.A., p. 360).

[12] "La doctrine physiologique, qui attribuait à un seul principe immatériel et intelligent les fonctions vitales ou organiques, come les actes d'une volonté consciente, put se rallier, jusqu'à un certain point, dans son principe et ses résultats, avec systèmes métaphysiques, où ces mêmes fonctions et actes se trouvaient placés sous la dépendence immédiate d'un moteur suprême. La doctrine idéologique, qui a reduit toutes les facultés humaines à la *sensibilité*, considérée comme effet générale abstrait d'une cause, ou comme propriété abstraite de son sujet d'inhérence, pourra s'allier et se confondre, à son tour, avec un point de vue physiologique, où cette même faculté sensitive attribuée à l'organisation matérielle y est étudiée dans ses mobiles d'excitations, ses circonstances et ses lois expérimentales" (*ibidem*, p. 324).

[13] "Ici la métaphysique, changeant de titre, se trouve transportée entièrement dans la physiologie, qui lui communique sa direction, sa méthode, ses instruments ou moyens d'analyse" (*ibidem*, p. 324).

[14] "Une philosophie toute systématique n'avait reconnu que deux grandes classes d'êtres: des machines ou de purs automates, et des êtres pensants. Elle ne reconnaissait donc point de science intermédiaire entre la *physique* ou la mécanique et la psychologie, ou entre la science exclusive des phénomènes de *l'esprit* ou de *sens interne*, et de tout se qui ne ressortait pas immédiatement de ce *sens intérieur*. . . ." *Rapports du physique et du moral de l'homme*, cit., E.A., p. 92.

[15] Description of Stahl's doctrine recurs throughout Maine de Biran's works. We note here the works in which this description is most extensive: *Mémoire sur la décomposition de la pensée*, cit., E.A., pp. 319–323; on the deriviation of Stahl's doctrine from Descartes' thought, *De l'aperception immédiate*, Vol. IV, E.A. (Paris: Vrin, 1995), p. 111; and on physiology's immersion in metaphysics, *Mémoire sur les perceptions obscures*, cit., p. 12; *Observations sur les divisions organiques du cerveau considérées comme sièges des différentes facultées intellectuelles et morales; des rapports qu'on peut établir entre cette sorte de division et l'analyse des facultés de l'entendement: examen du système du docteur Gall à ce sujet*, in *Discours à la Société medicale de Bergerac*, cit., p. 50; *Rapports du physique et du moral de l'homme*, cit. pp. 28–38, pp. 92–95.

[16] Sergio Moravia notes that "Stahl admitted the existence of an immaterial *force* in the human body: but then he attributed to it the activity of the organic functions. Thus his doctrine loses any trace of the difference existent between vital movements and conscious and voluntary movements, between corporeal acts and those that live in a different, *intérieur* space, . . ." (Sergio Moravia, *Il pensiero degli idéologues. Scienza e filosofia in Francia* (1780–1815) [Florence: La Nuova Italia Editrice, 1974], p. 511).

[17] *Rapports du physique et du moral de l'homme*, cit., p. 67.

[18] L'espace et le lieu, étant en effet, comme on a dit dans une école célèbre, les *formes* de nos intuitions *sensibles*, nous sommes naturellement et presque invinciblement enclins à revêtir de ces *formes*, toutes nos idées ou notions même les plus intellectuelles. . . . Puisque les sensations et les images se distinguent donc entre elles avec une netteté proportionée à la séparation ou circonscription naturelle des organes qu'elles affectent . . . , pourquoi ne chercherions-nous pas à imiter en quelque sorte la nature elle-même, ou à suivre la voie qu'elle nous indique, en fondant une division naturelle ou hypothétique, réelle ou probable, de sièges organiques attribués à des facultés diverses?" (*ibidem*, p. 60).

[19] Cfr. In this regard *Rapports du physique et du moral de l'homme*, cit., pp. 38f. For Hartley, the will was contained in this central nervous system (cfr. *ibidem*, p. 54), while Bonnet, remaining tied to the Condillacian concept of sensitivity, based consciousness only on the object and pure immediate sensation (cfr. *ibidem*, pp. 54–55).

[20] *Observations sur les divisions organiques du cerveau considérées comme sièges des différentes facultés intellectuelles et morales: des rapports qu'on peut établir entre cette division et l'analyse des facultés de l'entendement: examen du système du docteur Gall à ce sujet*, cit., pp. 47–81.

[21] *Rapports du physique et du moral de L'homme*, cit. The writing on Gall was read to the medical society of Bergerac in 1808, while the *mémoire* in question was from 1811.

[22] *Rapports du physique et du moral de l'homme.*, cit., p. 59.

[23] "Willis entre autres a voulu assigner dans le cerveau des districts séparés a des facultés intellectuelles diverses en logeant l'imagination dans le *corps calleux*, la mémoire dans *la substance corticale*, le jugement dans le *corps cannellé*" (*ibidem*, p. 69).

[24] For more complete coverage of this theme, cfr. Sergio Moravia, *Il pensiero degli idéologues,*, cit., pp. 510–516.

[25] Sergio Moravia notes that "the Biranian polemic regarding the *physiologistes* was first of all due (we know) to the impudence with which scientists such as Willis, Stahl, Gall . . . claimed to reduce the autonomy and the irreducible complexity of *intérieure, morale* life to the work of simple *physique* organs" (Sergio Moravia, *Il pensiero degli idéologues*, cit., p. 513).

[26] Regarding the importance of this formula, which recurs often in the works of Maine de Biran, we recall the judgement of José Echeverria, the editor of the first edition of this *mémoire*, "en avril 1804 commence ce qu'on peut appeler la période *biranienne* de Maine de Biran. Son thème fondamental est la distinction de deux vies dans l'homme; d'une part, la vie sensitive ou animale essentiellement passive; de l'autre, la vie proprement humaine dont le sujet est le *moi*, défini come effort volontaire s'applicant à une résistance organique. Biran aime citer le formule de Boerhaave: *homo simplex in vitalitate, duplex in humanitate*" (*De l'aperception immédiate* [Paris: Vrin, 1963], p. 9).

[27] *Décomposition de la pensée*, cit., p. 329. Pierre Tisserand notes that "cette idée d'une vie purement affective n'est pas nouvelle. On la trouve dans la conception de l'âme végétative et de l'âme sensitive" (Pierre Tisserand, Introduction, in *Décomposition de la pensée*, E.T., Vols. III–IV [Genéva-Paris: Slatkine, 1982], p. lvi). However, the originality of the Biranian analysis is in its departure from the use of the term "âme" where Maine de Biran conceives of it as a presence of "forces productives ou d'espèces d'âmes qu'ils distinguaient des modes divers de mouvement" (*ibid.*).

[28] *Rapport du physique et du moral de l'homme*, cit., p. 108.
[29] *Décomposition de la pensée*, p. 330.
[30] The existence of these states is amply documented in the various works of Maine de Biran and in the *Nouvelles considérations sur le somneil, les songes et le somnambulisme*, in *Discours à la Société médicale de Bergerac*, cit., pp. 82–113.
[31] *Mémoire sur les perceptions obscures et sur les impressions générales affectives et les sympathies en particulier*, cit., p. 44. Maine de Biran's judgement on this theme is interesting. He considered it the border between psychology and physiology: "Le sujet [that is, the theme of the obscure perceptions] . . . se trouve placé, pour ainsi dire, sur les confins des deux sciences qui embrassent tout l'homme; il appartient à la physiologie qui considère cet être mixte comme simplement vivant et sensible, et à la psychologie qui considère comme intelligent et pensant . . ." (*ibidem*, p. 10). In addition, this theme raises the question of the influence of Leibniz's thought on Maine de Biran, an influence and relationship already examined by important studies. Among the most authoritative is the essay by E. Robef, *Leibniz et Maine de Biran* (Paris: Alcan, 1925); Michelangelo Ghio, "Biran interprete di Leibniz", in *Revue internationale de philosophie* LXXV (1966), pp. 117–140; Emilienne Naert, "Maine de Biran lecteur de Leibniz", in *Revue de Métaphysique et de Morale* IV (1983), pp. 499–513; Patrice Vermenen, "Les adventures de la force active en France: Leibniz et Maine de Biran sur la route philosophique menant a l'eclectisme de Victor Cousin (1810–1845)", in *Exercices de patience* VII (1987), pp. 147–168. Maine de Biran wrote a contribution on Leibniz for the *Biographie Universelle*, edited by the publisher Michaud in 1819 ("Exposition de la doctrine philosophique de Leibniz", E.T., Vol. XI, pp. 435–489), in which he summarized and explained Leibnizian thought. Regarding the problem of the obscure perceptions in this context, Maine de Biran affirmed here that Leibniz, in addition to conscious activity, admitted the distinction between perception and apperception, which "distinction conforme à toute notre expérience intérieure" (*ibidem*, p. 468), establishing the basis for another irrefutable observation on human life, that is that "(l'âme) à titre de force sensitive, . . . s'ignore elle-même; elle ne sait pas qu'elle vit ou sent. . . . Telle est la source des *perceptions obscures* que Leibniz attribue à l'âme humaine . . ." (*ibid*). Maine de Biran acknowledges such perceptions but does not attribute them to the soul, as will be seen.
[32] "Or, n'est-ce pas d'un semblable état de vitalité simple qu'il nous faut partir, si nous voulons assister pour ainsi dire à la seconde naissance du sujet pensant, et chercher les limites qui les séparent de ce qui n'est pas *lui*, ou de *lui*?" (*Décomposition de la pensée*, cit., p. 44). "Tout ce qui est du simple fait de l'organisation, uniquement soumis aux lois de la vitalité, et étrange pensée, rentre dans le domaine exclusif de la physiologie. Cette science expérimentale devra donc fournir un point de départ ou un préliminaire essentiel à l'analyse des facultés humaines" (*ibidem*, p. 330). Santino Cavaciuti defines this innate simplicity as "the Biranian subconscious" (Santino Cavaciuti, *Il problema morale nel pensiero di Maine de Biran* [Milan: Marzorati, 1981], pp. 121–148).
[33] "La physiologie entrera donc essentiellement dans une analyse ou une science complète des facultés humaines, qui embrasse également le facultés dans la conscience de leurs actes et les conditions ou instruments nécessaires à leur déploiment, comme les facultés sensitives prises dans les affections immédiates et l'ensemble des causes physiques qui les déterminent" (*Décomposition de la pensée*, cit., E.A., p. 331).
[34] René Lacroze states that "(l'inconscient) n'est pas dû à une simple défaillance de

l'attention à la vie; dans le domaine intérieur il ne correspond ni à un acte manqué, ni à une privation; il est *positif* en tant qu'il manifeste une certaine activité et *essentiel*, parce qu'il résulte de la présence de certaines déterminations dans les centres" (René Lacroze, *Maine de Biran* [Paris: PUF, 1970]).

[35] Pierre Tisserand emphasized this characteristic when he affirmed that "la vie animale ou l'inconscient est . . . pour Maine de Biran, une vie originelle, complète en elle-même, radicalement distincte de la vie consciente personnelle" (Pierre Tisserand, *L'anthropologie de Maine de Biran* [Paris: Alcan, 1909], p. 79).

[36] On this and other themes the study by François Azouvi, *Maine de Biran*, cit., is particularly illuminating, especially when the scholar affirms that "la science de l'homme . . . est possible à partir du moment où sont clairement distinguées les limites et les méthodes de ses deux principales branches: la physiologie et la psychologie" (*ibidem*, p. 83).

[37] *Décomposition de la pensée*, cit., p. 44.

[38] *Rapports du physique et du moral de l'homme*, cit., pp. 112–113.

[39] *De l'aperception immédiate*, cit., p. 59.

[40] Maine de Biran, *Journal intime*, Vol. II (Paris: Plon, 1931), p. 75.

[41] *Ibid.*

[42] *Ibid.* "Le point de vue de la conscience": in his study, *Maine de Biran. La science de l'homme* (cit.), François Azouvi begins from precisely this and other passages in which the philosopher speaks expressly of *point de vue*, indicating the exact point of view, the exact approach which every context of his anthropology demands. Thus here we find the point of interest for the science of man and his definition inasmuch as it is a "mixed" science, beginning from different *points de vue*. We are of the opinion that this interpretation does justice more than any other prospective to the complexity and richness of Biranian thought, because it places at its center that "science de l'homme" which Biran had as his main interest and the driving force of his meditations.

[43] "Philosopher c'est réfléchir", affirms the author in his *Journal intime* (Vol. I. Cit., p. 233). Michel Henry notes "ce qu'exprime Maine de Biran sous le mot de "réflexion", c'est la profondeur même de la subjectivité, sa "vie intime" par opposition à l'être trascendant en général qui est sans dimension intérieure" (Michel Henry, *Philosophie et phénoménologie du corps* [Paris: Puf, 1987], p. 19). Reflection is intrinsically inherent to the method of Maine de Biran, so much so that he is known as the philosopher of reflection and father of that which is defined a the "philosophie de la (French) réflexion". An important study on this theme is that of Gaston Fessard *La méthode de la réflexion chez Maine de Biran* (Paris: Bloud et Gay, 1938).

[44] *Essai sur les fondaments de la psychologie et sur ses rapports avec l'étude de la nature*, Vols. VIII–IX, E.T. (Paris: Alcan, 1932), p. 15.

[45] *Ibidem*, p. 177.

[46] *De l'aperception immédiate*, cit., p. 25.

[47] *De l'aperception immédiate*, cit., p. 57; cf. similar passages in *Décomposition de la pensée*, cit., p. 362 and *Essai sur les fondaments de la psychologie*, cit., p. 179.

[48] The discovery of motility is the first real revolution upon which the most interesting discoveries of Maine de Biran will be built, one after the other, leading him well beyond Condillac, who as mentioned above had initially influenced him. This discovery was due to his reading Destutt de Tracy's *mémoires*, about which Maine de Biran said "la

première lecture des mémoires précités [that is, those of Tracy] fit dans mon esprit une révolution dont il conservera probablement toujours les traces" (*Correspondance*, Vols. VI–VII, E.T. [Paris: Alcan, 1930], pp. 80–81.)

[49] *Ibidem*, p. 246.

[50] *Ibidem*, p. 240.

[51] *De l'aperception immédiate*, p. 106.

[52] *Journal intime*, cit., Vol. I, p. 107.

[53] *Correspondances*, cit., p. 279.

[54] *Essai sur les fondaments de la psychologie*, cit., pp. 207–208. This distinction is well underlined by Bernard Baertschi, who affirms that in a first moment the body "est senti 'en masse' " and only in a second moment does a kind of localization intervene (Bernard Baertschi, *Les rapports de l'âme et du corps. Descartes, Diderot, Maine de Biran* [Paris: Vrin, 1992], pp. 114ff.).

[55] Michel Henry dedicated a very interesting essay precisely to the "subjective" nature of the body, underlining how the fact that "la détermination de l'être de la connaissance ontologique, qui est celui de l'ego, (est) être même du corps" (Michel Henry, *Philosophie et phénoménologie du corps*, cit., p. 79.).

[56] Xavier Tilliette, "Nouvelles considérations sur le Cogito biranien", in *Revue de Métaphysique et Morale* IV (1983), p. 438.

[57] "Appendice sur les deux rapports simples d'existence personnelle et étranger", in *Décomposition de la pensée*, cit., pp. 219–226.

[58] François Azouvi, *Maine de Biran*, cit., p. 219.

W. KIM ROGERS

THE RECIPROCITY OF HUMAN ORGANISM AND CIRCUMSTANCE

An Ecological Approach to Understanding the Actions and Experiences of a Human Organism in Its Environment

The goal of this essay is to think through the characteristics of human action and experience with the aim of repositioning them, rather than resolving all difficulties. I have not set forth to demonstrate the superiority of an ecological approach over past views, but simply to show how it can illuminate these areas of human life.

J. Bennett's statement in *The Ecological Transition* may well serve as a starting point for an ecological approach in philosophy as well as the behavioral and life sciences: "Ecology . . . expresses, in a single word, the idea of all components of a milieu in reciprocal interaction with each other" (1976: 163). Roger Ames, working in collaboration with David Hall, has suggested a similar starting point using the concept of polarity as that is treated in Taoist sources. "Polarity," Ames writes, indicates the nature of the relationship that obtains between two or more events which are "correlatively related, each requiring the other for adequate articulation . . . as a necessary condition for being what it is" (1986: 324). "A polar explanation of relationships gives rise to . . . a world of 'foci' characterized by interconnectedness, interdependence, openness, mutuality, indeterminateness, complementarity, correlativity, coextensiveness . . ." (1986: 325).

An ecological approach means that one takes the reciprocity between living beings and their environment, the polarity and mutuality in their interaction, as the basic framework for all of one's studies. Each sort of life is viewed as a particular, permutable system, involving all of its components in reciprocal interaction with each other. On the other side, let us recall the dormancy of living beings when tension between them and their circumstances is completely relaxed – i.e., when "nothing happens."

This study is concerned with the interactions of human organisms and the affairs they experience in their circumstances, which are seen from an ecological perspective as the mutual complementing and completing of each in and by these interactions. Our life is, as Ortega said,

A-T. Tymieniecka (ed.), Analecta Husserliana LXVI, 225–240.

a matter of "what we do and what happens to us" (1969: 36). Living in a given circumstance is not just a matter of the human organism existing beside some affairs. Rather, it is the existing of oneself and these affairs in tension and of one's negotiations with them, of their activities inciting one to action and one doing something about them.

Two actional ranges belong to the human organism's life (cf. Goldstein and Scheerer 1941: 1–7). In the first, one begins and ends one's actions at the prompting of the affairs now present in one's environment. One's activities are directed by the claims made on one by one's experiencing the sensory impressiveness of a present affair or group of affairs, or the contextual "belonging togetherness" of some set of affairs, or some sequence of events in one's circumstances. Here one is, qua actor, a client and a subordinate of one's circumstances. For example, an activity in which this first actional range is found would be the familiar experience of driving down a highway, thinking about some other matter while proceeding to deal with the traffic problems at hand. This type of action is sometimes called "automatic." The actions of very young children fall almost – if not entirely – within the range of this type of action, which is included to a greater or lesser extent in all of the human organism's actions.

In the second actional range, the human organism's actions are self-initiated and self-directed; one puts oneself into action. One starts one's performance, or renews it after interruption, and one initiates changes of activity. One stops doing one activity in order to begin another. The actions belonging to this range of actions are what might be called executorial, in contrast to the "automatic" sort described above.

There is, for the healthy human organism, always more than one action which could be attempted in one's circumstances. The human organism as executant will thus have to determine what it will do. But to act in this manner necessitates anticipating the conjoining of one's actions and one's experiences in one's circumstances in a certain way. The executory range of human actions thus involves the relating of one's possible actions to possible affairs.

Possibilities are acts and affairs which are conditionally doable or findable. They are first given one in and through certain patterns of readinesses effectually present in the human organism's bodily predispositions and concomitant actions (Bransford et al. 1977: 435) and later augmented by one's experience of life and one's imagination as one experiences in one's negotiations with one's circumstances some of one's actions and some affairs, as the border of a range or ranges of other activities and affairs not done or not here yet. These acts and affairs can be the "same" but can also be "otherwise" than they are at present, are able to be changed or replaced by something different, so that something new can begin to be done or encountered by one.

It should be noted that when one's acts fall within the first actional range of "automaticity," then no range of possibilities will be given to one, since one only does something about just those affairs then present.

A human life could not be based on "automatic" activity alone. It presupposes both actional ranges, the "automatic" and the executorial, and where the executivity of one's life cannot originate with oneself, as in the case of a very young child, then it must be present in the actions of others around one, such as one's mother. However, many actions originally belonging to the executorial range of human activity can "automatically" be done later.

From the first hours of life the human organism shows that it is capable of imitating the actions of other human organisms, and so of matching (in part at least) experiences which are correlated with or produced by the activity of others. Imitative actions fall within the range of "automatic" activity and are included in the behavior of many forms of animate reality besides the human.

In repetitive and routine activity, the human organism imitates his own previous actions and order of action again and again. Consider, for instance, the child who, when learning to write, continuously performs the movement of the hand and wrist by which a diagonal is made, until he becomes tired or something interrupts this activity. Imitation of both others' actions and one's own remains a primary way by which the human organism seeks to adequately negotiate with one's cricumstance throughout one's life.

Habitual activity arises when the human organism gives heed to a particular type of affair in any region of being as requiring a specific activity from oneself, namely, imitating "the same activity as before." One's habitual actions remain wholly within the first actional range, although the activity which is repeated may originally have been of the executorial sort. In so far as such comparisons can be useful, a habit may be described as an acquired instinct. The forming of these habits accompanies the composing by an individual human organism of one's own milieu of expected, "normal" contents, since being able to repeat the "same activity as before" necessarily involves one's encountering the "same" affairs.

Still, whatever one attempts to do, whether in an "automatic" or executorial manner, whether continuing to act as before or doing something else, whether in imitation of others or by habit, one may or may not succeed in doing it, for the physiological-psychological conditions of life (this body in this circumstance) may or may not allow it. These conditions provide the affordances for some affairs happening to the human organism and for the human organism to do something about these affairs. They also set for the human organism, along with the sort of community – with its history – in which one lives, the limits of one's life with those affairs belonging to the region of sensorially met and immediate realities in one's circumstances.

As disclosed by contemporary research in such areas as anthropology, archeology, history, and clinical psychology, the different manners in which human organisms live show that particular human organisms and communities of human organisms in specific circumstances only act in some of the ways that they are are capable of. Each of their actions reflects a particular restriction which they set upon what they can do, which also tends to restrict what they will experience. Thus, recognition must be given to the extraordinary parsimony of particular human actions and experiences.

Each action is a self- or socio-limited response of the human organism to the inciting presence and activities of the affairs one meets, in so far as one essays a certain kind of life in one's circumstances. That is, definite requirements are placed by one and/or one's community upon oneself, to be satisfied through negotiations with circumstances. One's actions then involve putting the temporal "distance" of specific exigencies between oneself as needing to deal with the affairs one meets, and oneself later when that will have been done.

The human organism has exigencies in the sense that, for instance, one needs to get to work without delay, more specifically needs to find the key to a door which will admit one into a laboratory, and even more specifically, one needs to examine a given specimen of possibly cancerous tissue under a microscope to which one will gain access through the opening of the door.

But aren't there natural exigencies? Of course. The physiological-psychological conditions of the human organism's life which ground the interactions of human organisms and affairs manifest themselves through the human organism's actions in connection with a certain set of needs. For instance, as an animate, bodily reality, one needs food to eat, air to breathe, water to drink, an environment which fits one's capabilities for action.

However, in contrast to the animal that always says "Yes" to life, the human organism can say "No." One is in Scheler's words, "the ascetic of life" (1961: 55). The human organism alone can repress any of these vital needs, or decide if it would be better to live no longer and suppress them all. Thus even these natural needs are effective within one's life only in so far as they become exigencies deriving from one's essaying a certain kind of life in one's circumstances.

It is here that Kurt Goldstein has made an important contribution to human studies. Through his clinical studies of brain-injured patients, Goldstein uncovered the role of ordered behavior, that is, of preferred patterns of acting and experiencing in human life (1963: 340–366). Not all actions, of course, are preferentially ordered, but much disorder appears threatening to the human organism. (If the human organism's actions or his circumstances appear to us as disorderly, we speak of their being "wild," and a "wilderness" always remains at the fringes of human life.)

Some of these preferences are to be found in the human organism's bodily predispositions, others are invented by the individual or acquired through participation in the life of a community. By the diversity of their preferences, which thus set restrictions upon their actions and experiences, the diverse particular forms of individual and communal life have been established.

Each of the human organism's acts must accord with its character; this is just the kind of human life which one essays in one's circumstances. That is, an act which accords with the unique, variegated and multi-leveled programming by the human organism of one's life in and through one's own preferred manners of acting. These preferences in their content and in their hierarchical employment will vary from one particular human organism and one particular circumstance to the next, but they will also reflect the "life program," that is, the culture of the historical community of which one is a member.

The human organism thus tends to follow certain regular patterns of action, that is to say, it tends towards developing one's own uses as well as adopting common usages, through which persistence one's character in preferred ways of acting is established and actualized. Nevertheless, one could at some succeeding moment modify this character, or even exchange it for a radically different one (or, as in the case of one type of mental illness, add on other characters) and so change one's preferred patterns of action. That is, human life should be seen in terms of the change or persistence of preferred patterns of individual and communal actions with regard to a circumstance which itself is thereby being more or less changed.

Gibson has noted (1966: 59–62) that all of the human organism's experiences depend upon the body as the original, primary framework within which one as actor is related to affairs (although the body is not as stable as he sees it, for significant changes in body size and sexual activity occur in the first twelve or so years of life, and other alterations occur through illness and old age).

Further, as Merleau-Ponty said, the body, in certain forms of behavior of the human organism in his circumstances has a certain hold on these circumstances (1962: 353–354). One's actions simultaneously specify one's own bodily powers and those of one's affairs. The experience of moving through one's circumstances simultaneously specifies the successive movements and postures of the mover as well as arrangements and rearrangements of one's circumstances (Mace 1977: 52).

The orders of sensorially met affairs in one's circumstances – for instance, their divisions into up and down, front and back, at the side, of their scale, the small and large, and the provinces of the manipulable and non-manipulable, the traversable and non-traversable, and so on – are correlated not only with the preferred patterns of his actions but with the human organism's upright posture, his height from the ground, the length of his limbs and extremities, his forward-looking eyes, and the movement of hands and feet.

There occurs through the human organism's activity, and especially its capacities for locomotion, a simultaneous zonal distinguishing and relating of intra- and extra-body affairs. In and through this duality of zones of encounter one experiences one's bodily sited-ness and one's installation amongst other situations (one is in this place, not that one). One can also experience changes of situation: of another place, another side, a different perspective, of proximity or distance, of approaching, receding, all of which also involve a reference to one's situation as something which is mobile. Hence the importance of an infant utilizing the capacities for movement of his or her bodily organs, learning to crawl, and so on, as their bodily maturation permits.

Affording the conditions for one's meeting of particular, immediately appearing affairs, but never themselves then as such mediating the affairs one meets, the energy spectrums there extend, correlated with the human organism's senses. Affairs are not encountered through the simple reception of the sensory components of experience: colors, sounds, smells, tastes, etc., or through the combining of such. Rather, such affairs are encounterable only on the basis of one's heeding sensorially given information presenting invariable distinctions and similarities (or, the unvarying difference of their presence or absence in certain varying situations).

This process of detection and differentiation of invariables begins even before one's birth and continues to increase in specificity of discrimination throughout one's life. The variety of invariable distinctions and similarities of observable unchanging contrasts and connections with which one's senses are capable of resonating, provides the human organism with the potential wealth of sensorially met affairs (Gibson and Levin 1975: 13, 15).

What one encounters sensorially is the affair itself, showing one or more of its multiple "faces." That affair which the human organism encounters in experience is not something "in" a mind nor "in" a nervous system (such as a representation or a stimulus). That is, it is not the effect of something which itself remains outside experience. One's experiencing is not a matter of a connection being made between "someone within one" and something outside one, but rather of the communication between oneself as a whole and as actor and some other reality or realities. It is a matter of address and response.

Each "face" of a sensorially met affair is encountered as something having a certain "thickness." There is about it that which is presentient of a latent platude; what is evident in experience appears as part of something more which is not evident, indeed, a "more" never wholly made evident. The "thickness" of an affair is something mediate only to be found through the immediate, that is, a surface. A surface reveals that something of an affair is concealed as such. One's explorations of sensorially met affairs always expose only surfaces, beyond which the "thickness" of things is still hidden. This is a type of experience of depth, but it is not the same as that found in the experience of one affair hiding another, which J. J. Gibson has discussed so well (1979: 76–86).

Further, sensorially met affairs may in experience be related to each other so as to suggest their connection to yet other affairs always waiting to be revealed, and thus for example, the supporting ground beneath our feet becomes "the earth," trees become "the forest," buildings and streets become "the city" (the "thickness" of a conglomerate of sensorially met affairs being experienced in terms of depth and the "nestedness" of all our interactions with affairs within other actions and other affairs on varying scales of duration and size).

Uexkull noted that the variety of types of affairs experienceable is equal to the variety of types of activities the human organism engages in to carry out in his particular circumstances (1957: 49). The human organism's patterns of actions are correlated with affairs' activities, that is, the preferred patterns of one's action are correlated with the counter-patterns one meets in one's circumstances. Each particular affair encounters acts, of course, controlling the kind of action one takes towards it, and each action likewise controls the kind of affair encountered.

Affairs can be experienced as undergoing certain kinds of changes, such as increasing, decreasing, aging, growth, and decay, while still remaining the "same" persisting affairs. In fact, the "persisting" elements or structures on a particular level of experience are themselves changing on other scales. Further, among the affairs one encounters there are events, that is, the activities of oneself or others and the changes they make in the one's circumstances, of more or less duration and occuring at slower or faster rates.

Past experiences "set the stage" (Bransford et al. 1977: 434) for subsequent actions and experiences through the human organism's having acquired patterns of readiness for encountering and articulating distinctive temporally ordered sequences as well as spatially ordered structures and thereby providing for one's recognizing the sameness as well as the uniqueness of information in one's present experience. This "stage-setting" experience of life enables one to direct one's further actions and explorations of one's circumstances by anticipating certain kinds of encounters with affairs belonging to any region of being, and to recognize in later experiences both when certain information is available and when it is absent. As one then expects to find some kinds of information rather than others, at times the affairs one encounters give occasion for feelings of surprise or disappointment.

Additionally, how the human organism experiences an affair in any region of being is influenced by his use, that is, by his persisting in preferred patterns of action. These preferred patterns of actions operate as selecting and accentuating controls upon the human organism's experience of his circumstances, so that one may encounter the "same" affairs over and over again in specific contexts – "same," in this instance, meaning that differences between affairs are ignored unless those differences go beyond the more or less fixed restrictions set by one's preferred patterns of action. These constantly reappearing affairs, or better, "types of affairs" that comprise the human organism's milieu, are the experience of a "normal" reality.

For the human organism to be capable of adequately negotiating with circumstances, some of the affairs one encounters must be the "same" as or match others one has encountered before, thus providing, by the consistency of affairs through successive experiences of one's circumstances, for the consistency of one's actions. Among the experiences of the "same" affair is to be included its absence (or in the case of sensorially met affairs, also the hiddenness of it or some part of it) as for example, in the case of the nickel for which I search on top of my desk only to finally become aware of its not being findable there at the moment. The very young child, it should be noted, must learn through experience of life to distinguish the absence of an affair from its non-existence.

Having the "same" affairs to deal with, one can then imitate one's previous actions, follow familiar routines, and perform the "same" actions again. Heeding the invariant features of the self-same, persisting affairs should not, as Gibson noted, be confused with heeding the invariant features which make different affairs similar. "Invariants over time and invariants over entities," as he put it, "are not grasped in the same way" (Gibson 1979: 249).

The experience of these activities and changes – in which something that persists must be met also – is entirely a matter of scale since events within one landscape and the landscape itself overlap or are "nested" in other events and landscapes at varying scales of size and duration.

Scaling is found, too, in what persists and what changes within the layouts of the affairs behind to the side, and before one, which approach and recede and transit the visual boundary between vague and distinct as one moves through one's environment. Here one should think of such experiences as driving a car down a road or walking along a street, or hammering a nail into a board. Some of the changes experienced always follow a certain pattern or order and some are idiocratic, depending on the random distribution of unique characteristics or the combination of characteristics of the particular portion of the environment through which one is moving or which is actively changing. But there are persisting patterns also as regards the sorts of randomness and the frequency of distribution of these unique characteristics to be encountered on every scale of experience.

Note that in experiencing scaling, one does not go from whole to part or vice versa, but from one whole to another whole relative to the size or distance of the affair or event, starting from the mobile situation of the experiencer.

At times, an affair may be encountered as a vague or doubtful something, and one cannot yet determine what to do about it, except possibly to inquire. One may then actively explore one's circumstances in order to make more information available, that is, not just move, but move so that it shows more distinctions and connections. Usually one seeks some "characteristic shape," the one which in terms of one's expectations, gives the most information. Such exploratory activity may be done just to get more information, but it is frequently done so one can accomplish something more in negotiations with one's circumstances.

The human organism's present experiences modify in some part such patterns of readinesses as one has acquired, which in turn allows new information to become available which in turn will modify those patterns of readinesses, etc. In this way, one always carries all of one's past experiences with one. (But this is very different from the way in which these same experiences can become available for recall.) Thus "present experience *depends upon* previous experience but cannot be reduced to memories of such experiences" (Bransford et al. 1977: 435).

Depending upon what the human organism's exigencies are, so an affair's characteristics come to have importance for one by affording one as actor with aids or impediments. Here one should remember Uexkull's (1957) description of the "tones" which animals and human organisms perceive as belonging to the objects of their experience. Ortega (1961: 199–200) and Gibson (1979: 127–143) have both also described the human organism's encounter with an affair in similar terms. The affair is experienced as being able to favor or frustrate, assist or resist one's actions. That is, the affair is encountered as an independent power or operation that in specific ways does or does not help one actualize that unique variegated and multileveled life program, one's character, which one essays in one's circumstances. By one's actions one interprets affairs and these affairs in turn interpret one's actions, which are then successful, partly successful, failures, and so on.

Accompanying and inseparable from every human organism's activity are feelings, emotional responses by which the varying relationships of actions and affairs are given positive or negative significance. Further, if in the course of negotiations with one's circumstances one experiences that one is capable of satisfying one's exigencies, then one responds by a feeling about oneself, a mood of well-being, or if one finds oneself incapable of adequately negotiating with the affairs which one faces, then one gets into a mood of anxiety. When one's body is damaged or diseased, one must seek new, more limited kinds of action and experience adapated to its changed structure, if a mood of well-being is to persist.

One may attend to an affair, centering on one of many affairs in one's circumstances. However, when one takes other affairs into consideration, the affair upon which one's centering attention focuses is as close to one as oneself; in relation to all other affairs, they exist together at the center. "Here" at the center is the presence of the affair and oneself – "there" is everything else which one relates to this presence at the center.

One's central attendance by the human organism to an affair in any region of being is correlatively a perspectival structuring of some particular landscape within one's circumstances whose foreground and background expresses how other affairs are related to further actions to be taken, or which may be taken towards this affair. Those other affairs which have the most relevance to the action one will or may take towards this affair are experienced as the closest affairs, those not quite so relevant experienced as further away, and some so irrelevant as to be "over the horizon" and not experienced at all. (This perspectival structure is independent of the bodily organization of circumstances in terms of proximity and distance.)

To illustrate this, consider the case of the nickel for which I search the top of my desk, only to finally become aware of its not being there at the moment. It may be on top of a stack of books within a few inches of my eyes, but it is absent for me all the same because it is beyond the horizon of my attention which is centered upon the desktop. The nickel thus sought for prevents my perception of the nickel before my eyes.

Through this perspectival structuring by one's attention of a particular landscape, the scene in which one does something about an affair is already constituted for one's on-going activity.

Since living for the human organism means not merely surviving, for one needs to live well – to actualize the "best" human life – the human organism seeks conjointly with negotiations with circumstances to meet just such affairs as will enable him to act in preferred ways and so as to allow that, within a limited range of variability, a specific kind of "fit" between oneself and one's circumstances can be maintained, in terms of which one is capable of continued preferred activity. One thus tends to seek not just some order with which one's preferred patterns of actions can be coordinated, but a certain "unsurpassable" ordering of affairs in one's circumstances. That is to say, developing one's own convictions as well as accepting common beliefs about the organization of one's circumstances supporting a continuation of preferred ways of acting. One becomes convinced that one is confronted by a universe of tacit and potential order towards which one as actor may turn with expectations of encountering regularities in the relations of the affairs themselves which can be coordinated in experience with one's preferences, both innate and learned.

Continuing to act in preferred ways requires that the human organism's actions take place within a specific field of action that is capable of connection with any other fields of activity in which one and others may act. Such a possible combination of fields of action, being in all its parts a field of action for us all, makes up a particular "world." A field of action is composed of interlocking "plans," that is, of particular individual or common patterns of acting which are connectable in advance (in the "future") through one's anticipation of being able to do "thusly," followed by one next doing "so and so," and so forth in the course which one takes in negotiations with his circumstances. These fields of action prefigure, within his experiences of his circumstances, specific landscapes with their particular ranges of typical affairs and correlated ways of acting, which then have a mnemonic role suggesting these activities. Each landscape may be connected with others, varying widely in scope and content, and may be "nested" in yet other landscapes.

One orients oneself to and is able to find one's "way" through the plurality of affairs experienced in a landscape by giving a pattern to their placement in accord with one's use. Some affairs stand out as "landmarks" in terms of their importance and are connected in a re-experienceable sequence of places correlated with a particular repertory of preferred actions. Relying on the regularity of the position of affairs, etc., in acting one can follow a "familiar route." Examples would be keeping things on one's desk in certain positions, going the usual way to the grocery store and so on.

With one's character in a particular landscape one can always justify one's actions to oneself: it is in each instance "best" for one to do what one has done (although later, from the standpoint afforded by further experience of life, or in a different circumstance, one may have "second thoughts" about it).

Sometimes the human organism will not or cannot simply "carry on as before," or just do what one observes another doing; one does not know what to do, or what is the "best" thing to do in one's circumstances. Sometimes one needs to act before what has incited one has become more than a vague or doubtful affair for one, when the determination of one's action must precede obtaining adequate information relative to that action. Yet one may experience a problematic situation not as anxiety producing, but rather as a matter calling for inquiry and exploration (cf. Straus 1966: 166–187): a possible action or affair is to be sought which best fits within the "gap" of a specific ignorance between the human organism and one's circumstances or between oneself and oneself.

But an outcome of such an inquiry or "journey" of exploration may not be just a matter of finding something which fits in that "gap," since it, or, if one fails to find it, the very inquiring itself, may be an incitement to change the way the human organism negotiates with its circumstances. One may change one's "plans," revising the field in which one acts or exchanging it for another or even inventing a new one, or one may change one's circumstances through a new arrangement or alteration of a part of it. Thus, instead of finding what one was seeking, one overcomes one's problem as a problem by acts of invention or innovation, and so can have new possibilities of action.

An affair which the human organism may be genetically prepared to respond to at the earliest moments after birth, and which one may more frequently encounter than any other soon after birth, is (the face of) some other human organism, usually one's mother. One thus in most instances comes to recognize another human organism before one knows any other kind of affair. The presence of another human organism is manifest to one not just as a body but as the one who is doing something to or for one, as an actor (the experience of a non-human activity comes later).

One meets one's own body as one among many affairs in one's circumstances through later experiences, for though the nearest, it is still not the first affair one encounters. As an extremely young child one only connects one's body parts with oneself when one links their appearance with one's own activity (or resistance) towards another body in which one finds a different activity.

The other actor that one encounters may be of the same or opposite sex, and recent studies seem to show that even young infants are aware of this distinction in others. However, the sexual distinction is not to be connected with the body having certain parts only, not something that begins here and ends there. A human organism's sexuality, just as the other physiological, psychological conditions of bodily existence, affords him (or her) aptitudes for varying relationships and sets limits for the possibilities of his (or her) life with other human organisms.

A final note: as the preceding study has been limited to an exposition of one human organism's life, it is, of course, necessarily incomplete. I have presented here only a "skeletonal" human life to which must be added the "flesh" of his socio-cultural community life, thus completing this consideration of human life by an account of the interactions of human beings with other human beings.

East Tennessee State University

REFERENCES

Ames, R., "Taoism and the Nature of Nature," *Environmental Ethics* **8** (1986), pp. 317–350.

Bennett, J., *The Ecological Transition* (New York: Pergamon Press, 1976).

Bransford, J., J. Franks, N. McCarrel and K. Nitsch, "Toward Unexplaining Memory," in R. Shaw and J. Bransford (ed.), *Perceiving, Acting, and Knowing* (Hillsdale: Lawrence Erlbaum Assoc., 1977).

Gibson, J. J., *The Senses Considered as Perceptual Systems* (Boston: Houghton Mifflin Co., 1966).

Gibson, J. J. and H. Levin, *Psychology of Reading* (Cambridge: MIT Press, 1975).

Gibson, J. J., *The Ecological Approach to Visual Perception* (Boston: Houghton Mifflin Co., 1979).

Goldstein, K. and M. Scheerer, "Abstract and Concrete Behavior," *Psychological Monographs* **53** (1941), pp. 1–7.

Goldstein, K., *The Organism* (Boston: Beacon Press, 1963).

Mace, W. M., "James J. Gibson's Strategy for Perceiving: Ask Not What's Inside Your

Head, But What Your Head's Inside Of," in R. Shaw and J. Bransford (ed.), *Perceiving, Acting and Knowing* (Hillsdale: Lawrence Erlbaum Assoc., 1977).
Merleau-Ponty, M., *Phenomenology of Perception* (London: Kegan Paul, 1962).
Ortega y Gasset, J., *History as a System* (New York: Norton, 1961).
Ortega y Gasset, J., *Some Lessons in Metaphysics* (New York: Norton, 1969).
Scheler, M., *Man's Place in Nature* (Boston: Beacon Press, 1961).
Straus, E., *Phenomenological Psychology* (New York: Basic Books, 1966).
Uexkull, J., "A Stroll through the Worlds of Animals and Men," in C. H. Shiller (ed.), *Instinctive Behavior* (London: Methuen, 1957).

Marlies Kronegger.

EVA SYŘIŠŤOVÁ

DIE SPRACHE DES TRAUMES UND DER TRAUM DER SPRACHE

Beitrag zur Phänomenologie der Träume in den kritischen Lebenssituationen

Mein Beitrag, welcher der Phänomenologie des Traumes in den kritischen Lebenssituationen gewidmet ist, geht von den authentischen Aussagen und Selbstexplorationen der bedrohten Menschen aus.

Die Träume in ihrer symbolischer Sprache und Bosch-artigen Bildern sind gleichzeitig eine merkwürdige Antizipation der Grenzsituationen des Lebens (z.B. des unlösbaren Wahls) und trotz scheinbaren Unverständlichkeit ermöglichen einen fast hell-seherigen Einblick ins Wesen des erschütterten menschlichen Daseins. Solche Träume, vor allem in den Knotenpunkten des Lebens, in den Momenten der ängstlichen Bedrohung, sind nicht nur die verblassten Schatten der verborgenen Wirklichkeit wie manche alltägliche Träume. Sie tauchen in einer Verdichtung als "Destillate" (als Essenz) des realen Geschehens. Sie dringen in das innere Drama der Lebenskrisis, das nicht rationell ergreifbar ist, ein. Manchmal erscheinen sie wie die verborgene allgegenwärtige Zukunft, welche real kommt, von uns nicht erwartet, aber desto ganz unausweichbar, als "Fatum", als Ewigkeit, die nicht die Horizontalen der Zeit kennt. Vor allem die Träume in den Siatuationen der äussersten Bedrohung, wo sich der Mensch auf der Kante seiner Endlichkeit befindet, sind fast "ontologisch", sie greifen tief zum Sein der zerspatlteten menschlichen Existenz.

Als Beispiele hatte ich einige Träume ausgewählt, die mir zwei Personen mittgeteilt hatten, mit der Bitte um Rat und Hilfe in der Situation der steigenden Spannung und Angst vor der kommenden Zersplitterung oder Sturz des bisher harmonischen In-der-Welt-Sein. (Beide, Herr R.N. und auch seine Frau waren begabte Naturwissenschaftler, die niemals die psychiatrische Hilfe brauchten.) Jeder hatte mich mit seinen Träumen allein, unabhängig von seinem Partner ausgesucht.

Die symbolische Sprache der ganzen Serie der Träume war so deutlich und expressiv, dass sie selbst das Erschliessen des Wesens der Krisis ermöglichte ohne Zugabe irgendeinen theoretischen Skeletten.

A-T. Tymieniecka (ed.), Analecta Husserliana LXVI, 241–252.

Die Träume von Herrn R. erschliesen den basalen ürsprünglichen Gehalt des tatsächlichen unlösbaren inneren Konfliktes seiner Frau, ohne er die faktische Begebenheiten kannte. Seine Träume enthüllten die Bedrohung ihrer Miteinanderseins. Er schaute auch in seinem Traum dem schrecklichen Sterben seines "Mörders" zu, und begriff auch seine Unschuld.

Die Träume seiner Frau, die mit ihm 10 Jahre harmonisch lebte, waren in tiefer Verbindung mit seinen hellsehenden Traumbildern. Tamaras Träume durchdrangen zum Kern ihrer unlösbarer Wahl. Sie sah den vergeblichen Zweikampf zwischen zwei Männer, aber keiner von ihnen ist mit Sieg ausgegangen: "Nur eine Illusion bleibt: ein brennender Strauch. Kein Tat, kein reales Verrat ist gewesen." Sagte Sie. Sie aber begriff, dass sie dem Traum der Sprache verfallen ist. Das war die äusserste Erfüllung für sie. Ihre Träume sind zugleich eine Antizipation der zukünftigen Ausmündung der Lebenskrisis.

Aussage von R.N.

Ich stand auf einer Strassenseite. Auf beiden Seiten von mir war je ein Soldat mit einem Dolch und einer Pistole. Ich war nicht gefesselt und war scheinbar frei. Die Wache verfolgte unverwandt und gleichmütig jede meiner Bewegungen. Mitten auf der Strasse standen zwei Jeeps. Zwischen ihnen hingst du, mit gefesselten Händen und Füssen, die Füsse mit Stricken zu dem einen, die Hände zu dem anderen Jeep angebeunden. Beide Jeeps setzten sich in Bewegung, jeder in anderer Richtung, zogen dabei die Stricke an und drohten dich in zwei Teile zu zerreissen. Auf der anderen Strassenseiten, hinter dir und den Jeeps, stand eine Reihe bewaffneter Soldaten. Sie standen da, die Hände auf die Gewehre gestützt und betrachten mich gleichfalls unverwandt. Ich musste zusehen bei dieser Marter, musste zusehen, wie sie dich auseinender reissen wollten, und durfte mich nicht bewegen. Du konntest die schreckliche Pein nicht mehr ertragen. Ich habe dich nie so verzweifelt schreien hören. Ich konnte es nicht mehr länger ansehen. Blitzartig erfasste mich der zwingende Gedanke, dieser Marter eine Ende zu machen: den Dolch oder die Pistole einem der nebenstehenden Soldaten zu entreissen und dich zu töten. Begreifst du das – als wenn man zusehen müsste, wie sie dich langsam auf einem Scheiterhaufen verbrennen – die einzige Gnade wäre da, dich zu erschiessen. Aber ich konnte mich nicht rühren, ich war völlig gelähmt infolge der Machtlosigkeit und Verzweiflung. Dabei konnte ich das Gefühl nicht loswerden, dass eigentlich nicht du das richtige und eigentliche Objekt der Folter warst. Als wärest du nur ein Instrument zu meiner Folter, als wenn der eigentliche Grund meine Vernichtung wäre.

Ich überlegte fieberhaft. Auf einmal fiel mir ein, ich müsste – auch wenn ich wusste, dass das praktisch unmöglich ist – gleichzeitig beide Fessel, beide Stricke durchschneiden, um dich lebendig zu befreien. Ich wusste, wenn ich das nur auf einer Seite schaffe, dass dich der zweite Jeep wegschleppt. Ich konnte im vorhinein nicht abschätzen, mit welcher Geschwindigkeit er sich entfernen würde. Ich wusste auch nicht, ob ich das alles auf einmal fertigbringe, ob mir inzwischen nicht die zwei Soldaten von hinten eine Kugel in den Kopf jagen würden.

Dann wusste ich nur noch eines: dass ich etwas tun muss, dass ich dich nicht so lassen kann, dass ich das nicht mehr länger ansehen kann. Ich entriss dem Mann rechts von mir den Dolch und durchschnitt die Stricke mit erstaunlicher Geschwindigkeit. Dann sah ich nichts mehr, fing dich in meinen Armen auf, du warst ohne Bewusstsein, aber du lebtest. Dann wache ich auf. . . .

Das war kein Bild, das man leicht vergessen kann. Dieses Geschehen, bedrohend und genährt von einem unbekannten Angstgefühl, wie es R.N. schilderte, blieb irgendwo im Hintergrund seines Erlebens, zur Klärung aufrufend wie eine hieroglyphische Schrift. Oder noch viel dringender. Er fühlte, dass der Traum wahrscheinlich eine enge Verbindung mit einem bischer unbekannten und unenträtselten Lebenskontext bedeutet.

Im Verlauf des folgenden Monates beunruhigten ihn weitere Träume mit einer ausgeprägten lebensbedrohenden Thematik, die er so intensiv durchlebt, dass sie sich – wie es ihm vorkam – mit einem Zustand von künstlerischer oder wissenschaftlicher Intuition vergleichen liessen. In solcher Intuition klärt sich meistens aber das Problem auf, oder es kommt zur Formulation einer schöpferischen Idee, derweil Roberts Traumereignisse auch trotz ihrer riesigen Lebendigkeit dunkel und unklar blieben.

Die weiteren Situationen, die sich ihm in den Träumen darboten, hatten eine gewisse Ähnlichkeit in dem Grundthema, sie waren gebunden mit äusserster Lebensgefahr. Sie nahmen nur die unterschiedliche Gestalt an und hatten in bestimmter Weise auch eine wesentliche Anbindung.

In der Nacht wurde ich auf der Strasse von einem Raubmörder überfallen, der mich erschiessen wollte. Ich war Polizist. Ich war zwar mit einer Pistole bewaffnet, aber der Mörder wusste, dass ich nicht schiessen darf. Es gelang mir, ihn zu Boden zu stossen. Der Mörder hörte, dass jemand kommt, und warf seine Pistole weg. Dann wurden wir vor Gericht gestellt. Der Richter begnadigte ihn und erklärte: "Bei diesem Menschen war das der Fall äusserster Not . . .".

. . . Ich war in meinem Arbeitszimmer. Auf einmal sah ich, dass gegen die Wand der Wohnung rückwärts ein Traktor einfährt, als wenn er die Wand einreissen, mein Haus zerstören und mich töten wollte; ich fühlte mich zwar bedroht, aber ich hatte keine Angst vor dem Tode. Durch reinen Zufall, wahrscheinlich durch den Stoss entflammte der Benzinbehälter im Traktor. Ich sah, dass der Fahrer Feuer fing, auf einmal in einem Feuerstand. Sein Kleid brannte, sein Haar, seine Haut . . . es war kein Mensch mehr, nur eine einige Flamme. Dabei schrie er, nein, er schrie nicht, er heulte vor Schmerz. Ich wusste, dass ich ihm nich zu helfen kann, dass es ganz aussichtslos ist. Ich sah den Menschen wörterlich verbrennen, ich sah, wie er verkohlte. Die Formen blieben erhalten, er war verbrannt und dabei lebendig, seine Haut war überall grauschwarz, alles war verbrannt. Dabei sprach er – ich weiss nicht was – nur eines, nämlich, dass er die Hoffnung zu leben nicht aufgeben wollte. Ich wusste nicht, ob ich ihm die Wahrheit sagen soll, dass das ganz ohne Aussicht ist. Ich emfand die Machtlosigkeit, die Unmöglichkeit eines Ausweges. Es war die völlige Apokalypse. Ich weiss nicht, wo all das herkam, früher hatte ich nie solche Träume. Für eine Weile erwachte ich völlig zerschlagen. In mir war eine tiefe Verzweifelung über das vernichtete Leben, über meine Machtlosigkeit. Eine Weile lief ich im Zimmer herum, konnte mich aber nicht beruhigen. Erst nach längerer Zeit schlief ich wieder ein und träumte wieder als wenn es kein Ende haben sollte.

. . . Diesmal sah ich eine Frau – sie war irgendwo im Krankenhaus – die auf eine sonderbare Weise behandelt wurde. Die Ärzte sagten ihr, dass sie ein Jahr lang in einem Ölbalsam liegen muss. In diesem wohlriechenden Bad lag also die Frau mittleren Alter, das Gesicht in ein Halbdunkel abgewandt. Auf einmal sah ich, dass in diesem Balsambad, das die Frau retten sollte, von ihrem schönen Körper Stücke, oder besser gesagt zentimetergrosse Hautstreifen mit Fleischstücken abfielen, so dass die blosse Muskulatur zu sehen war. Ich war erschüttert von dieser Destruktion. Der Heilungsprozess war ja im Wesen Mord an dieser Frau.

. . . Im Verlauf der weiteren Wochen sah ich in einem anderen Traum, dass mir ein Stück Fleisch an meinem rechten Unterarm feht. Eigentlich blieb von dem Unterarm bis zur Handfläche nur der Knochen übrig, das Fleisch war abgefallen, abgefressen, oder abgefault. ich überlegte: es bleibt nichts anders übrig, als dass sie mir den Arm amputieren. . . .

Diese Träume hatte R.N. in einem Zeitraum von etwa 1 1/2 Jahre. Sie traten irgendwie buschartig und gleichzeitig in Explosionen auf, dazwischen lagen scheinbar ruhige Zeitabschnitte.

Diese Inselartig auftauchende Träume, welche in manchen Umrissen ein beinahe psychotisches Gepräge hatten, trugen nicht den Charakter einer geistigen Krankheit. R.N. verstand sie zwar nicht, manchmal beunruhigten sie ihn, aber sie beherrschten ihn nicht, wie Wahn, oder Halluzinationen. Er verfiel auch nicht einer unbeherrschbarer Angst vor der Bedrohung. Er konnte sie aber nicht einfach auslöschen und von sich abwerfen, dazu waren sie aussergewöhnlich und sonderbar wirklich.

Er nahm sie hin, wie er sagte, als einen unzertrennlichen Teil seines Lebens, als eine sonderbare "Naturerscheinung", welche seine Gesetzmässigkeiten hat, wie ein kleines, wenn auch apokalyptisches "Wurder", das mit der gleichen Aufmerksamkeit wie zum Beispiel die bischer wenig durchforschten Prozesse in den Kernen der Sterne, nachgeprüft werden muss.

Er hat aber nie erfahren, dass die Begebenheiten, die in den Träumen des letzten Jahres durchlebt hatte, eigentlich wirklich geschehen sind. Er wusste nicht, dass sie in einem zeitlichen und bedeutungsvollen Zusammenhang mit dem Geschehen, mit wirklichen Situationen, an denen er unbewusst beteiligt war, standen. Im Traum waren sie nur in eine andere Sprache übersetzt, ungewöhnlich bedeutungsvoll, symbolisch verdichtet und gekürzt.

Er konnte nicht unbeteiligt sein an dem Leben seiner Frau, mit der er so lange innerlich tief verbunden war. R.N. ahnte gar nicht, dass er eigentlich alles ganz genau wusste. Dass er bis zu Kern der unlöslichen Wahl, welche das Leben seiner Frau zerspaltete, vordrang. Er konnte nicht daran vorbeigehen, oder besser gesagt, es konnte nicht an ihm vorbeigehen, weil seine Frau einen untrennbaren Bestandteil seines Lebens und seiner Welt darstellte.

Auf den Gipfeln der Träume, welche plötzlich mit scharfer Helligkeit die in Dunkeln schlummernde Landschaft beleuchteten, erschienen vor ihm in konzentrierter Form kritische Situationen im Leben seiner Frau.

Mit der hervorragenden Intuition, ähnlich wie im künstlerischen Schaffen, sprossen diese Träume in der Nacht seines Unwissens wie Lichtstrahlen, wie Reflektoren, die in die Mitte des Lebensdramas gerichtet waren.

Robert N. schätzte in seinen Träumen Tamaras und seine Situation ab, und das mit einer Hellsährigkeit, die die Grenzen des objektiven Wahrnehmungsvernehmens und des logischen Denkens überschritt. Allerdings nicht jeder Traum ist mit so einer zugespitzten Empfindlichkeit ausgestattet. Nicht jede Traumbühne ist so wesentlich real. Es ist notwendig, dass man an dem Schicksal des Anderen teilnimmt, damit wir ihn auch ohne Worte hören und damit wir in den Knotenpunkten seines Lebens beinahe alles kennen, auch ohne Kenntniss der faktischen Situation.

Auch Frau T., genauso wie ihr Mann kannte sehr wohl den Wert ihrer menschlichen Beziehung. Diese Verbindung bestätigte sich ihr auch dann- oder gerade in den Situationen, in denen sie gegen ihre Erwartung und trotz ihres Bemühens, versagte.

Als ihr R.N. nach einiger Zeit von seinen sonderbaren Träumen erzählte, lernte sie die Situation zu sehen, und sie zu verstehen. Aber sie konnte damals nicht eine Entscheidung machen, weil sie nicht wusste – ja damals als Robert von ihrer Folter zwischen den zwei Jeeps sprach, die in voller Geschwindigkeit in gegengesetzter Richtung losfahren wollten, wusste sie noch nicht, was geshehen wird.

Sie sah nicht die Möglichkeit einer richtigen Wahl und niemand konnte das für sie tun.

Ungefähr vor zwei Jahren traf sie einen jüngeren Mann, Bernhard H., dichterisch und musikalisch begabt, welcher zu ihrem Mitarbeiter wurde. Beinahe ein Jahr lang schenkte sie ihm keine besondere Aufmerksamkeit, trotzdem sie gemeinsam mit ihm an einer Übersetzung arbeitete. Sie schätzte ihn, ehrte sehr seine Begabung, er war ihr nahe, aber nichts mehr.

Herr B. aber begegnete in seiner introvertiver Einsamkeit einen verstehenden Menschen, der ihm, wie or sagte, von dem Verstummen gerettet hat. Wenn B. mit Tamara sprach, hatte das Gefühl, als wäre er allmächtig, als wäre er mit allem verwandt, als wenn sich ihm alles näherte und öffnete. Er kehrte in dieser Begegnung zu sich selbst. Bernhard entdeckte in Tamara einen Menschen, den er bisher nicht finden konnte. Seine Nächte, in denen er tastend herumirrte, fanden keinen Hafen.

Die Dialoge mit B. erfüllten Tamara mit nie geahnten Staunen. Es war eine Einstellung der Spiegel voll Visionen von einer unbekannten Stadt auf der Erde, so vertraut nahe und doch plötzlich so fremdländisch. Sie wusste, dass auch sie beide konnten nicht aneinender vorgehen, die, die sie so bekannt waren, hingerissen durch die gemeinsame Sprache.

Bernhard wusste, dass diese Begegnungen aussergewöhnlich sind, dass sie den Menschen fürs ganze Leben brandmarken, dass man durch sie das Leben erwirbt oder verliert. Er kann zu Tamara mit neuen aufdringlicher Hoffnung ganz ausgeliefert, angenommen mit dem unbestrittbaren Wissen der menschlichen Unzertrennlichkeit. Aber gerade in den Augenblicken, wo es schien, dass es unmöglich ist, die Hände nicht nacheinander auszustrecken, verfielen sie dem Schicksal zwei einsamer brennenden Kerzen von denen Bachelard schreibt: ". . . wenn sie sich einander endlich näherten, bezaubert durch ihre Vergrösserung und durch das Emporflackern, konnten sie sich nicht vereinen, und jeder ihrer Flammen behielt ihre vertikale Energie, um die Schärfe ihrer Spitze zu bewahren . . .".

Gerade in dem Augenblick der Wendung, des so befreienden Bruchs, welchen Bernhard immer von neuen erwartete und hervorrufen wollte, in den Augenblicken des menschlichen Empfangens, von dem beide Wussten, schieden sie schnell und schweigend voneinander mit einem Händedruck, jeder in eine entgegensetzte Richtung der Stadt.

"Das würde niemand glauben, das ist nicht mehr menschlich", sagte Bernhard einmal beim Abschiednehmen. "Sie können sicher nicht leugnen, dass es mir gelungen ist, ihre Welt zu öffnen, aber ich fürchte, sie haben mich nie in ihre Lebenshistorie eingeschlossen . . . sie nehmen mich nicht historisch wahr . . .".

"Es war hier kein Tat" – erklärte mir Tamara, "nichts sichtbares, aber desto so tiefes und so allgegenwärtiges. . . . Das Gedicht vielleicht, das Unsagbare, Schwiegende. Die Sprache des Traumes . . .".

Das Wesentliche begriff Herr B. nicht: dass die Sprache, mit der man den anderen anspricht und gewinnt, das unschuldigste und gleichzeitig auch das gefährlichste Instrument sein kann. Er begriff nicht, dass Tamara ihn ganz empfing in der narkotischen Entzückung der Sprache, in der er sich gab, mit der er sie erwarb und immer wieder fesselte. Bernhards Sprache war für Tamara viel wirklicher als für Bernhard selbst. Sie vertrat nicht das Leben, sie war das Leben selbst. Mit all dem, was er sagte, wie er es sagte, mit all dem eroberte er Tamara und bemächtigte sich ihrer. Aber diese Blendung, Resonanz, diese Wirklichkeit in Bernhards Worten waren so gross, dass Tamara gar nichts mehr brauchte. Bernhard begriff nicht, dass Frau T. in diesen Dialogen das Leben eigentlich bis zu Ende durchlebte, ohne es nötig zu haben, dass es sich in ein episches Ereignis entwickelt. Bernhard war für Tamara viel weniger greifbar als Robert. Trotzdem aber durchlebte Tamara das Leben mit Bernhard im Traume der Sprache. Bernhard vergass, dass ein Zwiegespräch mit seiner tiefen gegenseitigen Stimmung an sich ein anziehender Mittelpunkt nicht überschreiten kann, auch wenn man zu ihm zurückkehren will oder muss und wenn man dauernd von ihm angezogen wird. Bernhard aber wollte Tamaras reale Lebensgeschichte und die Sprache nur als einen Teil des Lebens, aber das brauchte Tamara nicht. Er kannte nicht die Macht seiner eigenen Sprache.

Nicht einmal Tamara begriff in ganzem Umfang das Wesen ihrer Verzauberung durch Bernhard.

Ihre "Schuld" bestand in der Tiefe ihres Lauschens den Worten Bernhards und in der Tiefe ihres Gedächtnisses. Darin beruhte auch die Unmöglichkeit ihres Loslösens von B. Aber wie auch dieses Band stark war, oder weil es zu stark war, brauchte es keine andere Bindungen mehr. B. hatte recht, als er sagte, dass Tamara ihn nicht historisch nahm.

Seine Versuche eigene Gespaltenheit zu überwinden, die immer von dem Gelingen scheiterten, trieben Bernhard zu einem hartem und scheinbar rücksichtslosen Angriff. Er verlangte von Tamara eine unaufschiebbahre Entscheidung, deren Tamara nicht fähig war.

Damals, gerade in der Zeit, sah sie Robert gefesselt und angespannt zwischen zwei losfahrenden Jeeps. Er wusste nicht, dass er im Traum einer realen Situation von Tamaras nötiger und unmöglicher Entscheidung zusieht.

Sie musste und konnte nicht wählen. T. wusste, dass jede ihrer Entscheidungen jemandem eine tiefe Niederlage bringt.

Robert war in seinem Traum durch einen Raubmörder bedroht – aber der Richter erklärte in demselben Traum von dem Täter: "Bei diesem Menschen ist es der Fall einer äussersten Not" . . . und begnadigte ihn.

Robert sah einen Mann, der verbrannte, als er Roberts Haus zerstören wollte. Er sah einen Mann, der nur noch eine einzige Flamme war, er sah sein grausames einsames Sterben in diesem Brand, diese unbarmherzige Folter, diesen aussichtslosen Durst, den verkohlten Körper und das Gesicht. Am schlimmsten aber war, dass dieser verkohlte Mensch weiter lebte. Dieser lebendiger Tote hatte noch Hoffnung, diese unsterbliche, vergebliche, hoffnungslose Hoffnung. Sein schwarzgrauer verkohlter Körper hatte noch dauernd einen unlörschbaren Durst.

Tamara im Mittelpunkt dieses ganzen Geschehens wusste, dass hier von Schuld keine Rede sein konnte, auch wenn sich alle gemeinsam unzertrennlich durchschalteten, ein unbekanntes Gericht über sich vorbereiten. Sie begriff, dass Situationen existieren, in denen man nicht wählen kann. Oder solche, in denen jede Wahl jemanden bedroht.

Tamara, gepresst durch das Räderwerk der unmöglichen Wahl, sich vergeblich fragend, wo eigentlich die Wahrheit ist, – diese "Leprakranke" aus Roberts Traum, verfiel in der Zerrissenheit der tiefen Schwermut. Sie reiste ab und arbeitete einige Zeit ausserhalb der Stadt, in der beide Männer wohnten. Es war kein feiges Entkommen. Sie wusste, dass sie Zeit brauchte zum Abstand und Selbsturteil und das alle eine Zeitlang allein sein mussten.

Aber trotzdem waren alle unzertrennlich verbunden. Robert träumte, dass ihm der Arm amputiert wurde. Bernhard sah seine Einäscherung ohne Erlösung in seinem Gedicht. Tamara durchlebte in der Zeit ihrer seelischen Erschöpfung einige Träume, welche sich auf die Leinwand ihrer unruhiger Nächte in einem scharfen, wie künstlichen Licht projizierten und welche sie sich gleich nach dem Wachwerden mit völliger Klarheit vergegenwärtigen konnte. Einer von diesen Träumen war in auffälliger Beziehung zu dem Traum Roberts, der ihre unlösbare Wahl symbolisierte. Er betraf diesmal aber einen unlösbaren Zweikampf der Männer:

Träume von Tamara:

Sie schaute aus dem Fenster ihres einsammen Hauses. Sie war allein da. Es war eine klare, frostige Nacht, überall Mengen von unberührtem Schnee, alle Wege zu dem Haus waren zerstört, alle Spruen verweht. Kein Weg führte weder hinaus, noch zurück. Aus dem Schnee brachen nur gewisse Reste von Schienen hervor, die in verschiedenen Richtungen führten, als wären sie zufor in die Luft gesprengt worden, drunter und drüber, zerstört. Die Bäume und alles Lebendige sahen aus als wären sie in Glas verwachsen. Einige Lampen beleuchteten die öde Landschaft mit künstlichem Licht. Eiszapfen, die an den Rändern der Lampen wie wachsende und reifende Traube hingen, schufen auf ihnen Trugbilder von Barocklustern.

In einer Entfernung von etwa Hundert Meter war eine einzige Strasse zu sehen. Der Verkehr war abgeflaut, es war schon sehr spät. Tamara sah auf dieser Strasse auf einmal zwei Panzerwagen, die mit grosser Schnelligkeit aneinanderprallten. Es war etwas unerbittliches, mörderndes und gleichzeitig blindes in diesem Zusammenstoss. Die Wagen waren gänzlich deformiert – niemand konnte diese Havarie überleben. Aus dem Wagen sprangen aber zwei Männer heraus und stürzten mit ausdrucksvoller Aggresivität aufeinender, als wollten sie einender töten. Die Personen liessen sich nicht identifizieren, alles spielte sich auf der entfernten Kulisse der Fahrbahn in Zwielicht der Lampen ab. Auf einmal, inmitten des Kampfes, waren beide Männer von Erdboden verschwunden.

Auf dem Platz der Versenkung spross aus dem Schnee ein Strauch von glühendem Feuer hervor.

Nach der Pause einiger Tage sieht Tamara im Traum Bernhard, der zu Hause an ihrem Tische sitzt wie bei einem Festessen, und isst gierig rohes Menschenfleisch. Das Blut tropft dabei in eine Kristallschale ab. Die abgenagten Knochen legt er neben die Schüssel auf den Tisch. Die Schale blieb bei diesem kanibalschem Akt strahlend und unzerstörbar in ihrer vollen Schönheit.

Tamara wacht aus diesem Traum mit einem Gefühl von Angst und Unwohlsein auf. Es geht ihr das Ritual des Opferns durch den Kopf. Sie wurde der hoffnungslosen Sehnsucht und der blinden Eifersucht geopfert.

Tamara dachte daran, dass eigentlich nur sie, mit ihrer Existenz "schuldig" ist. Aber sie wusste, dass es absurd wäre. Es kam ihr falsch vor.

Und woran war Bernhard schuld? An seiner ausdauernden Hingabe und mit seiner Sehnsucht nach einer nie vollendeten Tat? An dem Entdecken eines nahen Menschen, das manchmal im Leben einem Wunder gleicht?

Für Bernhard waren die unendliche Abschiede unerträglich. Vergeblich wollte er in sich das Bild von Tamara töten. Er ertappte sich bei dem hartnäckigen Wunsch, Tamara möge infolge einer unheilbaren Krankheit verunstaltet sein. Er selbst erschrack vor seinen Gedanken. Die Schwere ihrer Wahl begriff Bernhard nicht. Ihr Schweigen un ihr Fernbleiben batrachtete er als Entkommen und Verrat. . . .

"Sie verdienen weder zu sterben noch zu leben", hatte er in einem von der letzten Briefe an Tamara geschrieben. "Ist es nicht ein etwas spöttischer, grausammer und unsinniger Zufall", sagte Bernhard, "der mir eine Befreiung aus der Einsammkeit zuspielte um Alles zu verlieren, kaum dass ich es sah und erleben konnte? Woher kommt inmitten des kaum geborenen Glücks dieser Zwist, diese tötende Fratze?

Er siedelte weit weg von der Stadt, in der sie einander begegneten, in der sie diese qualvollen "Verlobungen der Trennung" durchlebt hatten. Eine Versöhnung gab es für B. nicht, er blieb allein.

Was aber wissen wir – sagte Tamara zu sich – die wir uns gegenseitig tief lieben oder hassen, von den Bündnissen, die uns verbinden? Was wissen wir, die wir urteilen und wählen – wenn auch mit äussersten Verantwortungsgefühl – und doch unwissend, wie Hazardspieler, wer von uns in diesem Augenblick verspielt hat, und wer von uns gewinnt?

Robert befand sich unmittelbar in seinen Träumen in dem Knotenpunkt des Dramas von Tamara und Bernhard, den er nie kennenlernte. Desto waren sie miteinander verbunden und sprachen sich miteinander in dieser sonderbaren Traumintarsion an, ohne davon zu wissen. Seine Träume waren eine Lichsspur in den Kern eines Streites und eines Urteils.

Aber sie waren keine Antwort auf die Frage, welche Tamara so dringend beantwortet haben wollte, inmitten ihrer unlösbaren Wahl. Existiert eine Entscheidung, welche in dieser Situation nicht eine Schuld wäre? Wo ist die Bürgschaft eines gerechten Urteils in diesem Drama? Begegneten sich hier alle Drei mit einer Nötigkeit, die sie selbst herbeigerufen haben? Robert befand sich nicht in dem Räderwerk der Wahl. Wo war aber die Schuld in der zufälligen und doch so schicksalsschweren Begegnung von Tamara und Bernhard? Machte sich Bernhard schuldig durch sein kompromissloses Gefühl, das nie zu Erfüllung kam?

Wir wissen, dass der Richter in dem Roberts Traume den hinterhältigen Dieb Gnade erteilte mit den Worten: "Bei diesem Menschen handelt es sich um einen klaren Fall der äussersten Not . . ." War Bernhard nicht wirklich ein Mensch in grosser machtloser Not? War für ihn nicht sein nie gestillter Durst schon eine höchste Strafe?

Die Sprache der Träume entblössste das Wesen des Geschehens. Roberts Traum, weit empfindsamer als das Augennetz, war ein hinreissender scharfsinniger Zeuge, vielleicht war er auch Richter, der die Gnade seinem potentionellen Mörder gab.

In Tamaras Träumen war aber kein Sieger in dem Streit. "*Es blieb nur Traum von dem brennenden Strauch*" sagte Tamara zu mir.

Karlova universita
Prag

REFERENCES

Blakenburg, W., "Phänomenologie als Grundlagendisziplin der Psychiatrie," *Fundamenta Psychiatrica* 5 (1991), pp. 92–101.

Syřišťová, E., *Die imaginäre Welt* (tsch.) (Prag: Mladá fronta, 1974).

Syřišťová, E., "The Reality of the Imaginary World," *Raw Vision*, no. 2 (1989) (London).

Syřišťová, E., *Die gespaltene Zeit* (tsch.) (Brno: Bollingenská věž, 1991).

Syřišťová, E., *Der Mensch in den kritischen Lebenssituationen* (tsch.) (Prag: Karolinum, 1994).

Syřišťová, E., *Das Gedicht als Heim in der Heimlosigkeit von P. Celan* (Uebersetzung der ausgew. Gedichte in tsch), Vorwort, Epilog (Prag: Foundation "Bíilý havran", 1994).

Syřišťová, E., "Die Bilder u. Mitteilungen von der abgewandten Seite des Mondes," *Tvar* N. 17 (1996) (Prag).

BRUNO CALLIERI

THE CONNECTION BETWEEN PHENOMENOLOGICAL CULTURE AND THE CLINICAL PRACTICE OF PSYCHIATRY

In psychiatry, more than a century ago, the wave of Positivist thinking, which at that time stood in strange and awkward juxtaposition to the Romantic movement, allowed the great blossoming of experimental medicine (following the model of the natural sciences – Claude Bernard is our reference here) and this constituted an extremely positive factor for medicine in the later part of the 19th century. However, it permitted the entertainment of the rash suggestions of a reductionism that was easily erected into a system by doctors who were epistemologically unprepared and unready.

Therefore, *alienist* medicine was determined univocally by a naturalistic environment, of the worst Lombrosian kind, arriving above all at the act of objectification, at the reification of the *other person*, of the sick person, which approach ended by completely forcing the suffering person into the anonymity of a category of objects, both in an etio-pathogenetic and a nosological sense.

In this framework, the function of psychiatrists with immense culture – I am thinking here about Tanzi or Bianchi – remained an essentially a-dialogical function, for fitting everything into explanations positing a mind-brain, psycho-brain causal identification.

The famous Griesingerian thesis that mental illness is brain disease, which dates from 1863, became absolutized with the radical medicalization of all who were suffering psychologically or were characterologically disturbed.

In all this, from my point of view, an inhuman act was performed, which deprived a man of his liberty, a liberty which I, as a clinical psychiatrist who has worked for years in asylums, have not only always respected, but have also perceived present and vital, as an existential dimension, even in those people who were most deeply immersed in that which we call mental illness.

On the other hand, we cannot but recognize the fecund horizon of naturalistic psychiatry, which today has yielded important results in biological psychiatry, in neuropsychology, and which we cannot *a priori*

A-T. Tymieniecka (ed.), Analecta Husserliana LXVI, 253–259.

minimize or negate, just as we should take real care not to overemphasize them in an absolutely undeserved and unworthy manner.

Even if these fecund horizons of naturalistic psychiatry are recognised, we cannot however fall into reductionism, as has happened in the past, in that banal, dogmatic and equivocal methodology against which Ludwig Binswanger wrote in 1957 a small and admirable little chapter "Psychiatry as a Human Science", which, first translated by Giacanelli in the "Writings" of Binswanger, was retranslated recently in a perfect contextualisation from the difficult Binswangerian German by B.M. d'Ippolito.

This new translation is an outstanding achievement because it has brought to us directly, in this sad moment of the domination of biological psychiatry, the precise voice, even though synthetic, of Binswanger, which sees psychiatry as a human science, and not just a psychological and mental science.

Binswanger addressed the optimism that considered that the problems of psychiatry can be resolved or will be resolved only by following the naturalistic hypothesis.

This was another act of faith possible only if one is not aware that the human being is to be unilaterally characterised by "life" (*Leben*) or by his natural being and holds instead, as Binswanger did, that he is to be understood fully, that grasping human presence-in-the-world is necessary, as is grasping the human person's need to historicize himself.

Therefore psychiatry is basically a human science: it is the science of human existence. Yes, it belongs to medicine, but *not only* to medicine.

Human existence is not only nature, but also culture and history.

The Husserlian *Crisis*, the famous *Crisis of European Science*, is one of the fixed points, the cornerstones on which rests all the thought of the century we are about to leave behind. At the heart of Husserl's response to the crisis is the rediscovery of intentionality, of consciousness, the concept of world-of-life, or lived world, on which are based the lines of inquiry opening the horizon of psychology and psychiatry, which saw an admirable flowering in Europe dating back, more or less, to the Twenties and Thirties. And it was working with concepts from Husserl's *Crisis*, precisely the opening of the *inter-subjective* relationship and *embodiment*, that enables Merleau-Ponty to demonstrate the importance of the concept of the lived body.

The problem of the interpersonal encounter persists. There is even today a need, to perceive and gather the *alter ego*, on whose constitution the Professor of Göttingen worked so hard, perhaps in vain, in his last years. Here perhaps is the future development of psychology, which could see an unfolding similar to that which occurred in sociology as a result of the work of the great pupil of Husserl, Alfred Schutz, who is much too often ignored in the psychiatric world.

It is possible to perceive a bipolar tension between nature and existence, between the 'case' as objectified in biological parameters, in scales and scores, in extremely subtle biochemical pathways, psycho-immunological, neuropsycho-endocrinological systems, and written up in international journals, and the 'case' met *here and now* in its singular reality, which is unrepeatable and irreducible. As a doctor, as a psychiatrist, I must say of the single sick person: "he is depressed, he is a depressed person, he is *that* depressed person".

This allows us an anthropological understanding in the etymological sense of *anthropos*, as the *unum* of the *individuum*, as above all some thinkers declared in the Middle Ages (e.g., Roscellino, Abelard's teacher, for whom "nihil *est praeter individuum*"). This was in the eleventh century, which saw enormous penetration of the sense of individuality, a sensibility which was then submerged under the invading sea of scholasticism.

In this way, the depressed person oscillates between the neurobiological explanations of his condition and research that anthropologically understands it in its existential dimensions, between the configurations of Seneca and those of Panofski, between the positing of a saturnine temperament to the attempt to recover the irreductible individual personality.

Between these two positions, that of the objectivised case in its biological parameters and that of the case confronted in its clinical and irrepeatable reality, I believe that one can stake out a *tertium*, which is the psycho-analytic approach. This is an ambiguous approach in the noble sense of the word, one without doubt causalistic in its metapsychology, but always a more humanistic one in practice. In any case, as it has developed in the last ten to fifteen years, it is balanced between nature and nurture. All this recalls an irreducible ambiguity between the old concept of explanation (the *Erklären* of Jaspers) and that of understanding (the *Verstehen*).

In reality, this distinction, which has become almost a scholastic principle in classical hermeneutics and in the Stockholm school (which takes up and develops old suggestions of Gadamer), is slightly marginalized: and the latest psychoanalysis seems to appreciate this fact. Thus this analytic approach allows us to render more complexly and more flexibly the picture in which we must absolutely place ourselves, not as observers, but as co-protagonists together with the patient.

We are not observers, we are *co*-protagonists in an event which is *co*-existential, and in this sense one could claim that the thinking of Gabriel Marcel is still alive: "*esse est co-esse*" (being is being-*with*).

Therefore, the psychiatrist, as Tellenbach says, must be a chameleon of methods (of methods, one must be careful to say here, not chameleon like in principles) because he, in his daily practice, meets the other person in his ambiguity between nature and existence, but always as a partner, never as an object reducible to a clinical case.

Psychiatry must be transformed into a dialectic process, articulated on the essential knot of inter-subjectivity, which is the proper human way of being. This involves a recovery of the subject, founded ineluctably on the singularity, or, better, on the *onticity* of the person, as is realised as well in existential anthropology. In reality, this leads us to conceive of consciousness as intentionality. Even before Husserl, it was Brentano who, in 1850, not far from the university where Griesinger worked, said that consciousness is always intentional consciousness, that is, is always consciousness *of* something. This has a whole revolutionary meaning for psychiatry, when we say that the 'I' always puts itself in relation to someone or something.

In this way, personalism and the question of encounter create the problem of the ethical dimension of the psychiatric act, as a regaining of liberty, from which platform we can propose a sound theoretical plan on which to found the therapeutical treatment of *psycho-analysis.*

At root, the 'I', even in its widest sense, the metapsychological, does not exhaust the reality of human existence: and it is this which Heidegger expresses, in his *Introduction to Metaphysics*, particularly in unfolding the concepts of *Wirzeit* (the time of us), and *Ichzeit* (the time of the ego).

This constitution of the "Us" inevitably involves the constitution of the *other.* And this is an unavoidable option for psychiatry: the necessity of constructing a phenomenology of inter-subjectivity, one going beyond the phenomenology of the 'I', for which we all are indebted to Husserl.

This phenomenology of inter-subjectivity is waiting to be constructed, not only in a theoretical sense, but in daily practice, following a path bristling with obstacles, which is profoundly ethical, in as much as it involves wholly the existential responsibility of the psychiatrist.

In fact, as Maurice Natanson, an American philosopher, states in *Psychiatry and Philosophy*, inter-subjectivity remains a radical problem of philosophy for those who come after Husserl. And it must be added that no one has to confront this problem in such a peremptory way as the psychiatrist; and the body becomes the central intermediary of this inter-subjectivity, naturally not in the sense of 'having' a body as Fromm, following Marcel, stresses, but in the sense that the body makes the encounter with the other person possible, as an accessible object in as much as the body is an embodiment of its subjectivity.

In this field which is propaedeutic, in every psychiatric clinical procedure, 'I think' should be replaced by 'we exist', a 'we exist' within which the encounter reveals the other person to me, not as an object but as a concrete singular existence, as *that* existence. And this must be emphasized, because in daily living, the 'we' which the phenomenologist encounters most often is the 'we' of indifference, a sort of cold, opaque, empty 'we': the 'we' of 40 passengers pressed together in a bus, the 'we' of human relationships in the social environment, the impersonal 'One' of daily existence.

Understanding this necessity of relationship, the centrality of 'with' reveals all the depth of Buber's insight that the fundamental fact of human existence is man with man, that is, human relationships. Having understood this, we should widen our subject to include the problem of psychiatrist-patient relationship, in particular in the world of psychotherapy, with its dialogical dimensions, which are not resolvable in psycho-logy, but as Mario Trevi says, in anthropo-logy. Here we find the world of encounter, which, to repeat with Buber, "is always grace, mystery".

This is the great subject matter which Max Scheler deals with in his small but fundamental book: *The Idols of Self-knowledge*, and of which lacanian Marisa Fiumanò boldly declares: "the present time is the time of transference".

This is, basically, the time in which one realises one's own '*ipseitas*; in co-existential ways. Along this path we find the equally important research of Victor von Weiszäcker on interpersonal relationships, or rather, on the bipersonal situation, of which he gives admirable examples (the two woodcutters who saw through a tree together, where it is not possible for only one to use the saw; the two rowers, and so on).

Here in its entirety the problem of reciprocity and participation is expressed, which demonstrates that to exist is also to participate, in accord with the interpersonal foundation of psychological consciousness.

On that foundation, the encounter is not to be understood only and simply as a "Mit-sein", being-with (that is, as an indefinite series, anonymous and impersonal, of daily returns to the *other*, which we always meet), but as a being-with-the-other, faithful to the insuppressible dialectic structure of the encounter (*Begegnung, Rencontre*).

At this point, the objectivated case, where we started out from, is transformed into an *alter ego*.

The anthropological development which derives from this dialectic and which tends to wedge itself between two poles made up of the anguish of the single person and the passion of existence, leads us to examine again in a completely different light, new subjects of primary importance in all of psychopathology, which then becomes Human Science: restlessness, the awareness of guilt, solitude, envy, shame, courage, expectation, hope, nostalgia, love, and I would also add, the search for God.

Areas of research open up, which can provoke perplexity or even scandal for many psychiatric authors, those who very simplistically and triumphantly reduce all of man to man-nature. And so, a path which is worth taking, opens up. Looking to the future, here is an immense moral task for us all.

University "La Sapienza", Professor of Psychiatry
Rome

REFERENCES

Baeyer, W. V. and R. M. Griffith (eds.), *Conditio humana* (Berlin: Springer, 1966).

Binswanger, L., *La psichiatria come scienza dell'uomo*, ed. Bianca M. D'Ippolito (Florence: Ponte alle Grazie, 1992).

Callieri, B., *Quando vince l'ombra. Problemi di psicopatologia clinica* (Rome: Città Nuova, 1982).

Callieri, B., *Percorsi di uno psichiatra* (Rome: Ed. Universitarie Romane, 1993).

Cargnello, D., *Alterità e alienità*, 2nd ed. (Milano: Feltrinelli, 1977).

Di Petta, G., *Senso ed esistenza in psicopatologia* (Rome: Ed. Universitarie Romane, 1995).

Galimberti, U., *Psichiatria e Fenomenologia*. Pref. E. Borgna (Milan: Feltrinelli, 1979).

Herzog, M., *Weltentwürfe. Ludwig Binswangers Phänomenologische Psychologie* (Berlin: W. de Gruyter, 1994).

Maggio, G., *Psicopatologia e linguaggio* (Milan: Masson, 1991).

Peursen, C. A., "Phénoménologie et Ontologie," in H. J. Buytendijk (ed.), *Rencontre* (Utrecht: Spectrum, 1957).

Schneider, K., *Psicopatologia clinica*, 3rd ed., Italian translation by B. Callieri (Rome: Città Nuova, 1982).

Schutz, A., *Collected Papers*: I, *The Problem of Social Reality* (The Hague: Nijhoff, 1964).

Stanghellini, G. (ed.), *Verso la schizofrenia. La teoria dei sintomi-base* (Naples: Liviana, 1992).

Straus, E., *Psychologie der menschlichen Welt* (Berlin: Springer, 1960).

Trevi, E., *Il lavoro psicoterapeutico* (Rome: Theoria, 1993).

Uslar, D. V., "Von Wesen der Begegnung," *Zeitschrift für philosophische Forschung* **13** (1959), pp. 85–101.

Weizsäcker, V. V., *Der Gestaltkreis*, 11th ed. (Stuttgart: Thieme, 1943).

Wyrsch, J., *Die Person der Schizophenen* (Bern: Haupt, 1949).

Zutt, J., *Neue Wege zur Anthropologie* (Berlin: Springer, 1961).

E. V. ALTEKAR

THE DYADICS OF COMPLEMENTARITY
Towards a New Vision of Reality

INTRODUCTION

> A clash of two doctrines is not a disaster, it is an opportunity.

The above epigram aptly describes the dyadic epistemological features of complementarity. The notion of complementarity not only handles much more adequately the important epistemological issues of modern science, but also allows significant advances in overall knowledge as well. This principle not only integrates knowledge that lies at different levels, but also suggests how to produce knowledge and also how to reconstruct the whole of philosophy.

The aim of the present paper is twofold.

(1) On the one hand, it is very interesting to see how complementarity organizes antinomic discrete thoughts into a homogenous form. In other words, how to integrate paradoxical and opposing thoughts coherently is taught by the princple of complementarity. Here complementarity functions as a general epistemological principle in the light of which discrete and paradoxical phenomena are united. In other words, it is a maxim or a rule which unites manifold structures. This function of complementarity I call the "Philosophical Foundations of Complementarity". By digging inside it, we can explore this type of philosophy.

(2) But on the other hand, there also lies another subtle function of complementarity which tries to explore altogether new ideas and creative thought structures. Here complementarity not only integrates and relates opposing philosophies, but also generates new philosophy (ies). Taken in this sense, it may be called a philosophy of philosophy. This function of complementarity I call "Complementarity as a Foundation of Philosophy".

If the principle of complementarity is explored in this way, the age-old problems of philosophy, such as the object-subject distinction, the mind-body problem, the problem of determinism and free-will, the realism anti-realism conundrum, etc. could be seen in a different way.

A-T. Tymieniecka (ed.), Analecta Husserliana LXVI, 261–286.

The argument presented here will be structured, and the paper is divided into four distinct discussions which are as follows:

(1) WHAT IS COMPLEMENTARITY?
  (a) The Correspondence Principle;
(2) THE PHILOSOPHICAL FOUNDATIONS OF COMPLEMENTARITY
  (a) As a General Epistemological Principle;
  (b) Applications of this Principle in Quantum Mechanics, Biology, Psychology, Sociology, etc.
(3) COMPLEMENTARITY AS A FOUNDATION OF PHILOSOPHY
(4) CONCLUSION

Let us first briefly look at the meaning of the term "Complementarity".

1. WHAT IS COMPLEMENTARITY?

Complementarity is a way of knowing and interpreting a world that exhibits itself paradoxically. Supposing that reality exists in a paradoxical and antinomic way, how then can one know it legitimately? It is via complementarity that one can understand reality properly. Complementarity steers us successfully through to claim the truth. Two opposite objects, events or concepts, although they seem to be contradictory, should not be taken to be simply contradictory but should be understood as being mutually supportive and complementary aspects of a single reality. If we change our epistemic structure in the above way, certainly we shall have a better claim on knowledge. Complementarity first and foremost accepts the dual or paradoxical nature of the world as a basic assumption (e.g., the wave/particle duality in quantum mechanics) for epistemic argument. Thesis and antithesis both are required for the fuller understanding of reality. These are not opposite but supportive units of knowledge. These opposite parts are reciprocally related to each other. The most important truth that complementarity reveals is the limitations to our inquiry.

The term complementarity not only plays its important role in physics but also in other sciences, and in philosophy as well. The term itself is very delicate and complex, giving rise to numerous obscurities. It is necessary, therefore, to clarify this term in more than one way. To do this, the principle of 'correspondence' should be grasped first. Of course, there are other allied sources of the genesis of complementarity, but I will consider only the correspondence principle.

1.1. *The Correspondence Principle*

Niels Bohr is commonly acknowledged to be the author of the correspondence principle. However we meet, in literature, different opinions about the date of birth of this principle. There is also a huge debate whether correspondence is an empirical or non-empirical concept. For example, Feyerabend treats it as an empirical concept.[1]

According to the correspondence principle, the quantum theory of the atom and of its radiation passes asymptotically into classical theory – when the quantum numbers increase or, in other words, when we may neglect Planck's constant (h). This theory shows how classical theory functions as a limiting case of the quantum theory. Bohr defines correspondence in this way:

> It is the combination of features which are united in the classical mode of description but appear separated in the quantum theory that ultimately allows us to consider the latter as a natural generalization of the classical physical theories.[2]

In the first instance, correspondence shows the relation between classical theory and quantum theory (or the relationship between macroscopic and microscopic phenomena). Quantum theory, although radical in its nature, cannot avoid using classical terms. This principle connects the new theory of atomic radiation with the old one.

The correspondence principle is also important with regard to the limits of epistemic powers. It exposes the limits of classical physics and also our knowledge in general. It is via correspondence theory that one can judge how classical physics works as a limiting case in quantum mechanics.

The correspondence principle shows how one could not only construct a very advanced theory, but also identify the limitations of the original theory. By knowing the rules of construction one may know what the conditions are under which an old theory may be applied and under what conditions it may not hold. In this process there seem to be ingredients of mutual correction and self-correction. This mutual corrective process plays a very important role in scientific activity.

The above-mentioned point is important in another context as well. Correspondence can also be used as a heuristic methodological device that includes the principles of theory construction. For the construction of a theory we require two things: 1) the discovery of new laws, 2) the formulation of statements. The correspondence principle provides this facility. It is through correspondence that semantic terms can be described and interpreted.

And then the correspondence principle is useful for bridging the gap between theory and observation within the hypothetico-deductive system as envisaged by Karl Popper.[3]

Although the correspondence principle serves as a heuristic device, it is a temporary arrangement and not a lasting solution to a problem. The correspondence principle is not a complete model describing reality; rather it is an underdeveloped model, helpful in settling certain contingent or urgent queries.

Having seen the overview of the principle of correspondence, let us now switch over to the main arguments of the principle of complementarity.

2. THE PHILOSOPHICAL FOUNDATIONS OF COMPLEMENTARITY

The line of my argument here will first of all show how the principle of complementarity works in different disciplines as a general principle. Then I will work out the philosophical argumentation of complementarity by which it can be proven to be a general epistemological principle. Previously the significance of complementarity, as envisaged by Niels Bohr,[4] had been recognised. This principle is actually a response to a physical paradox at the level of quantum mechanics.

2.1. *A Response to a Physical Paradox*

At the sub-atomic level, nature behaves in an altogether different way exhibiting a peculiar tendency toward contradictoriness. For example, particles of light behave paradoxically here. Light behaves sometimes like a wave and sometimes like a particle. This can be demonstrated very well by the two-slit experiment.

As shown in Figure 1, particles (electrons or photons) are emitted from the source S towards a screen Y in which are two adjacent slits. The particles reach the photosensitive screen Z and darkening is observed on the screen showing the diffraction pattern familiar from classical optics. The highest peak (the greatest darkening) is at C. The peaks correspond to places where the straight line paths from A and B differ by even numbers of half wavelengths, and the roughs where they correspond to places where the paths differ by odd numbers of half wavelengths. Thus, it looks like the particles are really waves. Nature has a surprise in store, however. If the intensity of radiation is gradu-

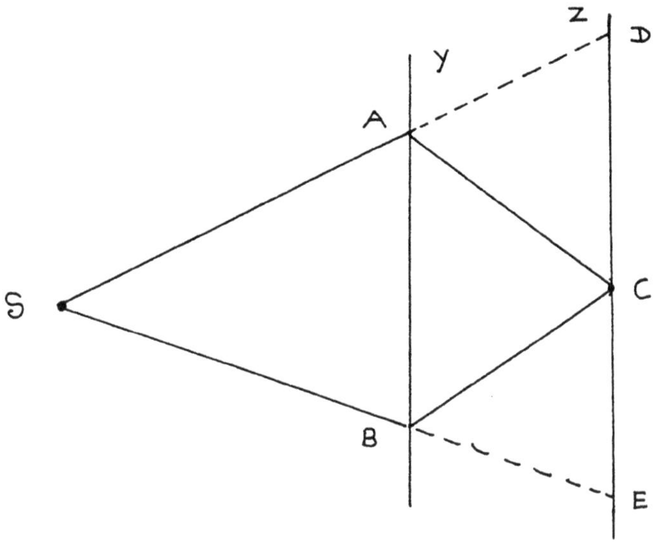

Figure 1. The two-slit experiment.

ally reduced, eventually single particles will be observed striking Z. It looks then like the electron or photon is really a particle. Classically, if the electron or photon went through slit A or B alone it would end up in the region of D or E. However, it is discovered that this happens only if the other slit is shut. This is puzzling because if it goes through A what should the matter of a slit being open or shut somewhere else (i.e., B) have to do with it? It would if the 'particle' were a spreading wave, but experimental objection is raised against this maneuver. If we get a scintillation or a blackening at a particular place on the screen Z, then we can say that the particle is definitely there. Until that moment, however, it does not seem to be definitely anywhere. In particular, we cannot suppose any definite trajectory between S and a place on Z.[5]

In conclusion then, we can observe the two phases of sub-atomic particles and so prove experimentally that they are opposite and mutually exclusive. However, this paradox can be successfully overcome with the introduction of the concept of complementarity.

It is not enough to merely solve the wave and particle paradox of quantum phenomena; it is also necessary to tackle the measurement problem at the sub-atomic level. In classical mechanics, there is no problem with measurement; if the initial velocity and the final velocity of a particular particle is known, a fruitful measurement is made. However, at the sub-atomic level, such type of measurement is not possible, it being difficult to access initial velocity. Moreover, a single particle cannot be isolated from the rest of the system, since it is an ensemble, and statistical laws cover measurement of the particle. In sub-atomic physics, the measurement problem is to be encountered when attempting to measure either electrons or photons. It is difficult to measure an electron. For its measurement two things must be known – its speed or momentum and its space or location. Quantum measurement poses the problem that simultaneous knowledge of momentum and location cannot be achieved with a particle like an electron. If the momentum is known, then its location cannot be known. Conversely, if its location is identified, its momentum cannot be known. Thus, these are mutually exclusive processes. Then the problem becomes one of how to know these two things in a meaningful way. Werner Heisenberg worked out a principle that shows the relationship between momentum and location. This principle is called the uncertainty principle.

2.2. *Heisenberg's Uncertainty Principle*

Any experiment in Physics, whether it refers to the phenomena of daily life or to atomic events, is to be described in terms of classical physics. The concepts of classical physics form the language by which we describe the arrangement of our experiments and state the results. We cannot and should not replace these concepts with any others. Still the application of these concepts is limited by the relations of uncertainty. We must keep in mind this limited range of applicability of the classical concepts while using them, but we must not try to improve them. This is the real paradox.[6]

Therefore, Heisenberg could on the one hand see the limits of the classical knowledge and on the other hand, feel the necessity of the old views and traditional concepts for quantum explorations. According to Heisenberg the new situation that occurred in the quantum world was altogether different from the classical view of the world. Here within the quantum theory the position and momentum of an electron cannot be precisely decided. Electrons absorb and radiate energy in discrete forms (or in discrete quanta) and further the electron's path is irregular. We cannot acquire reliable full knowledge of the electron because, if we try to measure the position of an electron precisely, we cannot at the same time, precisely measure its momentum and vice versa. If one is brought under control, the other is to that extent infinitely uncontrolled. Therefore, it is physically impossible to measure the position and momentum of an electron simultaneously. The relation between momentum and position is always uncertain. This physical uncertainty has been represented by the following formula:[7]

$$\Delta P \cdot \Delta X \geqq \hbar$$

where ΔP is momentum and ΔX is position and h is Planck's constant. This means that position and velocity are related not in any determinate manner but in an indeterminate way. This uncertainty is found under Planck's constant (h); h shows the basic or elemental indeterminateness on the basis of which infinite uncertainty may be deduced.

Heinsenberg's uncertainty principle explains the paradox of the two-slit experiment posed. From uncertainty there follow certain implications that are philosophically relevant and significant.

Among others, the foremost would be that out of the meeting of two different kinds of thoughts a new way of thinking emerges, that a new development would occur. This new development may be a revolution or at least a great change; thereby our whole conception of rationality and our thinking process itself may be transcended and take a broad turn and new direction. So when the Newtonian view clashes with the quantum view, a new turn in our very conception of science became possible and our very notion of rationality became broader and more comprehensive. Thereby our world perspective becomes richer in content.

Another consequence follows from the simple fact that when we try to determine the relation between the position and the velocity of an electron, our knowledge becomes increasingly inaccurate and uncertain, however well developed the method and sophisticated the instrument we use. Even probabilistic description cannot accurately measure the momentum and location of a particle. Now the problem is: despite our great efforts to measure the electron's position and momentum, why do flaws or errors in our observations remain? This question is related to another fundamental problem. This is the problem of the relationship between subject and object or between system and apparatus. The subject-object relation makes the problem of measurement increasingly complex.

That is because we cannot clearly draw a demarcation line between subject and object. The implicit problem in this regard is whether uncertainty exists in our observation due to the intrusion of the subject or whether objectivity itself is rotten or crude?

These two principles, the principle of correspondence and the uncertainty principle, respectively, compel us to think that mere physical technique cannot be of any help in getting rid of these paradoxes. Every time a question is posed to nature, it is paradoxically answered. Hence, ultimately, we must adopt some extra-physical way of tackling this problem. The concept of complementarity is one such way to be explored in overcoming the dilemma. Bohr used this concept for the first time in this context.

According to those advancing the concept of complementarity, there is a basic duality either in nature or in our thought, one that should not be merely taken as a contradictory fact, but which could be treated as a phenomenon involving complementarity. Bohr argues that a single picture gives us partial knowledge of a reality, whereas coupling together opposites gives us complete knowledge of that reality. This complementarity principle is, therefore, holistic or synthetic.

In place of a precisely defined conceptual model, the principle of complementarity states that we are restricted to complementary pairs of inherently imprecisely defined concepts, such as, position and momentum, wave and particle, etc. The maximum degree of precision in fixing either member of such a pair is reciprocally related to the measurement of the opposite number. Specific experimental conditions determine how precisely each member of a complementary pair of concepts should be defined in any given case. But no single overall concept is supposed to ever represent all significant aspects of the behaviour of an individual system precisely.

Bohr makes one bold claim with regard to the laws of quantum theory by stating that these permit or allow renouncing precisely defined concepts for complementary pairs of imprecisely defined concepts and models. Thus, he was able to prove that the use of complementary pairs of imprecisely defined concepts provide a possible way of discussing the behaviour of matter in the quantum mechanical domain. But a general point of view concerning the principle of complementarity goes further than this. For this assumption that the basic properties of matter can never be understood rationally in terms of unique and ambiguous models implies that the use of complementary pairs of imprecisely defined concepts will be necessary for the detailed treatment of every domain that will ever be investigated.

The main philosophical point that Bohr develops would be that the central feature of human knowledge is our distinction between subject and object, that the problems engendered by this distinction are logically similar to those which the instrument or system (field) distinction gives rise to in quantum physics, and that these can be handled by quantum description and the principle of complementarity. The movability of the partition between subject and object enables us, in some sense, to talk about ourselves. In many situations, for example, in introspection, experiences are inseparable from the circumstances under which they occur, and they are thus highly sensitive to the placement of the partition. Statements referring to different placement of the partition are complementary in the manner of statements made in quantum physics. The contrasts between complementary experiences are related to the differences in the conditions under which they occur. The richness of man's situation is rooted in the multiplicity of the conditions in which experience appears, or in the variety of ways of placing the object-subject partition.

The scope of complementarity is very broad. It is not restricted to a particular field of inquiry, but extends into other realms of systems of knowledge as well. Beyond physics, it is also operative in fields like biology, psychology, and the other social sciences. It is a general principle of inquiry. But a note of caution is necessary whenever its scope is extended. For, in each and every field, the sense of complementarity is different. For example, in micro-physics, we are dealing with antithetical components which ultimately belong to the same thing, i.e., matter, whereas in the case of biological phenomena, we are dealing with two different phenomena and philosophies so that it becomes very interesting to know how complementarity unites divergent philosophies together. So here materialism, on the one hand, and the phenomena of life, on the other, are necessary if we are to unite for the better understanding of living phenomena. Two opposite philosophies are deeply rooted in the conception of life.

After having taken stock of the complementarity in physics, I would like now to switch over to discussing the life-phenomena in the light of complementarity.

2.3. *Complementary Phenomena of Life*

According to Bohr when living organisms are placed under biological experimental conditions, they genuinely exhibit biological properties. The central problem in biology is that of how to know or understand "LIFE". The puzzle is whether a living organism is to be understood by reducing it to physico-chemical processes or whether it involves some higher and vital (teleological) principles. In the words of Ludwig von Bertalanffy, "Growth is undoubtedly one of the main problems of life. Indeed, growth has often been considered the central mystery of life".[8]

Growth, evolution, reproduction, metabolism, purpose, etc. are those biological processes which could not be understood simply in terms of reduction. Thus, a living organism is a conglomeration of the living and the non-living, of animate and inanimate components. The central debate in the philosophy of biology is whether reduction alone is sufficient for probing life-phenomena or whether in addition some anti-reductionist perspective is also required. In the philosophy of biology there are antithetical perspectives like mechanism vs. vitalism (this is the classic debate), reductionism vs. a systems view, structure vs. function, etc. which require interlinkage via complementarity. Let us briefly review these complementary pairs.

2.4. *Reduction vs. a Systems View*

The underlying philosophy of mechanism is the philosophy of Rene Descartes. According to Descartes reality is twofold: there is the body (or matter) and there is mind (or consciousness). All material objects or things are to be ultimately reduced to their indivisible parts or units. Explanation and knowledge of that respective object is given in terms of its basic units or parts. Due to this principle, we treat all living beings (including human beings) as being composed of two parts: body and mind. Hence knowledge of such living organisms is acquired by dissection, which approach is called reductionism.

But later on it was found out that the entire life-phenomena could not be dissolved into just material configurations. Recently, the limitations of reductionism have been pinpointed, for example, in the mysteries of mutations, neurophysiological processes, etc. which cannot be explained in the terms of reductionism only. An alternative model has been proposed, which can be called either a systems view or an organismic view.

There is another difficulty with the reductionist viewpoint. It has to do with the "structure" of living organisms. Mere structure does not exhaust all the facets and features of living; in addition, the concept of function is necessary.

2.5. *Structure vs. Function*

Structure and function seem to be opposites, but at the deeper level they are complementary and supportive of each other. For example, how may we learn about a cell? The cell has three phases or stages: birth, growth and decay. Now the important question is just how to acquire genuine knowledge from knowing its structure alone, since structure is a static conception. It gives a description of a certain stage of the cell at a certain point in time, but not the description of all stages.

The cell is a dynamic system, and in order to know it thoroughly it is necessary to know its dynamism through its functioning as well. Function throws light upon all the developmental phases of a cell right from birth to decay. *Both* of these components are mutually exclusive. For when we want to know the structure of the living being, we have to dissect the body, but when we do so, the living is dead. We cannot then know the organism's functioning and vice versa. Thus both components are exclusive; but they are necessarily related complementarity.

So from the foregoing argument it can be said that both reductionism as well as the systems view, if considered in an isolated way, cannot describe life phenomena fully and properly since each lays claim to half of the truth or reality. However, by thinking of these two alternative perspectives together, one can claim more reliable and better knowledge of the reality. Reductionism as well as the systems view are antithetical, but they essentially cooperate with and complement each other. It can also be gathered that when two opposite philosophies (e.g., materialism and animism) are coupled together, their claim to knowledge will certainly be far superior to that had when they are applied alone. The important truth lies in the process of complementarity in the sense that every antithetical or antinomous perspective serves essentially to show and identify the pitfalls and lacunae of its counterpart perspective.

Thus far it has been seen how complementarity works in physics and biology. It will be interesting to see its role in psychology and the other social sciences as well. Let us first look at the function of complementarity in psychology.

2.6. *Complementarity in Psychology*

The study of human behaviour may be undertaken from two points of view, from the point of view of an external observer and from the point of view of the experiencing subject. The first point of view may be roughly called behavioristic and the second phenomenological. Although these points of view tend to rigidly exclude each other, in recent discussions on the philosophical foundations of psychology, they are recognized to be more complementary than antithetical. This type of complementarity involves the most general orientation to psychological phenomena, but we may also note how the principle of complementarity works within the various particular systems of schools of psychology, e.g., psychoanalysis. For example, the work of such behaviorists as J. B. Watson, B. F. Skinner, E. C. Tolman or Clark Hull, on the one hand, and the work of phenomenologists like Brentano, Husserl, G. E. Muller, Karl Buhler, David Katz, Wertheimer, Merleau-Ponty etc., on the other is profitably studied in terms of complementarity. We then have more systematic and comprehensive knowledge about the human psyche.

Let us examine behaviorism and phenomenology very briefly.

2.7. *Behaviorism and Phenomenology*

2.7.1. *Behaviorism*

Behaviorism is a trend in modern psychology which is philosophically based upon physicalism, pragmatism and positivism. Behaviorism has been greatly influenced by Cartesian philosophy. It continues the mechanistic trend in psychology, reducing psychic phenomena to the reactions of the organism. Behaviorism identifies consciousness and behavior as the main units of which it calls the stimulus-reaction correlation. Knowledge according to behaviorism is entirely a matter of the conditioned reactions of organisms.

Behaviorism, in its most general sense, is concerned with the movement of the organism or its part in the external world. Behaviorism holds that only physical movement can be accounted for. Hobbesian theory gives us an important clue to the basis on which behaviorism came to be formulated as an attempt to interpret all mental states in terms of matter in motion.[9]

In 1914, J. B. Watson coined the term "behavior"[10] for the human act or human behavior which is extroverted and hence observable. Watson defines behaviorism as follows:

Psychology as the behaviorist views it is a purely objective, experimental branch of natural science which needs consciousness as little as do the sciences of chemistry and physics.[11]

Behavioristic psychology is thus a scientific study of human behavior eliminating all reference to mind, emotions, thoughts, feelings, etc. What Watson means is that all human behaviors are physiological processes that could be reduced to or be made explainable by the laws of physics and chemistry. Ivan Pavlov's conditioning experiments on dog behavior supported Watson's thesis.

Behaviorism was thus modeled after physics, especially Newtonian physics. But it is precisely because of this that behaviorism has had certain limits.

A. O. Lovejoy, in his essay "The Paradox of the Thinking Behaviorist", sought to prove that, "Behaviorism belongs to that class of theories which become absurd as soon as they become articulate". He further states that the behaviorist does make cognitive claims. One alleged contradiction consists in this, that awareness of things distant in time and space cannot be identical with movements of certain muscles or with bodily movements of any sort. For no such description of internal bodily events can account for such reference made to external objects.

Similarly, C. D. Broad makes this criticism against behaviorism. However complete the behavior of an external body answers to the behavioristic test for intelligence, it always remains a perfectly sensible question to ask, "Has it really got a mind, or is it merely an automation?" Here it is assumed that mental properties have been ascribed to actual physical behavior, yet a meaning cannot be derived from this, since intellectual activities cannot be identical with the physical and behavioral. There are other criticisms as well.

The above argument over behaviorism shows an important feature of complementarity's logic and epistemology. In quantum theory, we saw how knowledge of a corpuscle will drive us implicitly to embrace knowledge of the wave which is the counterpart of the corpuscle. Similarly, the limits of classical mechanics have been exposed by quantum phenomena.

In the same way behaviorism shows the following complementary pattern of indispensability and limitations:

(1) Necessity. Behaviorism has demonstrated its importance in psychological explorations. In other words, it has proved it is a necessary component for understanding man's psychological states, i.e., his psyche. The work of Watson, Skinner and others has shown that scientific study does explore mental processes more systematically and precisely.

(2) Limits. Behaviorism is successful to the extent to which man's external or extroverted actions are concerned, but it cannot grasp man's innermost mental structures properly. Here behaviorism is exposed in its limitations, which have been confirmed by studies revealing major gaps and defects in behaviorism.

By citing these two limits, it follows logically that we require an alternative scheme which takes note of the inner structures of man's mental life.

The view diametrically opposite to behaviorism is the phenomenological one. Although there seems to be only contrast between the two schools, at a deeper level they are complementary to each other, since both are concerned about man's psychological experiences. Behaviorism deals with external behavior whereas phenomenology deals with the inner experience of man. Hence, they are complementary.

After this deliberation, let us turn to the phenomenological account of psychology.

2.7.2. *Phenomenology*

As elaborately discussed in the preceding section, behaviorism is concerned only with man's external manifestations. Phenomenology, on the other hand, considers his inner aspects. Behaviorism shows how man responds to external stimuli and is constantly under the influence of his environment, whereas phenomenology is concerned with his inner state of consciousness.

a. *What is Phenomenology?*

Phenomenology believes that whatever is immediately and intuitively grasped about the life-world is a true reality. It is in essence the realisation that a true and pure form of consciousness prevails throughout our experience. It is the self-awareness that inner consciousness supports our experiencing the real world. From the phenomenological perspective the foundation of any inquiry – psychological, sociological, in either the natural or the social sciences – is in fact the person's understanding of everyday life wherein lies the source of meanings central to and implicit in inquiry.

Although the term "phenomenology" has a long history, it has gained popularity in philosophical and other literature mainly through Edmund Husserl.

The word "phenomenology" derives from two Greek words, phenomenon and logos. Phenomenon means appearance or that which shows itself. In psychology, phenomena are commonly defined as the data of experience that can be observed and described by the experiencing subject at a given time.

According to Husserl, phenomenology is a science of phenomena, of all the data of consciousness; or in other words, phenomenology is the science by which we come to know mind as it is in itself through the study of the ways in which it appears to us. Husserl's phenomenology attempts to answer one of the oldest and most fundamental problems of philosophy. What is the relationship between the objective reality existing outside the mind and the thought which we have of it? How are these two worlds, the world of thought and the objective world, to be fruitfully reconciled? The answer would be a) that philosophical inquiry cannot begin with anything other than the phenomena of consciousness because they are the only ones which are accessible to us, and b) only phenomena can reveal to us what things are. So here the endeavour is made to avoid the subject-object dilemma, which was a baffling problem for almost all previous philosophers. According to Husserl the only possible approach to knowledge of things is through the exploration of human consciousness. Thus phenomenology is nothing but a systematic way of fully exploring consciousness.

Husserl accords a central place to intentionality in his phenomenological explorations. For Husserl, consciousness is always consciousness of something. All the acts of consciousness are related to or point toward something. Knowing the content of consciousness means knowing the object itself. Since an object is accessible only in consciousness, the sole way to knowledge of things is through the examination of consciousness. The purpose of the phenomenological method is to make this examination effective, so that we can arrive at knowledge of things themselves without being entangled in solipsism.[12]

The most important epistemological strength of phenomenology is its scientific temperament. Phenomenological inquiry begins by bracketing a phenomenon or object in order to study it without any preformed judgements. The inquiry starts by suspending all assumptions and presuppositions. And when the study is completed, and only then, are the brackets dropped. This is what science actually demands. Here basic assumptions are suspended, therefore, no question arises about whether basic elements or basic principles have been left unexamined. Here inner consciousness is revealed to us; it opens us up for the examination of facts.

Phenomenology is, then, a very important component of the study of mental states. But it also has certain limitations. Let us list a few of them very briefly.

b. *Limitations of Phenomenology*

Like behaviorism, phenomenology also suffers from certain limitations:

(i) First there is the "Phenomenological Fallacy", the result of a logical mistake. The mistake declares that when a person describes his experience of things with their attributes like appearance, sound, smell, touch, etc., he is describing the literal properties of objects or things within his internal conscious screen, what modern psychologists call the phenomenological field.

All the properties of the object are said to depend upon the phenomenal properties of our consciousness. We first know the phenomenal properties of an object and thereby know the real properties of the object, it is said.

But actually in fact we do the reverse. First we know the real properties of the object (this has been proved by learning theory), and then on the basis of that, we derive the phenomenal characteristics. Thus, for example, we first see a green field and then comprehend the notion of "greenness". This is the criticism made by U. T. Place.[13]

(ii) Psychoanalytic phenomena cannot be understood phenomenologically since phenomenology mainly concerns itself with pure consciousness while psychoanalysis is concerned with unconscious aspects of the human mind. In comparing and contrasting phenomenology with psychoanalysis, major limitations of phenomenology may be identified. Paul Ricoeur has identified four such limitations.[14]

(iii) Phenomenological inquiry is vulnerable when it handles the problem of "Intersubjectivity".

Thus, while phenomenology has an important role to play in psychology, we must also note that the phenomenological point of view itself is not all sufficient. The limitations of phenomenology from the point of view of psychology as a discipline arise precisely in connection with its transcendental character. For Husserl, the other (person) is in a sense constituted in my consciousness, but the other is actually not the object of a subject but a co-subject.

On evidence, then, while psychology may be greatly enriched by the phenomenological point of view, it also requires principles other than purely phenomenological ones. In other words, the response of psychology to phenomenology must be framed in terms of the principle of cooperation.

What of the notion of complementarity in sociology?

2.8. *Socio-Culturological Dimensions of Complementarity*

Complementarity can be discussed at a general level overarching sociology and cultural anthropology, and it can be discussed on specific levels within the paradigms of sociology or culturology.

2.8.1. *Complementarity between Sociology and Culturology*

a. *Social but not Cultural*

Here one can argue that the sociological and the culturological aspects of human action are not contradictory but rather complementary. In the context of sociological inquiry, culture has always been treated as a part of the sociological studies, but as being secondary to sociological phenomena. But if we look at social life from the viewpoint of complementarity, this sociological interpretation is found to be inadequate.

b. *Cultural but not Social*

Social action can also be looked upon in terms of the meanings, norms and values which regulate social action. This is how cultural anthropologists view social action. Here entire social activity is explained through culture. Culturologists generally subordinate the social to the cultural and consider the science of human society to be a science with a narrower scope than the science of culture.

But I would say that this too is an inadequate outlook and that both of these viewpoints are required for the optimal understanding of society and culture.

c. *The Socio-cultural*

By treating either sociology as subordinate to culture or culture as subordinate to social reality, one cannot gain proper insight into human action. But if one can proceed in a coordinated way, by integrating sociology and culturology, one can definitely have a much enriched and more subtle purchase on knowledge of human action. If taken in this sense, these two aspects then belong to a single reality which can be called "SOCIO-CULTURAL". According to Nadel,

> Social facts are two-dimensional. Like any two-dimensional entity, they can be projected on the one or on the other coordinate, and so viewed under one or the other aspect. If we wish to find names also for the dimensions themselves, they seem suggested by the familiar words 'society and culture'. Society means the totality of social facts projected on the dimension of action. Mostly, when we speak of 'culture' and 'society' we mean a totality of facts viewed in both dimensions; the adjective 'social' . . . has always this double connotation. Nor do we possess a convenient term summarizing this twofold reality as such save the clumsy word "socio-cultural".[15]

For Nadel the word is clumsy, but for me it is not. On the contrary, it seems to me to be complementary. Both interpretations for social action are claiming the same thing but with different methodologies, conceptual schemas and articulations. Both approach a holistic point.

Such types of complementarity function at the general level of inquiry. But complementarity also functions within particular paradigms. For example, within sociology, the subjectivist interpretation of Max Weber, on the one hand and the objectivist interpretation of Durkheim, on the other, provide a splendid example of complementarity.

But let us see how complementarity is a general epistemological principle by citing some of its features.

2.9. *Complementarity as a General Epistemological Principle*

The previous discussions of how the complementarity principle can be applied in different disciplines have shown the power and fruitfulness of this idea. But the very fact that the principle can be illustrated in a variety of disciplines raises larger and more difficult philosophical questions.

The previous discussions have revealed that the principle of complementarity functions in individual disciplines in a dual manner. On the one hand, it is a response to the need to resolve certain antinomies and paradoxes. The wave-particle duality is a paradigmatic example of such an antinomy. So here it functions as a general epistemological principle, in the sense that it organizes or synthesizes opposite concepts, systems, theories and perspectives complementarily. On the other hand, it also functions as a growth promoting principle, in the sense that a paradox also opens the way for newer theories and hypotheses. The growth of quantum theory is a striking example of how growth takes place within a discipline precisely through the resolution of an antinomy. It is in this sense that the principle of complementarity may be called a principle of creativity. This latter view however, is better discussed in the next section; here the general epistemological features of complementarity will be discussed.

(1) The principle of complementarity is not a physical principle, nor any empirical reality; it is transcendental in its nature. Complementarity first of all identifies the two polar experiences of the same phenomenon which are needed to ascribe meaningfulness to it. In the absence of complementarity, these two experiences would have to be thought meaningless since they are opposite. But complementarity treats these experiences in such a manner that they no longer look vague and absurd. Complementarity accepts their co-present status, as being simultaneously true.

(2) The second important implication of the principle of complementarity is that it relates these two opposite experiences on its own level. Complementarity removes the threat of internal inconsistency between these two opposite manifolds of experiences and integrates them meaningfully under its own umbrella.

(3) Opposite experiences are synthesized through complementarity. It is through the principle of association of ideas that apparently discrete and disjoined sense-manifolds are united. Here a particular experience loses its own identity and becomes one with the complementarity realized. Thus, light cannot be called either a wave or a particle. This process is called the generalisation of a particular. When such a comprehensive integration is sought by apprehending complementarity, it functions as a rule of unity which determines all the manifolds.

Although complementarity is a vague, obscure and imperfect concept, it is necessary for the construction of a system of knowledge. Complementarity is a concept which is essential to solving the wave-particle duality; quantum mechanics is a telling example.

The form of complementarity is a universal which serves as a rule or maxim. For example, in various disciplines complementarity has shown how to coherently integrate different disciplines into a whole unified pattern. The complementary unity that we are talking about is a formal unity, which means formal unity of consciousness. Appearance must meet the necessary condition of the unity of apperception. Only then is knowledge possible. Only under such circumstances can we have a knowledge (of either an object or experience, etc.).

To conclude this section, we can say that complementarity presupposes a Kantian frame of reference and that within such a framework complementarity will be a rule, or a maxim which regulates the formation of concepts, hypotheses and theories. The rule of complementarity will function in a way analogous to the functioning of the categories of critical philosophy. It may be recommended that for Kant a category represents the rule of understanding which prescribes a mode of synthesis. The categories, therefore, are the forms of synthesis. Kant stresses that the function of the categories is to enable us to "read" phenomena as experience. The categories, therefore, are hermeneutic devices which constitute an intelligible experience of the world.

But there arise important problems at this level. One of them is that of how to relate the category of thought to the manifold of experience, they being heterogeneous? There might be another problem in the origin of the categories. How have the categories emerged? To make a synthesis of experience, categories are needed, but in order to synthesize the categories themselves, what is required? I think I will postpone the discussion of these questions until the end.

Let us now switch over to another feature or dimension of complementarity.

3. COMPLEMENTARITY AS A FOUNDATION OF PHILOSOPHY

The principle of complementarity not only organizes knowledge fruitfully but also opens new vistas to our understanding by transcending and transgressing antinomies and bipolarities. The principle of complementarity shows the limits of each term of an antinomy. Conversely, through these antinomous constructions one is simply not able to know the reality. That is to say that complementarity functions at such a deeper level that it cannot be empirically experienced or rationally understood. For it falls beyond observation and ratiocination. It is thematic in nature. This thematizing functions as a regulative and a constitutive principle of knowledge. It is regulative in the sense that it shows or prescribes the range of categories or concepts. But this principle is also constitutive in the sense that it shows how categories are content along with knowledge.

The principle of complementarity functions at such a deeper level that it, on the one hand, reveals the hidden powers and properties of the world and on the other hand, reveals the subjectivity of the individual. It reveals at once what underlies the world and the expression of self.

When the question of the application or function of complementarity arises, we have to turn to the Kantian programme of schematism. Just as Kant limited the role and function of categories to phenomena, so also the principle of complementarity should be limited to our knowledge of nature and not be extended to nature in itself. Such would be the general implication of a Kantian understanding of the principle of complementarity. But even within such a Kantian interpretation there is the possibility of raising basic and novel questions. For example, in so far as the principle of complementarity is understood as a metacategorical principle which prescribes the context of applicability of specific categories, it is somewhat analogous in its functioning to the schematism in Kant, for that schematism also is a rule which enables us to apply what is given in sensibility. Similarly and at a higher level, the principle of complementarity prescribes the applicability of concepts and categories.

So it is in the light of the Kantian frame of reference that the nature and function of complementarity is to be understood. But the implication of using a Kantian frame of reference is that its role must be seen as being epistemic rather than ontological. But this may create some delicate problems. Firstly, given the above Kantian understanding of complementarity, it must be obvious that it has to be *a priori*. A transcendental principle for Kant is not much concerned with phenomena in so far as such knowledge is *a priori*. Hence within a Kantian understanding, complementarity must be transcendental and *a priori*. The difficulty with this is that, as our previous discussions of the origin of the complementarity principle in modern quantum theory clearly suggests, this principle is the outcome of the history of a discipline. Niels Bohr, for example, was led to formulate the principle of complementarity on the basis of earlier developments in physical theories up to Max Planck. It is somewhat difficult to emphasize this historical context of complementarity within a Kantian frame of reference.

But a second constraint imposed by the Kantian model is even more serious, for the most important feature of complementarity is its growth promoting character. We have all along emphasized this feature, i.e., that it leads to novelty. It is difficult to see how a Kantian ahistoric a priori principle may explain creative novelty. Therefore, it appears that there may be certain constraints in the Kantian interpretation of complementarity.

Hence we may have to consider other possible philosophical frames of reference. One such model which may be explored may be called the Aristotelian-Leibnizian model. Here the basic understanding of complementarity would see in it the truth about things primarily and only the truth about our ideas as a consequence. In other words, it is because the world itself is complementarily structured that our cognition needs to be regulated by the principle of complementarity. The principle of complementarity, however, is not to be seen as a mere copy or reflection of things; it is to emphasize this aspect of it that the model is called Aristotelian-Leibnizian and not merely Aristotelian. For one of the important insights of Leibniz is that each monad cognizes other monads from its point of view and that such cognition is the result of the dynamic and spontaneous activity of knowing subjects. Knowledge for Leibniz is never a mere passive reflection but is an active contribution. But this constructive activity is founded on the nature of things. Hence within the Leibnizian framework, complementarity would be an activity of thought and not a passive reflection. At the same time it would also express the nature of reality. Hence within such a frame of reference, the principle of complementarity would be ontological too and not merely epistemic. Furthermore since the principle would be the result of an activity of the mind, it could well be said to be spontaneous. Thus, not only Kantian philosophy, but Leibnizian philosophy also can encompass the principle of complementarity.

It, therefore, appears that we have to deal with two philosophical models, the Kantian and the Leibnizian. Each model is needed for recognizing certain features of complementarity, and it is also clear that no one model is sufficient and that we need both. It seems that the basic situation of the wave-particle duality, which was the context in which the principle of complementarity was born, still weighs in our attempt to understand the meaning and the validity of this principle when we ask what is the significance of this principle and what role does it play in the economy of human knowledge. It appears that we have a duality of models: the Kantian and the Leibnizian. Perhaps at this higher level itself what we need to recognize is the functioning of the principle of complementarity. In the very attempt to understand complementarity, we may speak of complementarity not merely as a philosophical principle within a system of philosophy but also as a principle which relates different philosophical systems themselves. The history of philosophy itself may be understood under the guidance of the principle of complementarity.

From this angle, the principle of complementarity looks like the very basis or foundation of philosophy. In a similar way we can relate realism and anti-realism, determinism and free will, positivism and hermeneutics, etc.

4. CONCLUSION

The previous discussion demonstrates the functioning of complementarity as a maxim or a rule for organizing knowledge, on the one hand, and for producing altogether new structures and patterns of knowledge, on the other. This maxim (i.e., complementarity) is really a thematic principle which regulates and constitutes knowledge.[16] However, if we dig more, we find it hard to verify, testify to and properly conceptualize the presupposed foundation of the thematic principle of complementarity. But in the light of the dyadic features of complementarity, one can tentatively make the relevant argument.

The theory of quantum mechanics has taught us a lesson in how impossible it is to make a clear-cut demarcation between subject and object. This incapacity to make a clear distinction between the two has had a tremendous impact upon the age-old question of relationship between epistemology and ontology. It happens that the demarcation line between epistemology and ontology is unclear. The result of this is that all the definitions pertaining to epistemology and ontology are imprecise, defective and vague. Due to this imprecision, knowledge is not reliable and reality proper is not graspable. The whole of philosophy thus invites its being seen anew in the light of complementarity, begs for it.

Complementarity at least gives clues as to how to minimize the defects and lacunae within philosophy, if not eliminate them completely. Complementarity has two qualities which throw light upon the whole of philosophy. This can be shown schematically in the following way.

```
                  THE DYADIC FEATURES OF COMPLEMENTARITY
                                  |
                                  |
                                  |
            -----------------------------------------------
            |                                             |
            |                                             |
   UNITY (FUSION OR                                 TRANSCENDENCE
   IMMANENCE)                                             |
            |                                             |
            |                                             |
   -----------------                             -------------------
   |               |                             |                 |
SUBJECTIVE     OBJECTIVE                     SUBJECTIVE        OBJECTIVE
   |               |                             |                 |
   |               |                             |                 |
   |__STRENGTHS ___|                             |____ LIMITS______|
```

Figure 2. The dyadic features of complementarity.

4.1. *Unity*

Unity or fusion is sought between opposite phenomena or concepts via the principle of complementarity. Here, all the strengths and positive qualities of polar entities are admitted while their limits or lacunae are not considered. For example, when the wave/particle duality is considered, their lacunae or incapacities are not here taken into account, but how both contribute to the enrichment of the understanding of the phenomenon of light is considered. This type of unity is positive.

The Sāṃkhya system of thought offers something analogous. In Sāṃkhya Yoga, Prakṛti and Puruṣa fuse together due to their inherent capacities. Puruṣa has the power to activate Prakṛti and in turn Prakṛti in becoming active produces the world. This positive aspect of the duality can be identified due to complementarity. Similarly, Taoism is also to be understood complementarily. Yin and Yang are opposite in their manifestations but both are necessary for the realisation of the "Tao". Here, Yin and Yang complement each other. Similarly, here subject and object are to be understood as powerful unified forces which can be fused together complementarily. The reality which is claimed here is nearer to truth than when it is considered by isolating each aspect. Here complementarity guides us in extracting positive features out of opposed components or worldviews.

4.2. *Transcendence*

On the other hand, complementarity functions as a transcendental principle, in the sense that it shows or identifies limits or lacunae in each of the opposed viewpoints. The Copenhagen interpretation has shown that the phenomenon of light is not to be understood in terms of either 'particle' or 'wave' alone. A wave cannot transgress its limits to embrace or reduce the particle, or vice-versa. There is no point in sticking to either one of these interpretations. It is necessary to transcend them, just as Sāṃkhya Yoga grasps the incapacity of Prakṛti to produce Puruṣa as well as the incapacity of Puruṣa to produce Prakṛti.[17]

After having reached the state of transcendence, one can see the entire spectrum of philosophy as being holistic as opposed to being piecemeal. This is analogous to the insights of bootstrap philosophy or the Ecological Paradigm as envisaged by Geoffrey Chew and Fritjof Capra respectively. Bootstrap philosophy considers the entire universe as a web wherein everything is interconnected with every other thing. Nothing is seen as being primordial or fundamental. Everything is on par or equal. Similarly, the principle of complementarity treats rival claims as being on par. There is no distinction made between subject and object, for example because we have already transcended their limits.

But, at this higher level unity and transcendence also have to be thought of together complementarily. The principle of complementarity serves here as a critique of Philosophy itself.

If we will think of philosophical theorizing in the light of the principle of complementarity, we can suggest not merely resolutions of certain philosophical problems and oppositions but also envisage creative developments in philosophy itself. The promise of complementarity is the promise of creativity, and not merely the special scientific disciplines but philosophy itself would benefit.

University of Pune, India

NOTES

1 There is a debate between Feyerabend and Hanson on this issue; according to the former, correspondence is essentially empirical in its origination, while for Hanson it is non-empirical.

2 Niels Bohr, *Atomic Theory and the Description of Nature* (New York, Cambridge: Macmillan, 1961), p. 19.

[3] See Karl Popper's important works, *Conjectures and Refutations*, *Objective Knowledge*, etc.

[4] The quantum postulate, Bohr holds, gives rise to two distinct species of complementarity. First, there is wave-particle complementarity. Second, there is the complementarity of spatio-temporal descriptions and momentum-energy descriptions, which, according to Dugald Murdoch is 'Kinematic-dynamic' complementarity (see D. Murdoch, *Niels Bohr's Philosophy of Physics* (Cambridge: Cambridge University Press, 1989).

[5] J. J. C. Smart has very effectively discussed the two-slit experiment in his impressive book entitled *Our Place in the Universe* (London: Basil Blackwell, 1989), especially Chapter III on "Physics and Reality", pp. 65–78, under the rubric "Quantum Mechanics".

[6] See Werner Heisenberg, *Physics and Philosophy* (London, George Allen and Unwin Ltd., 1958) especially Chapter III, p. 46.

[7] In addition to the uncertainty between position and momentum as is shown by the formula (1) $\Delta P \cdot \Delta X \geqq \hbar$, there exist other sort of uncertainities, such as (2) the uncertainty relations held between energy and time $\Delta E \cdot \Delta t \geqq \hbar$ where E means the error in measuring energy and Δt is the inaccuracy in determining its moment of time. The more precise the measured value of energy, the less definite its dating becomes, another uncertainity (3). The principle of uncertainty states the inadequacy of the classical notions of particle and motion, as is shown by the following formula:

$$\Delta A \cdot \Delta y \geqq \hbar$$

Here, the conjugated variables are angular momentum ΔA and angular position Δy. Another kind of indeterminacy (4) is that between conjugated variables of momentum of inertia ΔI and the angular velocity Δw: $- \Delta I \cdot \Delta w \geqq h$.

[8] See Ludwig von Bertalanffy, *Problems of Life* (New York: Wiley, 1952), p. 136.

[9] Thomas Hobbes, *De Corpore*.

[10] J. B. Watson's behavioristic manifesto was promulgated in several lectures and in a journal article; please see his "Psychology as a Behaviourist Views It", article of 1913 in *The Journal Psychological Review*, No. 20, pp. 158–177 and his book *Behaviour* (New York: H. Holt, 1914).

[11] Watson, *Behaviour*, *op. cit.*, p. 27.

[12] H. Misiak and V. S. Sexton, *History of Psychology* (New York: Grune and Stratton, 1966), p. 411.

[13] See Place's article "Is Consciousness a Brain Process?", in V. C. Chappell's work, pp. 108–109.

[14] Paul Ricoeur, *Freud and Philosophy: An Essay on Interpretation*), Book III, "Dialectic: A Philosophical Interpretation of Freud", trans. Dennis Savage (New Haven: Yale University Press, 1970).

[15] S. P. Nadel, *The Foundations of Social Anthropology* (London: 1951), pp. 79–80.

[16] See Gerald Holton's works, especially his *Thematic Origins of Scientific Thought* (Cambridge, MA: Harvard Univeristy Press, 1973), *The Scientific Imagination* (Cambridge: Cambridge University Pres, 1988), and *The Advancement of Science and Its Burdens* (Cambridge: Cambridge University Press, 1986).

[17] See Dr. Sarvepalli Radhakrishnan's classic work, *Indian Philosophy*, Vol. II, rev. 2nd ed. (London: Allen and Unwin, 1929–1931).

DANIELA VERDUCCI

GIVING FORM TO LIFE

Processes of Functionalization and of Work in Max Scheler

1. THE AFFECTIVE INTENTIONAL FUNCTIONS

Precisely at the heart of his most mature phenomenological work, *Der Formalismus in der Ethik und die materiale Wertethik*, in the section "Fühlen und Gefühle", dedicated to showing the existence of an "ordre du coeur", an affective intentionality on the same level as cogitative intentionality, Max Scheler brings up "the class of the intentional affective functions". In doing so he refers to a type of operation for pure feeling which is analogous to that which Husserl, in *Logische Untersuchungen*, saw at the origin of cogitative activity and indicated as intentionality of the conscience.[1]

However, the use of this term to describe affective intentionality is not a simple concession on Scheler's part to the more fashionable studies of the epoch, which used functions in mathematics, psychology, biology and the history of philosophy. If anything, it is a contribution to these studies, since in this way Scheler demonstrates that this type of functional operativity is in use not only at the level of the biological and psycho-physical being, but also, equally, in the area of the ideal-axiological being. In particular, in each pure feeling of values, or whenever "we have the feeling of something" independently of the psycho-physical organization of our human species, intentionality of the conscience is expressed, even in the alogical field, through spiritual operations termed "affective functions".

In order to examine the intentional dimension of the sentimental sphere and its qualification in a functional sense, Scheler inquires into affective lived experiences of particular complexity. Simple affectives states such as "losing one's temper over something" or "enjoying a beautiful sunset" are not enough. In fact, in these states one is driven in the first place towards the cause of the feeling, gathered thoughtfully and by representation in the empirical events. Only in a second moment does the question arise regarding what gives meaning to the feeling, that is, regarding what makes the beauty of the sunset felt and what configures an empirical object as the source of rage.[2]

A-T. Tymieniecka (ed.), Analecta Husserliana LXVI, 287–295.

Vice versa, there are lived experiences involving not only the empirical affective state, but also a hearkening back to the a priori dimension of feeling, in which an immediate and immanent "feeling something" is given, so that this hearkening back does not need representative go-betweens. For example, when "we become sad for not having been able to rejoice in an event, as it merited",[3] it is the very feeling of sadness we experience, which sends us on to a different and better axiologic perception: that which characterizes the present affective state is, in fact, the sense of inadequacy compared to a quality of value, which our empirical feeling was not able to achieve, even through knowing it and being oriented towards it. Thus we are driven towards a level of affective experience in which "the feeling directs itself originally towards a kind of object that is its own, that is the values".[4] Here, every extrinsicality of relationship between feeling and its object is removed, nor is the psycho-physical organization of our human species relevant, since this dimension pre-exists our empirical self-reference to it, and it maintains itself even when we abandon it, leaving it in shadow. In fact, Scheler affirms: "This feeling has with its correlate-of-value the same relationship that 'representation' has with its object, that is, the intentional relationship". In addition, it constitutes a completed event in itself, which can find, from the empirical point of view, both "fulfillment" and "non-fulfillment", without its essential validity begin modified.[5]

The spiritual dynamic in which the affective intentionality expresses itself is first described generically by Scheler as "feeling's original self-reference, self-orientation towards an object, towards values";[6] but then he specifies: "(feeling is) an instantaneous movement in which something is given to me and 'comes to manifestation' ".[7] In other words, feeling draws its determination from that which meets it, since the latter turns out to be the appropriate termination of its movement. Thus intentional feeling qualifies itself as "movement determined by the end",[8] being induced neither by its spontaneity nor an extrinsic end. Rather, in pure feeling there is a sort of "entelechy" which drives it to realize itself in the direction of the values: that is, it manifests that it possesses the nature of pure "tendency" (*Erstreben*) and of being, therefore, all directed towards its own completion (*Ziel*).[9] In fact, Scheler adds, when we find ourselves in the presence of a behavior directed to value, the qualities-of-values contained in it seem to require, starting from itself, corresponding qualities in the reactions of response,[10] to the point that, as seen in the case of the lived experience, "I am saddened for not having enjoyed an event as it merited", we perceive next to the affective state, also the inadequacy of our affective reaction.

In the intentional relationship that functionally unites feeling with values, therefore, a complete equation between feeling and felt takes place. In fact, on the one hand, feeling-movement converges totally with the quality of value, while, on the other hand, the value completely pours itself into feeling. The unification of the two poles of the intentional relationship is so close that objective knowledge of feeling and felt requires our deliberate act of reflection.[11]

Unlike cogitative intentionality, which is pure intuition of essences, the intentional relationship between feeling and value completes an accomplishment of feeling. In this accomplishment of feeling, in external lived experiences as much as in interior lived experiences, we are met only by a quality of value.[12] In other words, the affective intentional relationship, rather than giving place to the intuition of the one or of the other essentiality, gives place to the axiological conforming of the spirit itself: in this way feeling enters in contact with values and brings them to manifestation, and for this reason Scheler hold that affective experience is more original than cogitative experience.[13]

Thus, the denomination of "affective intentional functions" suits the operations that connect feeling with values: in such peculiar dynamics, in fact, the two poles of the relationship express themselves in terms of each other, in the same way in which the continuous quantities are represented on the Cartesian axes by means of mathematical functions. The spirit which feels the value finds its full affective accomplishment in the same way that the value is entirely adapted by the corresponding pure feeling. There are neither excesses nor residues in this functional relationship, just as in the system of Cartesian axes each point in space has a corresponding numerical value, no matter how infinitesimal. In fact, the very sensory functions of seeing, hearing, etc. draw their model and their measure from that perfect ideal adaptation.[14]

In the case of the affective functions, therefore, that which is achieved is not an immediate and immanent vision-of-essence, but rather a unification of the movement of the spirit with the quality of the value. The phenomenological experience "in which is understood only as much as is given and in which nothing is given outside of the understood",[15] in pure feeling comes as adaptation of the spirit to the felt value, which disposes it, for example, to select knowledges in order to process them into a vision of the world, as can be shown, affirms Scheler, by analysis of the role-guide of the unities of feeling and of value in the formation of the vision of the world, expressed in language.[16]

Unlike Kant's synthetic functions of the intellect, or categories, Scheler holds that the affective functions in themselves do not open us to the axiological ideal world: rather, they represent the last part of a spiritual itinerary which starts out with spontaneous acts of love and hate, "discovers" the axiological horizon and is thus capable of respectively enlarging or restricting it. This itinerary continues with the acts of preference and deferring which gather the hierarchic order of the values. In fact, the affective intentional functions are a spiritual mechanism which is set off in the presence of an already defined axiological context; they are thus receptive of the *a priori* axiological dimension, already configured by the "ordo amoris" of each person or people: they neither created nor construct values, but simply enter in contact with them in the course of internal perception, of introspection, of the living affective relationship with the world, in the accomplishment of the acts of preference or deferring, of love or of hate, opening for us ethical cognition in its *a priori* axiological content and giving place to the axiological adaptation of the spirit, a source of every authentic praxis.[17]

2. THE PROCESS OF FUNCTIONALIZATION

Scheler draws on this functional peculiarity of the affective intentionality in *Vom Ewigen im Menschen. Probleme der Religion*, describing the process of functionalization as that which permits the history of the spirit and of culture. If in fact phenomenological experience completed mere ideas and in this exhausted itself, it would risk having no relevance to life, above all in the face of technological knowledge which seems to exhibit the capacity to produce substantial transformations. Vice versa, in the *a priori* dimension of feeling, the tendential movement of the spirit, directed towards its own end, is compared with the values configuring themselves not so much as essences unto themselves, but as ends to reach. In this manner an intentional praxis is determined which has "feeling-the-values" as its prototypical success, and its principal result consists in the spirit's acquisition of a more or less wide variety of adaptations.

On the other hand, according to Scheler, our spirit possesses neither innate nor congenital ideas: in its being is deeply rooted the capacity to acquire self-knowledge.[18] The spirit, that is, acts in force of its being and is fully rooted in a metaphysical totality and equipped to discover in it, bit by bit, the elements. In addition, from this is derived as much the content of "receptio" of a priori knowledge,[19] as the ontological qualification of the cognitive relationship.[20]

However, Scheler observes, if "our thought and our knowing cannot 'create', 'produce', 'form' anything", beyond the signs and the *entia ficta*, the "categories" are to be considered a pure Kantian invention, as long as they are understood as synthetic functional laws of the human spirit through which are coined, on a given material without order, the formal objective determinations on being.[21] The originality of the spirit and of its *a priori* patrimony remains next to that of being. The spirit, in fact, gathers in the content of the given a priori dimension as much as the a posteriori one, capable as it is of distinguishing the determinations of being-so (*So-sein*) – which belong to the pure spheres of the what-thing and of the essence – from the modes of existence (*Dasein*). In this way it shows that it hosts with things an authentic and originary relationship of being, establishing a contact which in and of itself is not mediated by the sensory organization of our body and which completes a totalizing and unitary intuition of the world. In fact, in terms of this intuition, the senses serve as analyzers, introducing the differentiations calibrated according to the degree of biological excitation, that is, according to the possible motor reactions, useful or harmful to the life of the organism which belongs to the spiritual person, as a subject of the intentional perception.[22] In conclusion, an opening to a spiritual sense is indispensable not only for gathering, in the *datum* itself, the a priori from which the objective cognition of a qualitative type is developed, but also for acquiring the a posteriori directed towards the cognition of the accidental and the quantitative.[23]

However, it remains to be seen how the spirit passes from one field to another: Scheler uses the intervention of the "functionalization" process precisely in this problem. There is, in fact, a simple law for the use of the intellect, directed towards accidental facts, by which the intellect is able to "determine them", gather them, distinguish them, intuit them, judge them according to essential nexuses. In this intellectual activity, the essence which the intellect comes to know as object of cognition (*Sache*) is transformed in turn into a form of thought about the objects, or analogously, the value, which opens itself in an act of love, becomes a form of love in which an unlimited number of objects can be loved. This also happens for that which has been the object of the will, which is configured as a form of wanting.[24] The cognition of essence in a larger sense, that is, not only cogitative but also affective and volitional, is the setting for a "formation" of the spirit, generalizing the dynamic of entelechial finalization which is in force for the affective functions in terms of values.

This functionalization process is a spontaneous one of the spirit, which we can recognize in the presence of complex lived experiences such as dissatisfaction, deviation from the law and objections to error. In these experiences, in fact, we note a comparative character which signals to us that there is in our lived experience the reference to cognitions of an essential type, that is, objective and universally valid: thus they have entered into functions even without having found themselves explicitly before the eyes of the spirit.[25]

In the functionalizing activity, the vitality of the spirit is manifested: nourishing itself with cogitative, axiological and volitive essences, it acquires and develops the variety of its forms of action, becoming even more capable of interacting with the other spirits and with extramental reality. Thus, both Spencer and Hegel were wrong when they, departing from opposite but convergent positions, delineated the evolution of human reason. The former, on the basis of the positivistic-sensate cognition theory, favoured the idea of a knowledge which is potentially congenital in the individual, which comes to light in the course of the evolution of the species by means of the organism's adaptation to the surrounding world. The latter, vice versa, on the groundwork of a rationalistic-constructive theory of reason and the spirit, represented the development of the idea according to a dialectical law which only takes into consideration the explication of what the idea already has in itself. In both cases, although for opposite reasons, the same result is arrived at: among the originary dynamics of the spirit no identification is made of a process capable of "conforming" to reason according to the multiple intuitions of essence, headed by its acts: therefore there is not even any true growth of the rational spirit itself.[26]

3. THE WORK PROCESS AS NON-INTENTIONAL ACTIVITY

The intentional and empirical praxes examined up to this point are not, however, the only types of action identified in Scheler's analyses, since work also presents an original mode of relationship in human action, in terms of the end to which it is inevitably oriented. In fact, if the intentional praxis has an intrinsic relationship of immediacy and immanence with the axiological end, and the empirical praxis figures as "fulfillment" or "non-fulfillment" of the latter's noematic pole, work activity exhibits instead an extrinsic *nexus* with the end, even though this end determines it.

In his *Erkenntnis und Arbeit* of 1926, Scheler defines work as a psycho-physical activity in which a given material is transformed, placing it as a direct alternative to cognition, which he considered as a spiritual activity directed at the ideal participation in things.[27] As early as 1899, in his essay *Arbeit und Ethik*, Scheler had examined the particular structure of work activity compared with other types of theoretical or practical active being, drawing on linguistic observations and using the psycholinguistic method adopted by Alexius Meinong and Christian von Ehrenfels.

In particular, Scheler had observed that the German verb "arbeiten", the linguistic expression of work activity, does not take a direct object, as do other verbal forms indicating production, such as the transitives "schaffen" or "erschaffen". This, therefore, demonstrates that the work does not have an intrinsic relationship with its own final end or the result of the action. In effect, from the point of view of pure work activity, it makes no difference whether the work is house-building or accounting: if the two actions can be indifferently indicated as work, it means that they possess the same essential structure and that this identity is maintained independently of the empirical ends to which the individual work processes are applied.[28]

At the same time (continuing the linguistic analysis), the activity expressed by "arbeiten" is not even intransitive in the absolute sense, since the verb is followed by indirect objects of place, introduced by the prepositions "in" or "an".[29] Work thus seems to occupy in the practical area a "sui generis" position: in fact, to the degree that it lacks an intrinsic relationship with its own end, it is not identified with "praxis", that is, with the moral intentional activity, the end of which is the process itself. On the other hand, since it is indirectly connected to its own end or result through the preposition of place, work cannot even be assimilated into "poiesis" or productive transitive activity, the end of which is external, according to Aristotle's famous definitions in his *Nichomachian Ethics*.[30]

In Scheler's analysis, work is thus neither exclusively practical nor totally poietic: rather, its distinguishing feature is its executive nature, inasmuch as it is directed at the material accomplishment and the transformation of the world. Work's structural specificity causes it to assume, therefore, a middle position between "praxis" and "poiesis", providing for their ideal dynamics the guarantee of consigning the effective results of a transformation of reality. For its part, from the opportune integration with the theoretical and practical intentional activities, work receives that indispensable finalistic orientation, which it lacks in and of itself.

4. CONCLUSIONS

This brief overview of the principal processes directed at practice has allowed us to demonstrate the close nexus which, according to Scheler, links the activities of the human being in an articulated unity. In fact, starting with the axiological opening provoked by the spontaneous movement of love, the spirit invests the world with its triple intentionality: affective, cogitative and volitional, giving rise, through the functionalization process, to its own formation and at the same time also disposing it to shape reality by means of work.

University of Macerata

NOTES

[1] "Dieses Fühlen hat daher genau dieselbe Beziehung zu seinem Wertkorrelat wie die 'Vorstellung' zu ihrem 'Gegenstand' – eben die intentionale Beziehung". Cf. M. Scheler, *Der Formalismus in der Ethik und die materiale Wertethik*, in *Gesammelte Werke* (from now on *GW*), Bd. 2 (Bern und Munich: Francke Verlag, 1954), p. 263. Future references to *Formalismus* will be indicated with the letter *F*.

[2] *F*, pp. 262–264.

[3] *F*, p. 264.

[4] "Das Fühlen geht ursprünglich auf eine eigene Art von Gegenständen, eben die Werte". *F*, p. 264.

[5] " 'Fühlen' is also ein sinnvolles und darum auch der 'Erfüllung' und 'Nichterfüllung' fähiges Geschehen" (*F*, p. 263).

[6] "Ein ursprüngliches Sichbeziehen, Sichrichten des Fühlens auf ein Gegeständliches, auf Werte" (*F*, pp. 262–263).

[7] "Eine punktuelle . . . Bewegung, in der mir etwas gegeben wird und 'zur Erschauung' kommt" (*F*, p. 263).

[8] "Zielbestimmte Bewegung" (*F*, p. 263).

[9] *F*, pp. 55–56.

[10] "Die in den betreffenden Wertverhalten liegenden Wertqualitäten fordern von sich

aus gewisse Qualitäten derartiger emotionaler 'Antwortreaktionen' – wie anderseits diese auch in ihnen in gewissem Sinne 'ihr Ziel erreichen'" (*F*, p. 264).

11 *F*, p. 264.

12 "Im Vollzug des Fühlens wird uns das Fühlen nicht gegenständlich bewusst: Es tritt uns nur eine Wertqualität von aussen oder innen her 'entgegen'" (*F*, p. 264).

13 *F*, p. 83, n. 2.

14 *F*, p. 387. In addition cf. M. Scheler, *Erkenntnis und Arbeit* (from now on: *EuA*), *GW*, Bd. 8 (1960), pp. 283–379.

15 *F*, p. 70.

16 *F*, p. 265.

17 *F*, p. 87 and p. 89.

18 M. Scheler, *Vom Ewigen im Menschen. Probleme der Religion* (from now on: *EW*), *GW*, Bd. 5 (1954), p. 195.

19 *EW*, p. 197.

20 *EuA*, p. 203.

21 *EW*, p. 195.

22 *EW*, p. 196.

23 *EW*, p. 196.

24 *EW*, p. 198.

25 *EW*, p. 198.

26 *EW*, pp. 200–201.

27 *EuA*, *Zusätze*, p. 448.

28 M. Scheler, *Arbeit und Ethik* (from now on: *AuE*), *GW*, Bd. 1 (1971), p. 168.

29 *EuA*, p. 168.

30 Aristotle, *Nichomachian Ethics*, VI, 1140b 6–7, in which he affirms: "In fact, the end of production (*poiesis*) is other than the production itself, while the end of action (*praxis*) is not: acting morally well is an end unto itself".

S. V. KOMAROV

THE CONSCIOUSNESS-CORPOREALITY PROBLEM

Science is a system of true statement about reality. This means that each science contains inside itself definite methodological premises and ontological assumptions concerning the structure of the reality and cognitive processes which make scientific statements intelligent. These fundamental assumptions and premises are the philosophical basis of scientific cognition.

The real goal of psychological study is to explain the genesis of the human psyche, and to describe consciousness as a phenomenon of living subjectness, as a matter of living conscious experience. The achievement of that goal also depends on the introduced ontological conceptions of the nature of consciousness, which determine the methodology of its study. So, analyzing the philosophical basis of psychological knowledge is rather important for solving practical problems of psychology.

The study of consciousness-corporeality is a pressing issue in modern psychology. It is a question of its spatial-temporal arrangement, but not a problem of consciousness of spatial-temporal relations, nor a problem of consciousness as a perceived corporeality. The latter are problems of our perception of the temporality of consciousness and existence themselves, the problems of understanding the internal synthesis of various phases of perception, of the conception of consciousness as time (*Zeitbewußtsein*), and the constitution of the corporeal experience in consciousness. But here we are dealing with understanding consciousness as a particular self-existing reality, that has definite temporal and topological characteristics as phenomena in objective reality. Such an understanding of consciousness is rather problematic for psychology, because it presumes the refusal of prevailing naturalistic and epiphenomenalistic conceptions of the nature of the latter.

The problem of consciousness has a long and distinct history in philosophy. Philosophy from the very beginning regarded consciousness as a problem of ontological, existential-practical premises and the basis of human existence in the world, including the explanation of psychological phenomena. The aim of this article is to show that the problem of a corporeality of consciousness can be defined and adequately solved

A-T. Tymieniecka (ed.), Analecta Husserliana LXVI, 297–321.

only within the frames of the ontological understanding of consciousness, and the tradition of its transcendental-phenomenological analysis in European philosophy. This tradition presents complex and internally contradictory intellectual formations in the history of philosophical thought, but we now put these contradictions aside and regard the latter as a definite common experience of understanding the phenomenon of consciousness. Only within the frame of this tradition can the problem of objective method in psychology be solved, and the corresponding constructions and objectual means can be produced to imagine consciousness as the phenomenon of objective reality, with a quite definite corporeality and spatial-temporal dimension.

I

Psychology as a science is closely allied to definite philosophical premises that classical New-European philosophy states to substantiate the process of cognition. We mean the fundamental idea of a *rational structure of observation*. The rational structure of observation is a theoretical construction on which every science is based, because it is the condition of objective analysis of phenomena. Indeed, one can explain the origin of knowledge of the surrounding reality only by separating the acts of obtaining that knowledge from its contents. Those acts themselves of obtaining the information about the external world, i.e. perceptions, having once happened, occur "unbreakably", in the sense that, having received the perception, we cannot understand how it has occurred, and all that we have is its contents, the perceived. The solution to this problem, when stated thus, can be achieved only if in the "givennesses" of the consciousness there is a certain all-embracing element which possesses the property of immediate authenticity, of givenness thru itself. The separation of such an element of our perceptions is indeed the philosophical operation called *cogito* that was first introduced by Descartes, but has become the element of thinking of all the New-European philosophy right up to Kant and Fichte. By separating *cogito* acts from all empirically given acts of perception, we can distribute the whole totality of conscious givennesses between those which can be rationally controlled and reflectively reproduced, and those which cannot be controlled and reproduced in such a way. Having excluded the latter as empirically accidental, classical philosophy thereby sets the abstraction of *observation* or the abstraction of *understanding*. To put it another way, it is assumed that we can cognize the world to the extent that we are able to in accordance with the uncontrolled and independent actions of the world upon the natural human reflective apparatus and their equivalents – controllably reproducible formations of consciousness.

This implies that the observation is not considered to mean any act of perception of external phenomena, but a special procedure of cognizing mind, which lies in the fact that within the frames of such a constructed observation we are catching the object to the extent that we, together with this object, are catching those acts with whose help it is given to consciousness. Then the object is considered to mean such contents of consciousness as can be separated from the states of a subject (on the basis of a *cogito* operation) that does not depend on them, and thereby can be considered by us as occurring in reality, but not in our mind.[1] Then we deal with the rational structure of the object. Otherwise, the *cogito* operation necessarily presupposes its "ontological double" – the operation of objectivation: the object as the phenomenon, independent of a subject's thinking and features of an individual psyche, reveals itself only within the frames of definite procedures of thinking.

Now one can attribute to the objects of observation only such properties as fit the criterion of observability, i.e. reflective doubling, independently of the real conditions of their realization. The possibility of objectification presupposes that a certain event in the world occurs as if twice: first, like a certain natural spontaneous event, "imprinted" by influences of the feeling and cognizing structure of a human, and second, repeating to infinity its reproductions in his (her) consciousness, now controllable. This receives the status of objective description: any theoretical claims on the formations of consciousness must be resolved to fit this criterion. In particular, there is a well known and scientifically fundamental differentiation between essence and phenomenon: science deals with essences of phenomena, i.e. structures of reality, stable and independent of the conditions of their perception, whereas phenomena appear like "visibility" or "the visible", i.e. like formations that do not have their own objective status.

This structure of observation, which is specially developed by classical philosophy and then approved in the real process of cognition, gives not only the possibility of understanding the reality – thinking in terms of the essences of things, but the possibility of its *misunderstanding*. Really, the rational structure of observation creates a space of cognition that does not depend on occasional psychic features of the perceiving apparatus, the mood of the observing subject, his (her) individual characteristics, and so on. The introduction of the concept of a physical body (and other fundamental abstractions of experimental science – particles, ideal gas, etc.) as a phenomenon that is fully opened to the external observation and does not contain hidden essences, is a definite organization of the observing consciousness, which determines the synonymity of concepts like 'spatiality' and 'objectivity', 'spatiality' and 'materiality', 'objectivity' and 'externality', 'subjectivity' and 'internal perceptibility', 'externality' and 'materiality', etc. This abstraction (phenomenon) is opposed to another fundamental abstraction of fundamental science – the abstraction of 'consciousness'. The category of 'consciousness' now may be applied only to phenomena that have no spatial dimensionality, and can be represented and interpreted only under conditions assuming their reflective doubling (internal observation). And these concepts cannot have any other contents within the frames of such organized thinking. In fact, the classical differentiation between 'soul' and 'body', between 'consciousness' and 'matter', etc., is the conclusion drawn from the analysis of this rational structure of observation.

So we cannot understand phenomena of psyche and consciousness like we understand physical phenomena: we understand *by* consciousness, but we do not understand consciousness itself. This impossibility of an objective description of consciousness in its whole empirically vital appearance is well recognized in philosophy. Rational psychology as it was historically formed inside classical philosophy – in theories of Locke, Leibniz, Wolff – was in effect not a theory of psychology with a precise meaning; as we know, it appears here as part of a theory of cognition, schemes of which represent rationalizations of a vital (psychological) cognitive process, just admitting its reflective reconstruction. The comprehension of it is the basis of Kant's thought, that the soul cannot be given to us like physical bodies are, and so rational psychology can be just a normative discipline, but in no case an experimental science. To do this one must recognize, in the definite organization of consciousness – rational structure of observation – the status of the ontological 'givenness', that is, to identify the 'givenness' of consciousness with its factuality, and to recognize the status of objective phenomenon of this identification. The objectificating operation being applied to the problem of psychology and the description of living consciousness constitutes the identification of psychological reality and the forms of its givenness to the cognizing subject, who himself is the result and definite formation of consciousness and psyche.

In fact, that has occurred in the history of psychology: the rational structure of observation is one of the theoretical premises of psychology as an empirical study. Really, classical psychological theories are known to be built upon the identification of consciousness with the concept of 'subject' that presupposes not only passive reproduction of human psychological states, but their absolute transparency and understanding for themselves. Consciousness is identified with mentality, with reflective doubling of mental contents, and procedures of internal observation and understanding introspection, based upon it, appear to be the principal means of analysis of psychological processes from D. Hume to J. Mill, W. James and W. Wundt. The latter, having developed the methodology of 'objective' analysis, in fact proceeds from the same paradigm of understanding consciousness as internal meaning. All experimental psychology, such as psychophysics and psychophysiology according to the complementary principle, presupposes the explanation by introspection and understanding of psychological subjects as a constant of a rationally reproducible set of mental states.

Dualistic dissection of reality and recognizing the substantial character of the world of human consciousness and subjectivity turn to opposite tendencies in the field of methodology. On the one hand, in philosophy *naturalism* begins to dominate with its interpretation of the psychical as substance, ascribing to the psychical all 'corporeal' characteristics as distinct from the characteristics of physical reality. On the other hand, in psychology an *epiphenomenalism* that refuses its own ontological status of phenomenon of psychology, begins to dominate. The discussion of the method of psychological studies, that started, in fact, from the very moment of constituting psychology as an experimental science (in the last third of the nineteenth century) was based upon the former recognition of the independence of the physical world and the psychical world. The differences in substantiating its own method of psychological studies as opposed to scientific ('physical') research originate either from recognition of the latter as mediated experience and the former as immediate experience, or from the distinction between 'external and 'internal' experiences as dealing with different realities. Nevertheless, the classical definition, "Consciousness constitutes the sum of states cognized by us", remains the basis of the solutions offered.

The successive conduct of the subject of psychology with its 'doubles' leads to a crisis in modern psychology. The recognition that the psyche itself cannot be observed separately from the language of its phenomena and expressions, leads to the appearance of conceptions that are based on the recognition of the complexity and ontological 'density' of consciousness, i.e. the assumption inside consciousness that such links and formations are inaccessible to classical procedures of reflection and controlled rational reproduction. Consciousness therefore has lost its substantial character, turning to mental formation, determined by the phenomena of existential non-mentality – for example, by 'the unconscious' in psychoanalysis, or 'the structures of social practice and language' in Marxist psychology, by its own formations in actual genesis in Gestalt psychology, etc. The refusal of naturalistic and quasinaturalistic imaginations by the human consciousness and psyche has accompanied those theories by introducing new methods of consciousness-data analysis and its secondary rationalizations by their objective-referential interpretation and reduction to initial 'objectivenesses' of consciousness and by the analysis of consciousness in its actual, individually integral states.

The origin of modern conceptions of consciousness is connected to the understanding that with respect to consciousness, we deal with such cognitive situations when the object of studies by its own existence shields itself in its phenomena, givennesses and formations, and, in fact, closes the possibility of its direct objective observation. Any psychological processes – sensations, perceptions, impressions, emotions, etc. – can be objectively represented as happening in objective reality just because they are already represented thru their realization. Consciousness comes out as reality, inseparable by means of external observation and physical research from the observation of conscious life and meanings.

It means, in fact, recognizing that consciousness has its own ontological status, as a kind of special reality that demands coordination in a single regulated and logically homogeneous experience in the description of cognition of definite ('physical') phenomena and the phenomenon of consciousness itself. Or in other words, it demands the coordination of consciousness and its contents.

Subjectivity (consciousness) is a certain objective event and a phenomenon of reality, and as such it is 'a body' that possesses all the spatio-temporal characteristics of existence. This means that the problem of a 'corporeality' of consciousness is conceptually the problem of searching for consciousness description as a self-existent reality, the problem of interpreting consciousness, objective existence without coordinating with the states of cognition's subject itself and its psychophysiological apparatus. In fact, this is the problem with the objective method in modern psychology.

II

So, the consciousness-corporeality problem deals with its ontological status in the structure of existence. Nevertheless, classical New-European philosophy, solving the problem of cognition, just proceeds from an ontological understanding of consciousness as some objective, vital empirical event and unsubstitutable act in existence. Here we are dealing with that meditative experience that was first discovered by Descartes and developed by the transcendental-phenomenological tradition of European philosophy.

Indeed, the nominalistic late Middle Ages trend leads to the philosophical overturn in views on understanding the nature of consciousness thru its individualisation and psychologicalization, and its attempt to treat thinking and spirit as subjective spiritual phenomena, and to treat knowledge as contained in consciousness, in psyche. Within the frame of such formulation, the problem of strict and precise knowledge is bound with the understanding that between the reality its representation in cognitive thinking cannot be withdrawn by purely theoretical and speculative postulations on the identity of the structure of mind and nature, the similarity of rationality and the structure of existence, etc. In fact, beginning with Bacon, in philosophy there has existed a distinct difference between knowledge itself and its existence, that is the difference between object contents of thinking and thinking itself as an act or state of a cognitive subject. This difference has its own experimental bases: indeed, we cannot, moving in a single observational experiment from the object of influence upon our organs of sense, come to the contents in which that object "has been reflected", "has been imprinted", or has been understood, etc. It is as if acts of cognition had two connected poles, and carrying out a 'physical' phenomenon at one pole simultaneously carries out its "impression" or "observation" at another pole, although there is only one objective action![2]

So if we are interested in knowledge as something that really happened, as an empirical and experimental phenomenon, then the question of criteria for its authenticity comes up with the discovery of this experience, where knowledge is realized as a phenomenon having really ontological status. It is unified in some way in classical philosophy independently of its factual and textual interpretation, in the works of Descartes and Spinoza, Locke and Leibniz, Kant and Fichte. Really, philosophy, asking the question of conditions revealing the soul or thought, distinctly differentiates those conditions – form or event of thought or knowledge – from the contents of this event: they are unnoticeable and go to the background, when the contents have already happened. This determines the formulation of the question concerning the possibility of cognition as a question of ontological premises for its fulfillment in specific acts of perception, impression, observation.

Being so formulated, the problem can be solved just in case such actions are admitted in the structure of the cognitive act, in which the cognitive act itself shall coincide with its understanding – in a unified act of its fulfillment and realization. In other words, the inseparability of the contents of a cognitive act from its existence as the action of the world itself shall fix, and also identify, thinking and existence as events of the world itself. Classical philosophy, starting from Descartes, just introduces the concept of 'self-consciousness' in its existentially ontological, but not logical, comprehension: *cogito* is neither the procedure of understanding, nor a philosophical abstraction for understanding; it is a really happening phenomenon of consciousness, i.e. a certain event, whose nature is not epiphenomenal, but really existential. These are actions in reality or actions of reality – because cognition itself is a certain vital event, actually displayed in the minds of living humans. *Cogito* thinking as a special experience (form) of consciousness is a definite real act (together with other cognitive acts) in existence, before and outside of which all arguments about cognition, phenomena of knowledge, and coincidence of structure of thought with reality described by it, have no meaning. What distinguishes it from other cognitive acts is its fundamental ontological structure: it is factuality and 'givenness' of me to myself, as if I discovered myself again for the first and each time; it is the condition which determines all of my existence and its possibilities. These are events-in-the-world of realization of me as an empirical subject, i.e. such a subject related thereto is a reason to tell about the fulfilled experience, the received imagination, the accomplished knowledge.

In other words, raising the question about coincidence of thought and imagination with cognized objects, the philosophy of Modern History doesn't reveal understanding, but objectively acting layers of consciousness, wherein there is a 'merging' (*Verwachsensein*), organic entity of existence and way of existing, an identity of cognitive subject and its activity; and this question conceptually fixes these in the appropriate theoretical construction of *cogito* or *Ichheit*.

This is the exposure of a real ontology of the world. Consciousness is understood as a certain, unconditional and absolute event-act, to which we cannot attribute any predicates or attributes, because we ourselves as empirical (psychical) beings, and all our thinking forms, and our whole world as conformist, cognizable, etc., are possible only after the accomplishment-determination of these events. All our experiences (revealings of 'I') are already inside this act (appearance of consciousness), and we cannot arrive at this by any effort of our will. So consciousness is understood as absolutely non-objective, non-objectivated, and an irreplaceable event in the structure of existence. In this meaning, consciousness is a movement or thought of God absolutely free of accomplishment or the will-to-existence (*energeia*, *potentia*). This act of structuration of the existing and of us as subjects is a single action, from which we cannot move to the source: our 'I' is just the idea of God. Then, realization of the happening is given by this existential "chronothopy" of conscious events: an attempt to think of subjectivity as a certain objective phenomenon, entering reality in such a transcendental way, represents this lasting experience of an appearance of consciousness (me in the world) and the world as synthesizing time and space and as the production of phenomena and artifacts. Time and space themselves are the results of this vital movement of God, are the condition of our possibility to perceive, feel, and experience something. And in this meaning, time and space are objective forms of reality, given in the act of their realization together with our appearance, and thereby already putting restrictions on all our abilities. In other words, *cogito* is a certain *thopos* of containing 'I', the properties of which are the transcendentality of free action, an existential comprehending (i.e. phenomenal) "chronothopy" of conscious phenomena, and restrictions imposed by these phenomena on their understanding, etc.

This is the real metaphysics of consciousness that is revealed by classical philosophy as a special meditative experience, that demands a special technique of meditation and thought. Consciousness is power, induced inside the world by our every step, and inseparable from us as living and feeling creatures: we do not see what we see and what consists of some trans-subjective, 'ideal', 'flesh and blood' view of ourselves. It is not just a property of the theory,[2] but the event, that is realized in all its apparency, but the property of existing, of realizing thought itself as an event existence.

Consciousness, comprehended thus, appears as *cogitatio*, i.e. not empirical, but transcendental or 'pure' consciousness. *The transcendentality* of consciousness, in philosophical terms, is the recognition of non-objectivity and the non-objectivability of consciousness, and the indication of its practical efficacy and of the "appearedness" of vital experience. Consciousness is not a phenomenon, but an action itself. Transcendentality of consciousness is a vital empirical movement: God has thought, the world has moved, irreversible processes have occurred, the world has crystallized in this motion, and the necessary connection with things has been stated, including necessary connections of consciousness, understanding, observation, perceived by us as empirical connections of psychological meanings, impressions, associations, etc. We constitute ourselves thus, having determined the fact of our occurrence as empirical creatures. In other words, classical philosophy's arguments about 'pure' consciousness cannot be understood in a logical context, because all these categories – *cogito, transcendental consciousness, 'pure I'*, etc. – are not ideas in the common sense, but characteristics of real ontology in the world of human phenomena. 'Pure' consciousness is not the indication of some empirically existing consciousness or psychologically revealing phenomenon. It is the recognition of factuality and the necessity of realizing, in the structure of existence, acts of consciousness that must be repeated every time and at every moment, stating and certifying me as an empirical subject. The 'pure I' is an event of the world that constitutes the organization of the world, its conformity, intelligibility, etc., in which there is a place for the individual 'I'. This occupation of its own place in the world is the necessary moment of existing of the world itself, without which it is not determined in the causal meaning. The causality of phenomena is being stated as a law after an interaction of phenomena has really occurred in the world; the same applies to causing in us, as living creatures, some states that are psychologically experienced by us as states of being understood, of being recognized, as states of thought. It is already impossible to penetrate into a 'pre-gravitational' state of linking global forces after their completion, because we ourselves have already 'occurred' by this very act. We have occurred as the empirical 'I'.

In this meaning, consciousness is a phenomenon, i.e. a special thing, that has no essence outside of itself. If in the world an event of thought, soul, knowledge has happened, and I appeared to myself in such a way, hence thereby in the world's structure, simultaneously with my appearance, there have occurred some irreversible changes and transformations, determining restrictions not only on the possibilities of the world, but on the possibilities for thinking about it. Consciousness, understood in its existential state, is a 'world point' where the line of the nature and of the cognizing idea itself intersect, along with the existentially practical determination of the nature's mode of action in any case, of the mode of cognition, and of how this mode – by its phenomenon of occurrence – is allowed by the world.[4] This is an act of statement and determination (predetermination, to be exact, because it presupposes participation of the living 'I') of existence itself, at the point of performing this act of consciousness. Here consciousness is the knowledge of what happened to knowledge in cognition, of the transformation of *living, expanded within real space and in real time*, the action of thought, imagination, correlation to objects outside it with the image, behind which the action itself occurs unseen. So the category of consciousness is just the indication of the non-substantiality of what happened – ascertaining the linkage between objective action, corporeal change and sensual imagination, i.e. the fact of understandability without imagination or interpretation in the objective meaning. It is the real binding of modes of thought and objects of reality, having expanded as an act of comprehension, but irreducible to its objective composition. Therefore, a phenomenological treatment of consciousness – as the experience of objective, sensual consciousness of the world as special object and as a real subject of cognition – is again and again repeating the ontological possibility of cognition itself as an efficacy.

That understanding of *transcendence and irreversibility (individuation)* of consciousness is the basis of understanding consciousness as a phenomenal existence through classical philosophy. We can say that all New-European philosophy is being built as transcendental ontology. Its concepts allow us to analyze consciousness in its ontological meaning and in its 'corporeality', without connecting it to states of psyche, and abstracting it from the human as a subject of the latter. The improbable abstraction of *cogito* or '*pure consciousness*', is really fulfilling nevertheless, because consciousness and thought are action itself, and the result and product of such living action is our empirical and psychological 'I'. The category of 'reflexion' or 'self-consciousness' produced by philosophy is, in its essense, a special tool for keeping *cogito* in the state of consciousness, for being *cogito* in these states, because concerning consciousness as an actual phenomenon, the description of conditions, under which it can 'appear' or 'be induced', is possible. All the differences between empiricism and transcendentalism are in reality the different paths to this transcendental *Ego*: to split the psychological experience from the transcendental experience is very difficult due to their merging, and in the artificial construction of reflexion, the realized experience is built as a set of modi – of sense, impression, imagination, etc. The description of these consciousness modi is the way of empiricism, with consciousness, existence, possibility and conditions as their aggregates.

When Hegel raises a question in his philosophy about the appearance of knowledge and shows that reflexion is really an empty concept, i.e. is understood as a phenomenon that shows itself, then he continues this tradition of transcendentally phenomenological analysis of consciousness. Really, his claim that the soul is emptiness is a receptacle (das Rezeptakulum) which includes everything,[5] is the same understanding of actuality of the consciousness phenomenon as Descartes'. The soul is actually always present, but not the whole soul. The whole soul is present in the full development of time, whereas in the imagination – *hic et nunc* – it appears in an empirical way as something solitary, and is understood by us as the aggregate of solitary impressions; they are the tatters that we take for its essence. Whereas it is the connection, the receivability of those separate experiences connected by us 'in time' (time itself is the changing of those experiences). It is this 'joining' in time (in our imagination already) that is known to be in the fundamental traditional representation of the soul's substantiality. We imagine it to be so, but the soul is not a thing, it is an action of *intro*-receivability, or connection of imaginations. This connection of the 'solitaries' and their existence 'at once', 'as a whole' and 'together' is that very idea of soul when we think it. It is in the moment of thinking of its idea, because here thought itself is an action of a soul. This action of a soul as thinking its idea is reflexion. Therefore Hegel's reflection does not appear as a purely logical act – the knowledge of action, which is included in knowledge itself, but as some experience of consciousness, as the 'work' of a living soul here and now. It is the action of subjectivity, in which there exists a non-differentiation between the identity of the existence of consciousness and the mode of its realization, to be exact, the form of its appearance, thinking. As a logical contentious act, it is a premise of cognition, situated not inside but outside it, i.e. it is really an ontological premise, given together with the cognitive acts themselves. As a step of thought, this statement by Hegel in terms of its real contents does not differ from Kant's statement, that transcendental actions of consciousness are validated not by the presence of a substance included in them, but by their practical disclosures – revealing, practical realizations of a logical (gnoseological) function of pure forms of reason with the contents of experience (a synthesis of contemplation).

Nevertheless, from the very beginning Hegel treats reality itself in a different way, because it is known to appear in his work as consciousness (spirit as such), viewing itself in the form of an object. Correspondingly, he puts the conditions of possible experience under the heading of a consciousness to be changed. So Hegel leads the analyzed understanding of consciousness by classical philosophy to its logical hypostasis: his identity of idea and existence does not establish an understanding consciousness as existence, but is a theoretically cognitive premise and the principle of analysis of historical forms of human cognition. From this point of view, reflexion really, factually – in acts of human thinking as reconstructed by Hegel – turns to reflexion about the logical contents of knowledge. Here consciousness first appears as an external relation of 'givenness' of an object of cognition, then as internally stressed self-differentiation in itself, and at last, as an identity in phenomena of humanized, spiritualized reality. Cognition, as a whole, appears as thinking by categories – 'thinking in concepts' – of its own forms and revelations, and finally turns to a 'canonical' ordering of objectivated forms and revelations in the process of revealing the cognition thus realized. So in Hegelian philosophy the problem of understanding is eliminated: as reason in real contents of knowledge lacks any means of objectivation, so the contents of reason can be achieved thru the usual reflexive interpretation of implicit cultural mechanisms of cognition. The problem of an actual presentation of consciousness is eliminated here.

This "logicalization" of consciousness or "ontoligation" of logic has allowed the spread of a vast development of historical forms of human cognition and culture, and thus has determined new horizons of philosophical thought. Nevertheless, such formulations of the question lead to the loss of the 'living' cognitive process, reducing it to either cultural reproductive mechanisms or a psychological interpretation of its individual revelations. The psychologization of consciousness in philosophy has happened. The recognition of the fact has been reflected in twentieth century philosophy's turning to the analysis of 'living' consciousness.

The specific feature of modern philosophy, independently of differences between trends, ideological orientations, language, etc., is that this philosophy is known to raise a question about the necessity of qualifying human existence as the decisive point of all philosophical problems. This also determines its relation to classical philosophy in formulating a consciousness problem: if classical philosophy and cognitive theory take the question of the personal existence of consciousness as a real problem, so modern philosophy treats just the question about the total significance of truths and the intersubjective nature of consciousness as a problem. Appeals to the analysis of the world of daily experience and affirmations that Kantian or Cartesian *cogito* cannot explain to us the essence of cognition as such, because they alone – as certain experimental acts – need to be explained, presuppose the development of the other path to a transcendental *Ego*.

Nevertheless, in spite of the fact that modern philosophy abdicates classical differentiations between 'phenomenon' and 'essence', but treats the experience of consciousness as a 'phenomenon', as 'vital forms', as *dasein*, where everything is immanent and acts of consciousness are already indistinguishable (*cogito* are not already specially picked out from the whole continuum of cognitive acts), in spite of all these facts, consciousness is still explained *ontologically and transcendentally*.

This difference in the ways of 'catching' living subjectivity is determined here by the change of the aim of philosophical thinking. Really, Husserl, faced with the problem of revealing the immanent structure of consciousness – the existence of 'pure phenomena', treats consciousness as a givenness in which everything takes place, and hence, is observed as a premise of philosophical work that we need not specially ground. So the task of phenomenological analysis is concluded in a revelation of an essentialistic structure of consciousness, irrespective of its existential contents. In other words, the analysis of consciousness presupposes ontological neutrality with respect to objects of experience themselves, because the 'pure' structure of conscious acts is reached irrespective of the presence of their objective reference. It can be reached by the totality of special procedures of work with living consciousness – the series of psychological, eidetical, and at last, transcendental reduction.[6] It is connected to the fact that the primary understanding of consciousness links up with the phenomenon of the world as a correlate of ontologically understood consciousness: immanent intentionality of consciousness means – in such a way, in this act of consciousness, in this experience – that the world as opening reality is not the thing that opposes any other object of cognition, but as we have already stated, the thing that we reproduce by the fact of our being conscious creatures. We never could have seen a thing, should the matter have been in sensual perception. Looking at a thing, we simultaneously create a context of perception for it, some reasonable horizon as a background to perceive it. This is our analytics of consciousness, a kind of articulation that anticipates each work of understanding, similar to the Cartesian *cogito* or the Kantian *transcendental apperception*. Thus, phenomenology demands the consideration of existence as a correlate of consciousness, that which we "think about" concerning properties of revealing consciousness. Phenomenological reduction is that procedure of 'stretching' this operation – transnatural and unnoticeable by us – of the 'analytics' of consciousness, to be precise, its intending: we cannot leave this work of consciousness behind (because we ourselves – psychologically experienced – are the result of this work), but we can relegate it to stable structures which can be seen thru specific and reducible meanings. That is why later in Husserl's works, the investigation of intentional structures of consciousness changes to procedures of construction, the *ideation of ideas*, free variations in fantasy. Nevertheless, this phenomenological reduction as the way to a transcendental *Ego* is based upon the recognition that consciousness is an ontological phenomenon, a transcendental act in existence. And in this meaning it repeats classical philosophy: in phenomenology the objectivity itself is constituted by a radical self-reduction of subjectivity to its own primary structures. In *Cartesianische Meditationen*, Husserl returns to *cogito* consciousness: transcendental reduction brings consciousness to the area of absolute consciousness, the structure of which is *ego – cogito – cogitatum*.

Transcendency in any form is a self-constituting meaning of existence within the limits of 'I'. Any conceivable meaning, any conceivable existence, whether it is named immanent or transcendental, is outlined by an area of transcendental subjectivity as constituting meaning and existence. It is useless to try to comprehend the universe of original existence as something outside the universe of possible consciousness, possible cognition, and possible obviousness.[7]

In fact it is the recognition that *ego cogito* is not just an abstraction, but a real state of affairs in the world – the phenomenon of transcendency of consciousness as its living existentiality.

Nevertheless, phenomenology's primary intention, to describe immanent structures of consciousness (intentional structures in their ontological neutrality), leads to the fact that this loss of essentiality of consciousness occurs because transcendency is really being rejected, and transcendental ontology is being built as eidetic ontology. It reveals at once how phenomenology turns to ontology (in *Crisis of European Sciences . . .*): the world is understood as just a correlate of consciousness, but not as a space in which, among others, a phenomenon of consciousness takes place as something existing and essential.

Thus Heidegger, have formulated the idea of his fundamental ontology, comes from the necessity of leaning on an experience of a subject's existence that supposedly is known neither to classical nor to transcendental phenomenology. He begins from the statement that the question about existence primarily 'sends' the asker to a certain experience, in which this existence is found to be 'accessible'. Existence demands its own method of revealing, which primarily and essentially differs from all forms of conceiving the existing: therefore it is as if it could be detected neither with the help of abstractions, 'essence-event (phenomenon)', nor with the help of the concept of 'phenomenon'. The point is the discovery of a certain experience of ingenuous givenness of existence, and the problem is to qualify the existence of the existing, which possesses the possibility of asking questions about its existence. This 'existing' entity is a human being, and the existence experience of the existing is existentia (*Da-sein*), i.e. special and always individual experience of existence of subject-in-the-world, of his non-spatial, but objectively reasonable *situ*-ation. Existence is being analyzed in connection with this, 'the existing': the existing is what in one way or another is determined thru existence, and vice versa, existence, as it is exposed to us, is a moment of *Da-sein* structure. Thus, according to Heidegger, ontology is possible only as a phenomenology of the description of existence, of *hic*-being. This point reveals the paradox of formulating the question in such a way, and phenomenological description gives way to hermeneutical interpretation: the objectness that must be described is non-objective *existentia*, and it is always undetermined. An act of existence happening is not determined if it has not happened; by the happening itself the world is finally determined and the connections between objects and phenomena are stated. *Existentia* is pure temporality; the possibility of 'the existing' being and not being itself is ex-istentia, i.e. the transcendency of existence by each step of its happening and objective description (the objective description is itself the act of transcending, because it is a real and objective event, an event in reality). Therefore, by phenomenologically describing Dasein as the possibility of existence, we discover its 'closedness', its apophaticity. This discovery of the apophaticity of existence as a back cover of the finiteness of '*hic*-being', is really the description of the original nature of existing. Fundamental ontology, therefore, is really the ontological basis for human existence. In this meaning, fundamental ontology is the expanded phenomenology of life, a constant transcendency of itself.

This analysis of the thinking steps in modern philosophy shows that the introduction of new concepts ('phenomenon', 'vital forms', existentials', etc.) and new procedures for consciousness analysis (phenomenological reduction, language destruction, etc.) is connected precisely to the change of aim and to the development of new methods of analysis of the *objective nature of consciousness*. In this meaning, modern philosophy as if from the very beginning, discovers an essentiality of consciousness, and from this point of view all the differences between classical and modern philosophy are not so serious, because they still share the principle that the ontological understanding of consciousness is a certain action in existence, and its existentality and phenomenality, but not epiphenomenality.

III

Such a philosophical understanding of the objective nature of consciousness is the foundation for a scientifically correct formulation of the question a corporeality of consciousness. *Philosophical analysis demands the understanding of consciousness as an objective action, irreversible and transcendental relative to its empirically fixed presentations.*

As an objective phenomenon, consciousness must be comprehended as the interaction between objective reality and the human body, materially expanded and fulfilling in reality itself. The problem is that in the universe, conscious acts possess irreversibility as objective phenomena; given this relationship, in reality consciousness performs transcendentally with respect to the language of internal experience, as the action of adhesion of the objective-practical, as a sensual cloth of corporeal feelings and performed meanings. This interaction appears as a natural objective action that has its own 'space' and its own 'time' (history), and that is by no means identical to its realization in the substratum of the human body – in the form of perceptive images and psychological experiences of meanings. Vice versa, it is necessary to understand the latter as real actional-semiotical and symbolic transformations of this objective action, and as such – *caused* by it. Therefore the consciousness-corporeality problem is simply the problem of the 'embeddedness' of the phenomenal matter of actionally symbolic acts of consciousness and psyche into the eventual row of historical, social and cultural reality.

Really, it is the irreversibility of the course of events, including the actions of subjectivity, that make this subjectivity real. In the real space and time of the conscious act there is material, living motion, practically expanded as an objectual position of our body in the objectual reality, and simultaneously contracted as a symbolic substitution of experience and sensual image. It is as if living motion itself had two poles – empirically fixed change of position in objective space among external objects, and causation by this very change of respective images, thru which this change is experienced. Therefore in reality we cannot, by a simultaneous action, trace a motion in the objective reality that generates this tracing. So the *corpus* of consciousness is not a human body, taken in its anthropomorphic structure, but is in reality expanded each time *hic et nunc*, a sequence and simultaneity of living motion of objective practical action of the individual and substantial outcome of phenomenal matter of living experience. A human body here performs just like a functional organ of these transformations and adhesions. This is the objectively practical 'body' of life consciousness, expanded and existing outside an individual head; this is the sole reality, including subjectivity, that is not an idea or spiritual essence.

This is a specific matter of living experience, conceiving of a body; it is that *thopos* of ideas and 'I' that classical philosophy pointed at.

Consciousness in all its modi, as inseparable by discrete and sequential observations in time, enters the world transcendentally, i.e. is produced and generated materially and is able to be a localized 'body' of actual adhesions, *hic et nunc*, of living motion (sensomotorical acts) to an individual's objects of practical activity. This actual adhesion of feeling to the 'physics' of objective practical situations *hic et nunc*, 'binds together' into a single junction of this psychic state and state of the external world, into a stable connection, presented and understood as meaning. *This stable connection is thopos, corpus, not internal, but an objective space and place of existing of these acts of consciousness.* Consciousness or subjectivity is expanded in time, as a substantial practical contraction of practically obtained connectivities of the objective situation itself in all its space, and of actual into ideal sensual images.

Really, this process presents some quite substantial action as a practical realization of symbolization: fundamental perceptive categories – space, time, motion, shape, color, position, etc. which serve as practical landmarks, as a fundamental practical activity of an individual, are contracted in sensual images, and thus hidden by this actionally symbolic work. These primary layers, being the fundamental living motion of perceptive experience of consciousness (being its biodynamic and sensual tissue) are contracted, and not simply covered, by other 'upper' layers of consciousness – sensual images, images of fantasy, imagination, etc.[8] This contraction is living motion itself, with 'sediments' in the form of states, internally experienced and described as such in intraspective terms. A human moves as if to the world of meanings and concepts, reflects to 'upper' layers of the world built by him, *consciously* operating with objective images, words, symbols and meanings. But those sensual transformations must always take place to perform the perception as an individually sensual act: all mental images that can be described in the language of internal feeling and determined now (i.e. after performance of the perceptive act) in terms of aim, such as reason, emotion, realized reflective acts, aimed action, etc., relate to this definite system of motions – 'adhesions, substitutions, transformations' – inside this lasting sensual experience as forms of existing and expression of the given perceptive structure. In this unique experience of motion construction, the human psyche has become the unique 'place' of a human body's transformation into an inorganic corpus of consciousness and meaning in its literal meaning. Vice versa, transindividual by nature, the field of meaning in described real objective actions of living consciousness begins to exist as something organic, inseparable from the subject's body. Therefore this living structure of objective action in the perceptive structure of a body is the action of production, or performing those stable mental formations of psyche – 'upper' layers, by which we mean the surrounding world and their own contents, but behind which 'contraction', 'crystallization', 'structuration' have performed on this sensual cloth of living interactions, and so our consciousness does not see itself and its actions. The extraction of these formations is the revelation of dependence of upper layers of spatio-temporal interactions and motions, characterized by intelligence and ideality, and it is a practically impossible operation, because they can be revealed most in this extra-temporal and extra-spatial reality of the ideal.

Those formations of consciousness appear to be such ideal models, whore ideality is in a 'presentation' in them of this stated connection of sensual experience with symbolic meaning. To be exact, they are themselves the 'presentation' of this connection, its substantial representation. This representation is phenomenality as existing; it is not a representation of something to somebody, but a representation as a form of existing. Ideality or phenomenality of sensual images of consciousness is the indication of their existing as 'presentation' of a substantial realization of consciousness, i.e. its transcendentality. The presentation reveal themselves as objectivities of consciousness that now possess their ideal space and time. Not the objective space-time of their realization, but a quasi-space and quasi-time of a presentation of understandability.[9] In this there is an elimination (*aufhaben*) of that living action of consciousness, that has turned and gone, contracting into these presentations. Meaning is that phenomenon of representation, 'presentation' in an ideal form of sensual image of passed living motion and production of perceptive structure. In this meaning, meaning is a *thing* of our comprehension. In this meaning our 'I' is also 'a presenting' or a model of understanding, but a model as a living lasting existence, related to the definite system of motions of living experience. This presenting has meaning while this experience of its conscious thinking is fulfilling; the understanding of that has found its mirroring in the thesis of classical philosophy about an idea as a soul's form of existing: the 'I' exists only in the moment of its thinking, i.e. as an idea of 'I'. In this meaning our empirical 'I' is just a product of 'pure I'.[10] It is just an indication of a factual symbolic character and ideality of images of consciousness as their modality and representativity of practical realizability of consciousness.

Therefore meaning is eventual in principle, it is what is thought in and together with what happened in the world. Meaning, understood as knowledge, is the event of thought's realization. And so meaning is practical, realizing, from the beginning each time, a 'connection' between objective contents of knowledge (which is contracted and compressed 'former' experience that took place before as an actual experience of knowledge – a statement of understanding) and the state of the sensual apparatus, i.e. the course of action of the reality itself. In other words, this is the phenomenon of transformation of states of objective reality into objects of living subjectivity. The way we understand, explain, etc., knowledge depends on how that *thought* happened for the first time, and since it always happens in such a way, determining the utmost possibilities of its existence. Something in the experience of an individual becomes a fact of his life, if it has any meaning, i.e. it is actually realized in a transcendental way as a connection between an objective situation and its 'sediment' in the sensual apparatus.

This happening now becomes a constant of consciousness, determining now, as a law, all the current experience of psychic and conscious human life. It is impossible now to get back to the primary state of absence of meaning, because in the meaning an objective scheme of the psyche's work as a natural process is presented in the ideal way.

Therefore, those 'adhesions', 'substitutions', 'structurations' are the practical work of the psyche and consciousness, the pure 'corporeality' of subjectivity, taken in its spatio-temporal entity and irreducible to its own psychologically experienced images. In other words, this is not an experience of consciousness, with observer-produced actions from outside, but the product itself, determined by them. Consciousness is a naturally acting force, and so each psychological process may be observed both from the side of its objects, and from the side of the consciousness itself. This shift to the act of appearance of consciousness is that very attempt of 'stretching' the consciousness field over those living transformations, over what it has already closed in the inseparability of the ideal. 'Splitting' these formations of consciousness, connected in the process of their realization with their products – images, understanding, construction, language forms, categories of thought, etc., appears to be a complex procedure of phenomenological reduction (but not in Husserl's variant). Each new remarking and shift of focus and periphery as a tool of reflexion presents a new experience of consciousness. Therefore each identification and differentiation, when a subject needs to implicitly identify himself with one pole of the consciousness field – the only condition under which they can be together in the single common perception of the object of action – is its 'transfer' into another state or structure, accompanied by its 'corporeal' reconstruction. Each structure of consciousness (imagination) corresponds to its own 'corporeality' or phenomenal spreading. All the abstract constructs like *cogito*, 'reflexions', 'ideation of ideas', 'epoche operations', etc., are tools of philosophical analysis, and work an individually objective and semantic cloth of consciousness.

This interpretation of psyche with the attraction of fundamental spatio-temporal and objectively semantic categories is a new experience of consciousness. But this use of special understanding models is not simply an apparatus of thought, with the help of which one can avoid the traps of screening one's own consciousness. The issue here is that independently of any method of philosophical analysis, our psyche works in this way, opening thru its quasi-objective formations the space of the possible world of consciousness, and further developing empirical psychological structures themselves, in it and thru it.

IV

Modern psychology connects the solution of the consciousness-corporeality problem and the possibility of its objective analysis with the refusal of any naturalistic interpretation of it and an understanding of the dual organization of phenomena of conscious life, that follows from the very nature of a human being as a feeling and living creature. Understanding consciousness as an existential, practical phenomenon in the structure of existence presupposes the introduction into scientific circulation of some fundamental items of its analysis.

First, it presupposes the elimination of premises of classical philosophy and psychology (in *cogito* theory) to take givenness of consciousness itself as a reference point. It is necessary to distinguish the *psychology* of consciousness and its *ontology*. Consciousness is not a psychological process in the classical psycho-physiological meaning, it is not one of the psychical processes, but some level at which all specific psychological processes are synthesized. To be exact, it is a section of their *givenness or existence*. Therefore all phenomena of the human psyche may be regarded as phenomena of consciousness, not because of their having become objects of consciousness, as is presupposed by the classical procedure of reflexion, but exceptionally because of their possibility of being regarded as natural functions or dimensions (modi) of consciousness.

Second, it means understanding the difference between consciousness itself and the language of its givenness or description. Consciousness as a naturally historical phenomenon is such that the method of realization eliminates conditions in which it lives and exists. The peculiarity of consciousness is that it is a phenomenon possessing the property of self-designation; in other words, as a specific object, consciousness is identical to its interpretation. Thus, for consciousness it is important to differentiate the 'language' of consciousness itself and 'language of the explorer' or meta-language of its description. But this means that each description of consciousness is its new experience.

Third, understanding consciousness as an objective phenomenon presupposes the possibility and necessity of analyzing it without the description in the language of internal experience, etc. This objectivity presupposes that we can describe such phenomena, whose natural revelations contain in themselves subjectively actional transformations of reality as sediments. The objective description of consciousness means in this case the possibility of its description regardless of changes in the psycho-physiological feeling apparatus, and generally regardless of the human individual and his body. It is a true description of consciousness in an 'inhuman way'. Its principal cause is the idea that cognitive theory must first refuse psychological qualities in its primary constructions, and introduce them later as phenomena of consciousness, of specific consciousness, but not as something naturally given. We can never objectively detail the sides of a psychological process until we include subjectivity or consciousness as a necessary element of the objective reality.

This understanding of consciousness as an objective phenomenon in existence presupposes revealing a *real ontology* of the world and comes out as a problem of *philosophical ontology*. This ontology must give the description of a structure of existence and consciousness as a free, non-causal, but non-eliminable and necessary element in existence, without which the existence itself is not complete. It presupposes the development of special theoretical constructions that must come out as 'pseudo-topological' concepts or 'quasi-objectivities' of consciousness, permitting objective discussion about consciousness, avoiding traditional interpretations and distinguishing between the real status of things and consciousness, and describing conscious phenomena in all their spatio-temporal organization.[11]

Only in this case, I suppose, it will be possible to enter an actual 'effect of the real' in psychology and to describe the human psyche as a living action and act.

State Technical University
Perm

NOTES

[1] See Chapter 1 of Descartes' *Principia philosophiae*, wherein he produced forms of thought about reality as independent of thinking and objective; but all that work is developed in the reality of thinking itself.

[2] "Although acting and being acted upon are often quite different, nevertheless the action and the exposure to action are always the same phenomenon that has two names, because it can be related to two different subjects". R. Descartes, *Passions de l'âme*. Oeuvres de Descartes publiées par Ch. Adam et P. Tannery. Réédition. Vol. XI (Paris: 1969–1974), 327. Quoted after Декарт Р. Страсти души.// Соч. Vol. 1 (Moscow: 1987), p. 482.

[3] This misunderstanding of the nature of the *cogito* and Descartes' argument can clearly be seen in the *Fifth Disquisitio of Gassendi*. See R. Descartes, *Méditations*, Oeuvres, Vol. VII.

[4] Cf. Spinoza's arguments in *Ethica* about the adequate idea of a thing as of a state of a body, understood as actual objectual motion along with the form of the thing itself. This understanding of thinking as a method of action of a body, i.e., of a spatially spread phenomenon, is the basis of his concept of substance.

[5] Hegel, *Encyclopädie d. philosoph. Wissenschaft*. Quoted after Гегель Г.В.Ф. Энциклопедия философских наук. Vol. 1 (Moscow: 1974), p. 123.

[6] E. Husserl, *Ideen zu einer reinen Phänomenologie und phänomenologischen Philosophie*. Quoted after Язык и интеллект (Moscow: 1995), pp. 14–49.

[7] E. Husserl, *Cartesianische Meditationen*. Hua, I, p. 86. Quoted after Свасьян К.А. Феноменологическое познание (Yerevan: 1986).

[8] The concept of 'layer of consciousness' is a concept of meta-language, the language of study and theoretical description, but not of natural language concerning consciousness. In the meaning of a proposed description of the actionally symbolic work of consciousness, the latter is a *tabula rasa*.

[9] Within the frames of understanding of the ideality of this time, the problems of *Zeitbewußtsein* are taken up in Husserl's *Zur Phänomenologie des inneren Zeitbewußtseins*.

[10] Cf.: "There is self-consciousness, generating the imagination **I conceive**, which must have the possibility of accompanying all the other imaginations and be the same in every

consciousness, therefore this self-consciousness cannot be accompanied by any other (imagination), so I also call it the **primary apperception**. I also call its entity a **transcendental** entity of self-consciousness to mark the possibility of apriori cognition on the basis of this entity. Really, diverse imaginations that are given in certain contemplations would not be **my** imaginations if they did not belong all together to self-consciousness alone; in other words, being my imaginations (although I would not realize them as such), nevertheless they necessarily must conform in self-consciousness, because otherwise they do not all belong to me". I. Kant, *Kritik d. reinen Vernünft*. Quoted after Кант И. Соч. В 6-ти т. Vol. III (Moscow: 1964). pp. 191–192.

11 M. K. Mamardashvili, [The Classical and Non-classical Ideals of Rationality] (in Russian), (Moscow: 1994), pp. 80–89.

JARI KAUPPINEN

DEATH AS A LIMIT OF PHENOMENOLOGY
The Notion of Death from Husserl to Derrida

Si la possibilité de ma disparition en général doit être d'une certaine manière vécue pour qu'un rapport à la présence en général puisse s'instituer, on ne peut plus dire que l'expérience de la possibilité de ma disparition absolue (de ma mort) vient m'affecter, survient à un *je suis* et modifie un sujet.
– Jacques Derrida, *La Voix et le phénomène*[1]

Und der Tod? Monaden können nicht anfangen und nicht aufhören.
– Edmund Husserl, "Monadologie, Beilage XLVI", *Zur Phänomenologie der Intersubjektivität, Dritter Teil: 1929–1935.*

1. THE QUESTION OF DEATH IN HUSSERL

Can death be a theme for thought within phenomenology, even the "phenomenology of Life"? What Husserl writes about death is only some cryptic remarks about death, dying and birth.[2]

In his few short illustrations involving life and death, death is contrasted to the infinite life of the transcendental ego. In contrast to this infinite life is the possibility of my empirical or natural death. In *Ideen* (§ 54) there is the famous remark about the paradoxical possibility of a psycheless (*seelenloses*) ego, after Husserl had sketched the imaginative possibility of the whole of humanity's non-existence in a description of the phenomenological attitude. Humanity does not exist any more, neither myself nor my fellow human beings (*Nebenmensch*); nature and the psychic would be annihilated (*vernichtet*).

Further on, when Husserl characterizes the transcendental-apodeictic ego as immortal, he speaks freely of the immortality of that Ego (*Unsterblichkeit des transzendentalen Ich*). For Husserl the empirical ego is mortal (*Husserliana* XI, pp. 377–381). At a later stage he describes monads as being immortal because monads have no beginning (*anfangen*) or end (*aufhören*), (*Husserliana* XV, p. 609).[3] For them death is like unconsciousness (*Unbewusste*), a primary ground for consciousness

A-T. Tymieniecka (ed.), Analecta Husserliana LXVI, 323–348.

(*sedimentierte Untergrund des Bewusstseins*) and a dreamless sleep that gives birth to the subjectivity: being before birth, in death, and after death (*ibid.*, p. 608). When a human monad dies, it could sink into absolute sleep (*versinkt in absoluten Schlaf*), (*ibid.*, p. 609). Thanatos is traditionally called the twin brother of Hypnos, and sleep seems also to be one of Husserl's favourite analogies for death. For example, in elaborating the themes of sleeping and waking, he asks whether death is not the brother of sleep? (*Der Bruder des Schlafs*). (*Husserliana* XXIX, p. 338)

The most interesting remark is Husserl's saying that "The Death of Other is the death constituted first" (*Der Tod der Anderen ist der früher konstitutierte Tod*).[4] This is surely in contrast to Heidegger, who insists on the primariness (authenticity) of one's own death in *Sein und Zeit* and even later in the *Beiträge*. Thinking on death usually follows or accords with Heidegger and with Husserl. There is Heidegger's obvious critique against Husserlian subjectivity; but as recent Husserl research has clearly shown, Husserl had a theory of transcendental intersubjectivity.[5] And one can even say that in the chiasmatic interplay Heidegger describes between mortals and immortals as well as earth and sky, he thought about death on the level of transcendental intersubjectivity, if one's own death bears the mark of subjectivity. Another question is that of the primordiality of the subject, because for Husserl the constitution of the ego is also primordial, even in the case of transcendental intersubjectivity.

Husserl himself criticizes Heidegger in "Die Anthropologische Welt, Ende August 1936", when he continues to argue that in the phenomenology of absolute originary evidence (*Evidenzquellen*) death is what differentiates (*ausscheiden*) the transcendental ego from self-objectification as Human (*Husserliana* XXIX, p. 332).[6] This is the authentic (*echten*) phenomenology grounded (*begründeten*) by transcendental reduction. He also says that death is not experienceable (*Den Tod kann an sich niemand Erfahren*) and that it is only experienced in others (*an anderen, ibid.*).

Yet then there is also the question of the anthropologizing of phenomenology. What status does the other(s) have? This leads us to the *Cartesian Meditations* and to the question of intersubjectivity when the possibility of subjects' subjectivity constitutes intersubjectivity thought of as the possible absolute subjectivity. If we believe that Husserl had a theory of transcendental nonsymmetric intersubjectivity, then the other would be transcendental although it will still be constituted by my transcendental ego, my transcendental life being wrought by the death of the other. Death, like birth, then, is a limit concept for phenomenology.[7]

I try further on to elaborate the meaning of the notion of death in phenomenology through an explication of Jacques Derrida's work. If the notion of death is a limit for phenomenology, then this certainly explains why death could not be described in phenomenology. That is to say that it is not that Husserl simply did not think the death but that the nature of phenomenology cuts death off from the scope of phenomenology. Yet we can say that Husserl opened the way to thinking about death. The post-Husserlian "phenomenology of being mortal" could paradoxically be called "the phenomenology of death" because there was so little to do with the notion of death in Husserlian phenomenology proper.

The focus here is also on pre-predicative experience, sometimes also called "transcendental aesthetics", because death could only be manifested at a pre-phenomenological level. Death surely breaks the transcendental reduction, but the notion of death may not be grasped without such a reduction (of natural and empirical death). Death comes to mean something symbolic, through a system of symbols, i.e., the system of language. Through language we can inquire into the possibility of "my death"; the possibility of intersubjective language is also the way to the others' deaths and to the space of ethics.

2. DERRIDA ON HUSSERL AND LITERARY OBJECTS

Jacques Derrida's critical exfoliation of phenomenology in the book titled *Speech and Phenomena, and Other Essays on Husserl's Theory of Signs* (*La Voix et le phénomène*) is strongly opposed to Maurice Merleau-Ponty's *Phenomenology of Perception*. According to Derrida, perception does not exist and the thing in itself (*la chose même*) hides and escapes infinitely (*La Voix*, p. 117; *Speech*, p. 104). Derrida is interested in the forms of non-perception. This also avoids the positions taken in the *Psychologismusstreit*. Should we, 1) think about phenomenology in psychological terms and affirm that the noema is purely psychological or 2) hold that a noema is not perceptual or psychological but purely conceptual. That the noema is conceptual does not mean that this proposition is constituted without a concept or a reference to a psychological origin or genesis as a structure. Psychology as escape from psychology constitutes the "hard anti-psychological core of phenomenology" – as Derrida urged in his first article "Genèse et structure et la phénoménologie".[8]

The subject of Derrida's doctoral thesis back in the fifties was the ontology of the literary object.[9] Derrida then developed the ontology of the literary work of art up to a point at which it may rupture the phenomenological reductions. In advance, one could also note that Derrida's *oeuvre* continues the major problematics of French phenomenology in the forties and the fifties, if we agree that already then importance was given to framing a phenomenological theory of imagination and language. For the "Cartesian transcendental egologico-phenomenological" Husserl there obviously was very little space for a theory of language and imagination. Besides there was in the French reception of Husserl already the "anthropologized" problem of being-in-the-world. These questions touch the problematics of pre-predicative experience, but I treat them subsequently as background or context for Jacques Derrida's work on Husserl. Sartre and Merleau-Ponty can be mentioned as Derrida's targets in developing a stronger phenomenological theory of language. He was questioning the limits of phenomenology and not just trying to solve the obvious contradictions between the phenomenologico-transcendental ego and the world. One could even say that the Sartrean problematics of the transcendental field of the ego metamorphoses first into the question of the transcendental field without a subject (Hyppolite) and second to the question of the possibility of a transcendental language and then to the constitution of language as writing in the pre-predicative world (Derrida). This recalls Eugen Fink's famous critique of Husserl that the meaning of language as a transcendental reality must change after the transcendental reduction. If we were to say that there is a contradiction between the Husserlian transcendental ego and the world, then we could neatly say that Derrida's efforts in creating a non-concept or undecidable "infrastructures" (like *différance*, which is not a concept or a word and always differing and deferring) try to solve obvious difficulties without any Hegelian dialectical correction of intra-phenomenological difficulties.[10]

Derrida developed his critique in *La Voix* around the possibility of "my death". "My death" is for Derrida the possibility of speech and saying as well as of all semiological systems. It also means the possibility of radical non-perception. The possibility of my absence constitutes, according to Derrida, the core of the Husserlian "*lebendige Gegenwart*" (*La Voix*, pp. 60–65; *Speech*, pp. 54–58).

Derrida's argumentation in *La Voix et le phénomène* as well as in *De la grammatologie* is typically Heideggerian with death opening temporalization as the impossibility of possibility within the dialectical movement of instants in the retention-now-protention structure.[11] One could even say that the great themes of *De la grammatologie* and *La Voix* are taken from a few pages of Heidegger's *Einführung in die Metaphysik*.[12] For example, he writes about the Living Present in *De la grammatologie*:

We must then *situate*, as a simple *moment of the discourse*, the phenomenological reduction and the Husserlian reference to a transcendental experience. To the extent that the concept of experience in general – and of transcendental experience, in Husserl in particular – remains governed by the theme of presence, it participates in the movement of the reduction of the trace. The Living Present (*lebendige Gegenwart*) is the universal and absolute form of transcendental experience to which Husserl refers us. In the descriptions of the movements of temporalization, all that does not torment the simplicity and the domination of forms seems to indicate to us how much transcendental phenomenology belongs to metaphysics. But that must come to terms with the forces of rupture. In the originary temporalization and the movement of relationship with the outside, as Husserl actually describes them, non-presentation or depresentation is as "originary" as presentation.[13] (*Gramm*, pp. 90–91; English transl., pp. 61–62)

From Derrida's early approach to bridging phenomenology and epistemology one could develop the two following consequences.

(1) In *De la grammatologie* he asks in what way is time founded by an interiorized apriori, as a prior temporalization that does not follow the model of the exterior determination of linear space.

(2) In *La Voix* the living consciousness is contrasted to the external system of signs, for instance to the Saussurean structural semiology or symbolism as unfulfilled and empty intentionality. Symbolism is the possibility of expressing signs, the possibility of expressing and repeating non-existing things and stating facts of death (non-presentation).[14] A famous example of symbolism (as opposed to "sense certainty") from Hegel's *Phenomenology of Spirit* is cited: "When I Spent, death speaks in me" or "death is present in my language".[15] Thus, Derrida says of transcendental life and the possibility of "my death":

To think of presence as the universal form of transcendental life is to open myself to the knowledge that in my absence, beyond my empirical existence, before my birth and after my death, *the present is*. I can empty all empirical content, imagine an absolute overthrow of the *content* of every possible experience, a radical transformation of the world. I have a strange and unique certitude that this universal form of presence, since it concerns no determined being, will not be affected by it. The relationship with *my death* (my disappearance in general) thus lurks in this determination of being as presence, ideality, the absolute possibility of repetition. The possibility of the sign is this relationship with death. The determination and elimination of the sign in metaphysics is the dissimulation of this relationship with death, which yet produced signification. (*La Voix*, p. 69; *Speech*, p. 54)

In the Husserlian phenomenology of the living consciousness even the silent re-flective (wordless) monologue is based on the living flow of constituting consciousness. The enunciation "I am dead" becomes meaningless, although it can be understood as empty symbolism, one having no existing point of reference (*Sinn* or *Bedeutung*) or intention, a non-fulfilled intention. The concept of language, according to Derrida, is in itself contradictory to the consciousness constituted by a living flow of reflection, with the sign I, which is no longer there (*Je suis* being *Je suis mortel*). In *La Voix* this means a critique to the foundation of phenomenology in so far as phenomenology is determined by a transcendental apodeictic self-foundation.[16] Yet a parallel case which Derrida uses as one of his epigraphs in *La Voix*, is the famous Edgar Allan Poe short story *The Facts of the Case of Mr. Valdemar*. Mr. Valdemar speaks there about his sudden decease: *And now – now – I am dead.*[17] As indicated, here is the possibility of the coexistence of death and language.

In his later texts, which are not primarily concerned with phenomenology, Derrida thinks of the phenomenological *epoché*, for example, as a reduction to "phantoms", the (im)possible return of the dead at a phantasmatic-imaginary stage.[18] But in *La Voix* Derrida asked whether "the sign is originally wrought (*travaillée*) by fiction" (*La Voix*, p. 63; *Speech*, p. 56); the next step was to also ask: in what way does a literary work of art exist? What if we are no longer able to conceive of a literary work as an essence in a phenomenological stratum, as a material object or as an intentional object of consciousness in a subject-object dichotomy (as an object for its creator/author and a receptive subject)? In any case, such could already be understood on the basis of Husserl's "Ursprung der Geometrie" (*Beilage* III, to § 9a in *Krisis*), which Derrida translated, the introduction of the same being his first published work. Husserl (*Husserliana* VI, pp. 367–369) says there that objects range along three levels of ideality and that the literary work of art is a "bound object".[19] Husserl reserves the highest ideality for mathematico-geometrical objects, the infinite idealities that are "ideas in the Kantian sense" (*Ideen*, § 143) and which are constituted originarily (*Urstiftung*) on the teleological horizon (*Endstiftung*). At the lowest level of ideality, there is a stratum of material but having an empiricial sign or a sound with the same identity (e.g. lion) in consciousness. At the second and higher stratum there is a sign or symbol having different expressions (e.g., "lion" and "Löwe"), but the same repeatable and intentional meaning. The first two levels are the levels of bound (*gebundene*) objects. For Husserl this kind of object exists from its constitution as an intentional and ideal object. It is also bound to the world and culture in contrast to the non-presentable and only once existing and but reproducible infinite mathematico-geometric idealities.

For Derrida, phenomenology does not give an answer to the problem of the object nature of literature by virtue of the strange doubleness of literature's essence in being. Still he says in "This Strange Institution Called Literature" that we must start from Husserlian phenomenological reflection on the literary object as a bracketing and also a "putting phenomenology in crisis".[20] The opened space of literature is not a priori but historical, and we could ask on what conditions it also refers to institutional-political decisions about literature. In Derridian language literature is situated ontologically at the level of becoming, with the future, in that sense that literature means perhaps "nothing-in-being". Literature or writing does not exist, or if it exists, it does not exist on the level of an entity.[21] This explains why in Derrida's account of understanding and signification we can determine meanings in certain contexts (the problem of the indeterminacy of meanings) but the horizon of contexts must remain ever undecidable. In any case, the space of literature opens itself towards the future and as a structure of meanings it always remains open for fresh interpretations. The horizon is determinable, that is, it can be decided, but the horizon of constitution cannot be specified because of its structural openness. Derrida even says, that since there is no intrinsic criterion for the literariness of literature, which is but formalistic designation for a sign system or the essence of its being, this means that all criteria are strictly historical, extrinsic, contextual and institutional.[22] This is the same maneuver which Derrida deploys against the limits of Husserlian phenomenology: the inside ("there is nothing outside") always refers to the outside, to an absolute outside, like the other or death. Writing as an inscription always becomes ex-scription also, with the absolute limit being an absolute outer realm.

We find something in *De la grammatologie*. There Derrida in his determination of the being, postulates besides the pole (*polos*) of language another origin for language in the north, namely writing, which is the absolute pole.[23] The north in *De la grammatologie* means Jean-Jacques Rousseau's thinking about the absolute of death and work in contrast to the passion which is the origin of language (*Gramm*, pp. 309–326; pp. 216–229). Rousseau says in his *Essai sur l'origine des langues* that language is born out of passion, an affection which means life, music, and which constitutes the south pole, contrast to that other pole of writing and the dead letter.[24] This absolute death, writing, means also that which is absolutely outside all that which constitutes the other pole of language and life. One could say in addition that there is a chiasmatic turning of language between these poles.

3. TRANSCENDENTAL AESTHETICS, AUTO-AFFECTION, AND THE PASSION OF ORIGIN

Now will review the first chapters of Daniel Giovannangeli's important book *La Passion de l'origine. Recherches sur l'esthétique transcendantale et la phénoménologie.*[25] Giovannangeli has been specializing in phenomenological aesthetics and deconstruction from as early as 1979 in his book *Écriture et répétition*,[26] where he offered the "first systematic interpretation" of Derrida. The quest for the origin is the main question here, the passion for origin and (à la Rousseauistic) the origin as a passion. Passion also refers here to passivity, as in the Husserlian passivity in activity, as in the synthesis of passivity and activity (*La Passion*, p. 136). Giovannangeli is interested in searching out certain remains that constitute transcendental aesthetics, which are not representable (*imprésentable*) and are not present in phenomena. This is, according to Giovannangeli, the question of post-Husserlian phenomenology: is such a phenomenology without phenomena, operating with the invisible even possible? In Lyotardian language this could very well be "Anesthetics"; other references made here are to Michel Henry, Gérard Granel and Merleau-Ponty (*La Passion*, pp. 19–24). Giovannangeli's question is about the space of death: what remains after life or what is an experience of undecidability? (*La Passion*, p. 12, pp. 103–104) He does not regard Derrida to be any longer a phenomenologist or a structuralist as in his earlier book *Écriture et répétition*.

In the first chapter titled "*Affection et auto-affection*" he notes that Husserl formulated his own project as transcendental aesthetics.[27] Husserl also presented his own work as a radicalization of the Kantian project, and to confirm this Giovannangeli begins his reading straight from *The Critique of Pure Reason* and with the notion of self-affection, as developed in § 8 of the second edition (B 68–72).[28] He reminds us that Kant proceeded using a symmetrical analogy (§§ 2/4, §§ 3/5) of time and space when in *The First Critique* (§ 2–4) he starts with the concept of space's metaphysical and transcendental situation (*Erörterung*) and then continues (§§ 4–5) with the concept of time's metaphysical and transcendental situation. In the *Critique* (A 21–22/B 35–36) Kant says that in transcendental aesthetics time and space are *a priori* the primordial forms of intuition (*Anschauung*), time being the condition for internal intuition and space that for external intuition. The foundational constitution is the form of sensibility as *aisthesis*. In traditional language, we are speaking of the conditions of possibility, namely, the conditions of the possibility of an experience. Further on in the first edition of the *First Critique*, Kant proposes the famous transcendental schematism with the notion of three syntheses of transcendental imagination (*Einbildungskraft*, A 118–119) as the method for transcendental aesthetics. Giovannangeli explains the Kantian spatio-temporality:

L'intuition spatiale contient de simples rapports: l'étendue, le mouvement, les forces motrices. Le temps contient, lui, des rapports de succession, de simultanéité, de permanence. C'est qu'il n'est plus, comme pour Aristote, la mesure du mouvement. Aristote, en somme, définissait le temps par rapport au movement et faisait reposer la permanence du maintenant sur la permanence du mobilité. Kant renverse ce point de vue. Il ne subordonne plus le temps au movement; il en fait la condition de celui-ci. (*La Passion*, pp. 15–16)

We will go no deeper into Kant's argumentation than the crucial point of Giovannangeli's analyses. They roughly consist of what Derrida argues with reference to Husserl and Kant about the possibility of transcendental aesthetics (in both the Husserlian and Kantian senses). Derrida was at the time developing the notion of inscription as the habitation where the aesthetic world is always already situated. In a short section about transcendental aesthetics he says in the last chapter of *De la grammatologie*:

Indeed we say of inscription *in general*, in order to make it quite clear that it is not simply the notation of a prepared speech representing itself, but inscription within speech and inscription as *habitation* always already situated. Such a questioning, in spite of its reference to a form of fundamental passivity, ought no longer to call itself a transcendental *aesthetic*, neither in the Kantian, nor in the Husserlian, sense of those words. A transcendental question on space concerns the prehistoric and precultural level of spatio-temporal experience which furnishes a unitary and universal ground for all subjectivity, and all culture, this side of empirical diversity, as well as the orientations proper to their spaces and times. (*Gramm*, p. 411; English transl., p. 290)

The Kantian model of temporality then is interesting only as a modification of an inner time-consciousness that precedes spatiality as an a priori. Deconstruction, then, should say that temporality is always already open and defers to space, but does not follow the determination of spatiality. The problem is, then, the origin of time as the origin of a temporal consciousness which is based on the living moment. It is not represented but presented; it is repeating and representation that is the *hysteron proteron* to original presentification. The infinite regression of the imagination is interrupted by the force of living consciousness as an absolute in the flux of consciousness. Giovannangeli refers to the fifth chapter of *La Voix*, where Derrida discussed Husserl's distinction (in *Inneren Zeitbewusstseins* § 16) between primary remembrance (retention) and secondary remembrance (reproduction). There the difference constitutes this relation as one between two modes of non-perception (and not adopting the Husserlian distinction between perception and non-perception). Primary and secondary remembrance must be in relation to nowness and to each other as relations to nowness, and in addition they are constituted by an irreducible absence that Derrida calls the trace (*La Passion*, pp. 22–23; *La Voix*, pp. 72–73; *Speech*, pp. 64–65). Derrida continues his argumentation referring to the twinkling instant or the blink of an eye.

As soon as we admit this continuity of the now and the not-now, perception and non-perception, in the zone of primordiality of common to primordial impression and primordial retention, we admit the other into the self-identity of the *Augenblick*; nonpresence and nonevidence are admitted into the *blink of the instant* (*clin d'œil de l'instant*). There is a duration to the blink, and it closes the eye. This alterity is in fact the condition for presence, presentation, and thus for the *Vorstellung* in general; it precedes all the dissociations that could be produced in presence, in *Vorstellung*. (*La Voix*, p. 73; *Speech*, p. 65)

What is important here is the deconstruction of the event of the *Augenblick* and the identity of nowness; but also this difference in identity opens the space between re-productive imagination and past presence. Death as an anticipation could be in the becoming and not just as the re-produced imagination in the mode of no-perception.

At the end of *L'Origine de géometrie* Derrida gives his first example of *différance* when he does not accept a dialectical – or, in the language of *Le Problème de la genèse dans la philosophie de Husserl*, a dialectics of dialectics and non-dialectics – solution to the aporias of Husserlian inner time-consciousness.[29] Derrida thinks about the founding instant of the living moment: if there is no retention preceding it, the moment must preserve in itself retention (as well as protention) as such. That is why the origin of time consciousness is open to the past as well as the future as possibilities. If there is present the living consciousness, past presence (pro-tention) and the becoming of future presence (re-tention), so there must be the intentional present instance (*main-tenir*) of preserving intuition, which is in itself constituted by the deferral of temporality and differing of spatiality. This analysis refers to the Kantian linear and rational model, which follows the exterior model of a line, the line of writing as a universal teleology. According to Derrida in *De la grammatologie* Husserl was more right than wrong to say that phenomenology does not follow the model of (natural) sciences, but in the analysis of constitution there is a certain mathematical model which retains its prerogative. This leads, for a moment, to the "new transcendental aesthetics".[30]

A new transcendental aesthetic must let itself be guided not only by mathematical idealities but the possibility of inscriptions in general, not befalling an already constituted space as a contingent accident but producing the spatiality of the space. [. . .] Now if one lets oneself be guided by inscription as habitation in general, the Husserlian radicalization of the Kantian question is indispensable but insufficient. We know that Husserl reproached Kant for having allowed himself to be guided in his question by ideal objects already constituted into a science (geometry or mechanics). To a constituted ideal space a subjectivity constituted (into faculties) corresponded naturally. And from my present point of view, there is much to say on the concept of the *line* which so often intervenes in the Kantian critique (*Gramm*, p. 411; English transl., p. 290).[31]

Although Derrida speaks here of spatialization, this must be very familiar to readers of the second part of *Sein und Zeit*. The structure of Heidegger's argumentation resembles Husserl's elaboration of space and time, but there are differences because he does not follow a Kantian model. The spatiality of space does not follow the temporality of time or vice versa. Heidegger points to the constitution of being-in-the-world before any Kantian *a priori* temporality and spatiality. By death he means here being unto death. We must dwell here on the question of death and auto-affection – death escapes phenomenological self-presencing – before going on to Heidegger. In *La Voix* Derrida made some brief remarks about Heidegger when he introduced the notion of auto-affection (*La Voix*, pp. 93–95; *Speech*, pp. 83–85). Auto-affection deconstructs and at the same time makes possible self-presence; for example, in the problematic structure of touching/being touched, it could roughly be said that seeing in auto-affection the origin of *Dasein's* temporality follows the model of Husserl's *inneren Zeitbewusstseins*. This does not mean that Heidegger just turns upside down the Husserlian theory of time-consciousness as Michael Theunissen says in his *Negative Theologie der Zeit*, but that Heidegger repeats the analysis of the *Urimpression* as auto-affection.[32] If the temporality of *Dasein* bears the sense that *a priori* means temporally before, then it is without a foundation except as a difference; there could be no authenticity (*Eigentlichkeit*) of *Dasein* if there is no originary founding point for *Dasein's* temporality.[33] The original temporality of *Dasein* must remain a possibility over against the "fallen" or vulgar temporality of *Dasein*. This was at stake in Derrida's early deconstructions of Heidegger *Les Fins de l'Homme* and *Ousia et Grammè*,[34] the latter especially.

Derrida says in *La Voix* that auto-affection "produces sameness as self-relation within self-difference", "sameness as the non-identical", and a transcendental subject as a movement of *différance*. Derrida was then deconstructing the phenomenological voice's self-closeness in hearing and understanding (*s'entendre parler*) with reference to the world, space, the body, the external, etc. (see also *Gramm*., p. 146; English transl., p. 98). This in addition means that auto-affection is the condition of self-presence. And after that, Derrida confirms, we must admit that "no pure transcendental reduction is possible"; still it is necessary to "pass through the transcendental reduction" in order to get these results (*La Voix*, p. 92; *Speech*, p. 82).

> However, as soon as one takes the movement of temporalization into account, as it is already analyzed in *The Phenomenology of Internal Time-Consciousness*, the concept of pure auto-affection must be employed as well. This we know is what Heidegger does in *Kant and the Problem of Metaphysics*, precisely when he is concerned with the subject of time. The "source point" or "primordial impression," that out of which the movement of temporalization is produced, is already pure auto-affection. (*La Voix*, p. 93; *Speech*, p. 83)

4. A BRIEF INTRODUCTION TO THE "PHENOMENOLOGY OF DEATH"

Death is usually said to be something that cannot be phenomenologized. Death becomes something which constitutes the limit of phenomenology, and also a condition of possibility for phenomenology. Derrida has in several ways pointed out that death cannot be thought within phenomenology. Death is something unrepresentable, which means that it is something which is neither here yet nor here anymore to be represented, nor is it a *Gegenstand* of Husserlian consciousness (in noematic-noetico analyses). Derrida's argumentation is based on post-Husserlian phenomenology, about which we can even say there is nothing more typical than doing a phenomenology of mortality. Given the impossibility of that I will subsequently refer to this kind of thinking as the "phenomenology of death" in quotation marks.

The expression "phenomenology of death" is here borrowed from Françoise Dastur. She uses the expression in the context of Hegel's *Phenomenology of Mind*. In her *La Mort*, in describing the notion of death, she says that death can never be an object for Husserlian phenomenology.[35] But we can just as well ask whether the task for phenomenology is not then to go beyond these limits (essence, representation). After that maneuver the term "phenomenology of death" would encompass the non-phenomenologizable. We could unfold a quasi-phenomenology, to use Simon Critchley's recent characterization of Derrida's style.[36] This impulse is confirmed in recent phenomenology in the special number on being born and dying in *Alter. Revue de Phénoménologie* 1/1993: *Naître et mourir*. There Natalie Depraz and Jean-Marc Moullie ask how it is possible to proceed with a phenomenological description of death.[37] According to them, death is difficult to apprehend because it not an experience, an intuition or evidence, but is something non-given that yields perception (p. 109). In addition, death and birth could be thought in terms of passivity – as in Husserl's *Analysen zur passiven Synthesis* and *Ideen II* – as something that the ego is not active in.[38] The hour of my death always arrives at a time when I am late getting there.

The term "phenomenology of death" is then used in an unsymmetric way vis-à-vis the term "phenomenology of life". For example, when such terms as *zoè* or *bios* are deconstructed, death becomes a limit or boundary for *bios*. The classic example is that of an autobiography completed just at the hour of death. The auto-bio-graphy changes to a hetero-thanatography wherein the possibility of saying *I am* or *I live* depends on saying *I am dead.*[39] This is also the possibility of narratology.

So we have to ask seriously, then, can there be a phenomenology of death? Then we would have something like a thanatologized phenomenology that a) is different from a phenomenological point of view as a phenomenology of death, if we try to distinguish it from theology and medical or psychiatric thinking (e.g., thanatology) and b) gives an account lying between the phenomenologies of death and Husserlian phenomenology. A stimulating excursus about the meaning of death and the philosophical positions taken in the debate over the meaning of death is the work of Philippe Ariès. In his short book *Western Attitudes toward Death: From the Middle Ages to the Present* he sketches a symmetrical descent between four historical stages in the attitude towards death.[40] 1) First death in Antiquity or the Middle Ages meant "we all die", i.e., all men are mortals; 2) with the Renaissance emphasis on subjectivity came the concept "one's own death"; 3) in the Enlightenment death was considered also as the other's death or "thy death"; 4) and in modern society death has been medicalized and hygienized; it is "forbidden death". I will not go into the possible argumentation against Ariès in historical or sociological perspective, about why or in which way the scene of death is now forbidden, and whether there is any correspondence between the attitudes towards living and towards death (lifestyle/deathstyle)?[41]

In his *Apories* Derrida characterises Ariès' work as an onto-phenomenology, a regional ontology, that avoids as beyond a limit the questions of "What is death in general?" or "What is the experience of death?" or whether death "is".[42] He inverts the question of "my death" as a possibility by asking "Is my death possible?" as a question of singularity.[43] The question of my death is described as an aporia, as an impossible experience or as an experience of impossibility, or the difference in-between. The question of the limits of phenomenology is, then, a question of a border since death *par exellence* is a border (*finis*). This means in addition, to elaborate the question of the death as a limit in a certain tradition, one that is bio-anthropo-thanato-theological as Derrida calls the Western tradition. We have as well to re-mark thinkers in this tradition in relation to bio-anthropo-thanato-theological thinking.[44] Derrida asks why death is such a limit for all these thinkers? And why a Human death?[45]

Phenomenology should make its analyses before these historico-genetic problems enter in. One cannot think about the genesis of death because the concept of genesis implies death as some kind of telos. However, the second and third stages Ariès delineates seem to be very familiar: the concept of my death is a very Heideggerian way to express *Dasein*'s temporality and "Thy death" seems to be close to a Levinasian analysis in which the death of the Other (*autrui*) has priority over my own death.[46] It is just that these stances also seem to be the Human Situation and not just historical phenomena. The phenomenology of death should regard these things before any historical or social determinations. This also means that we cannot posit any cultural or natural origin for death although we can think of death as a universality that escapes all these representations.

Dastur writes a chapter in *La Mort* about the phenomenology of being mortal. She shows clearly that man became a finite being, *Endlichkeit*, a limited creature in relation to the infinite and transcendental creature (God). This is the meaning of finitude, being mortal. Then we could ask about this creature's relation to death and its temporal being. Is there just my own death (as being in the world) or is there the others' death (or even co-dying) as being? *Dasein* is in the phenomenology of death, the first candidate for a "transcendental finite". In *Sein und Zeit* temporality opens death (or death opens temporality) as a re-solution and other modes in being in the world (*Entschlossenheit, Er-scholssenheit, Ent-decktheit*); (*Sein und Zeit* §§ 45–46, § 59).[47] The openness of *Dasein* is constituted from closing (*Schlossen*), but as a possibility it is in relation to *Un-ent-schlossenheit*, the decision being related to the authenticity and the opening of the primordial truth (*alethêia*) in relation to the voice of con-science (*Gewissen*) and the possibility of its own being (*Seinkönnen*); (*Sein und Zeit*; § 60).[48] For Heidegger, the basis of certainty (*Gewissheit*) is death: my death is certain as a foundation of certainty; even in everyday inauthentic speech (*Gerede*) there is a certainty: "They" say, "It is certain that 'Death' is coming" (*Sein und Zeit*, § 52).

For Emmanuel Levinas, the second member of the school of "the phenomenology of death", the central problem is, not just my death as the possibility of an impossibility and the impossibility of a possibility, but the death of the other.[49] The death of the other is ontologically prior to my death. He starts thinking on death by referring always to Heidegger – over a long span from *Le Temps et l'autre* (1947) to *Dieu, la mort et le temps* (1993). Levinas points out that death in Heidegger is constituted in relation to states of mind (*Stimmung*).[50] But he does not consider the ownness of one's death (*Jemeinigkeit*) as the only mode of being towards death.[51] Levinas also thinks about the anguish over death, but he turns it around as he points to being-for-the-other.[52] The Levinasian concept of intersubjectivity is not a critique of Heidegger; it springs forth from the Husserlian egology and "the experience of other" in the *Cinquième méditation cartésienne* wherein is described the transcendental ego's constitution of the community of other transcendental egos from the intentional subjectivity of a subject as an objectivity. The third thinker of the phenomenology of death would be Eugene Fink, for his great book *Tod und Metaphysik*.[53] In all these phenomenologies – in so far as we can call them phenomenologies – death is an absolute master, a limit that cannot be crossed. In *Apories* Derrida named three thinkers of death (Freud, Heidegger and Levinas) who think of death as an absolute limit, as a thing beyond the pleasure principle, as the possible fourth prominent thinker of death, Jacques Lacan would say.[54]

The question still is, how can we speak about death, what is death? What is dying (*mourir*), an experience as present presence, before it is imperfect (deceased, died) or perfect (have died)? It can be expressed as a relation to the other in the self that dies. Death can also be the nearness in distance. For the "phenomenology of dying" we have in addition the work of Maurice Blanchot wherein death could be an experience of impossibility; death is not the presence of a subject but the being of a divided subject (*il y a*) and the neuter. The narrative subject disappears and shifts from *je* to *il* as a neutrum, pointing to *il y a* as a subjectless presence.[55] In the work of Blanchot there is a distinction between infinite dying as an impossibility and finite death as a possibility, and this should be re-read as Derrida points out in *Apories*. His new reading of Blanchot is given in his piece "Demeure" wherein he elaborates as an example Blanchot's *récit* or short *L'Instant de ma mort*. There the moment of my death continues from and in this instant infinitely (*désormais toujours en instance*).[56] Death also means the space of literature, which is the opening of the space of death as Blanchot used to say in his book *L'Espace littéraire*.[57] One could also speak of the origin of art, as an absolute passivity, before the opposition between the active expression, creation, passive perception and reception as aesthetical disinterest, as the music of Orpheus ascending from the other night, the night of night.[58] The synthesis between the active and passive is in itself a passive synthesis as a movement to the absolute outside as Levinas usually says. But now we must also turn back to Heidegger, whose thinking about death points to the outside as well as the inside.

5. DEATH, TESTIMONY AND TRANSCENDENTAL SCHEMATISM: HEIDEGGER

One main reason for the interest here in Kant has been the importance of the phenomenological interpretation proposed by Heidegger in *Kant und das Problem der Metaphysik*. There he tries in the chapters about transcendental imagination (*Kantbuch*, §§ 24–35) to frame a method (*schemata*) for transcendental aesthetics.[59] Transcendental imagination (*Einbildungskraft*) would combine Being with time, as a third grounding possibility between meaning (*Sinn*) and apperception (§ 27). Heidegger analyses, then, three pure syntheses (Apprehension, Reproduction, Recognition), the intratemporality of *Einbildunsgkraft* (§ 33), and proceeds to time as pure auto or self-affection as the time character of finite self (§ 34). This also means the finitude of man. The question of imagination or fiction in this context is important because imagination is the force that gives life to phenomenology, as Husserl says in *Ideen* (§ 70). Imagination has a crucial meaning for Heidegger – imagination could even give meaning to Being if we follow a suggestion made by John Sallis.[60] What are, then, the conditions that constitute imagination or the faculty of imagination? Must we say, like Rousseau, that imagination is the primary of human abilities because it can invent itself? And is death something that constitutes imagination? Or could it be that we could yet think of death as imagination and imagination as death?

This is what Maurizio Ferraris proposes when he writes about death and imagination in a book called *Mimica. La lutte e autobiografica* (1992).[61] He argues that for Heidegger the faculty of imagination makes possible discourse about death because there is no originary origin of the originary self or ego. Ferraris continues the interphenomenological argumentation – in favour of Levinas and with and against Heidegger's argument for the priority of the I. He asks if is it not possible that imagination makes possible the notion of the I, in that sense that the I (*io*) is founded by the imagination when we seek the apriority of the transcendental categories of space and time. Then the imagination would be a priori because it gives the conditions of the possibility of the I (*l'immaginazione che offre la condizione dell'io*). Is the "I" possible,

Ferraris asks. He also proposes that the death of the other is possible before "my death", before the possibility of my authenticated death and, ultimately, the Heideggerean imagination. Here Ferrari offers a deconstruction of Heidegger that is analogous to Derrida's deconstruction of Husserl's sign theory.[62] That is to say, Heidegger repeats, in his argumentation, the same movements (between interiority and exteriority) that Husserl made in his theory of signs, introduced in the first *Investigation* of *Logische Untersuchungen II*. The Derridian deconstruction, proposed in *La Voix*, goes without saying for Heidegger. Heidegger starts from authentic death, inwardness, as Husserl starts from the silent monologue (reflection after reduction) in a consciousness having no outer expression. Since signs are also indications, as outer signs, the point that Derrida makes is that Husserl should prefer first to seek the nature of a sign before making this classical distinction between the inner and outer functions of signs. This Derridian manner is to make an absolute inside consisting of an absolute outside. The outside always corrupts the inside, subjectivity being already constituted by irreducible trans- or intersubjectivity.

> Does not this "dialectic" – in every sense of the term and before any speculative subsumption of this concept – open up living to difference, and constitute, in the pure immanence of experience, the *divergence* (*écart*) involved in the indicative communication and even in the signification in general? (*La Voix*, p. 77; *Speech*, p. 69)

Ferraris claims that Heidegger in speaking of the possibility of my death also excludes the outside as the possibility of the other's death before me as well as the Other. This includes also Heidegger's treatment of the notion of imagination. This is not only to say that the model of the others' deaths is the model for my decease but also to decide that the imagination as thrownness and schema building (*Entwurf*) is built on the originary temporality of *Dasein* and that imagination is not the origin of temporality (or subject, inside, outside, etc.). Ferraris tries to work out a solution between Heideggerean inwardness (so called "mysticism") and the space opened to the other in Heideggerean text.[63] Here also is the grounding of reason as a *Herzengrund*, in contrast to phenomenological intersubjectivity and other as corporeality.[64] Giovannangeli as well noted the motive of a exterior or strange voice in Heidegger (*La Passion*, pp. 88–89). *Dasein* is opened in its core to the outside, to the other as it carries the voice of a friend.[65]

6. CONCLUSION: FROM TRANSCENDENTAL AESTHETICS TO THE SPACE OF ETHICS

In Husserlian phenomenology transcendental aesthetics forms the area of pre-predicative experience. This means that it is the level of primordial data or *aisthesis* which grounds the "higher" level of Judgement. In its Kantian version, in *The Critique of Pure Reason*, a short chapter is devoted to Transcendental Aesthetics before Transcendental Logic is taken up. One could even say, with Tran Duc Thao in his *Phenomenology and Dialectical Materialism*, that the Husserlian project of transcendental phenomenology was divided between these two strata, between transcendental aesthetics and transcendental logic.[66] In Husserl, transcendental aesthetics, sometimes also called the "aesthetic world", was primally discussed in few short remarks made mainly in *Formale und Transzendentale Logik* (*Slusswort*, pp. 256–257) and in *Cartesianische Meditation* (§ 60). Transcendental aesthetics describes something pre-scientific and foundational, which like the notion of *Geos* founds the possibility of a horizon for the human being in the *Lebenswelt.* This is what we have thought on here also, under the name of death. This pre-predicativity, the pre-phenomenological, means the constitution of the horizon in a horizon of constitution. In the Husserlian context we mainly sketch *aisthesis* as a foundation for philosophical reflection. Reflection was immediately founded by inner "living experience" (*Erlebnis*). In the flux of experience the moment of the Living Now (principle of principles) of the living consciousness (*lebendige Gegenwart*) is like a self-predication as self-presence. The ground could be also expressed in the terminology of givenness (*Gegebenheit*), which grounds consciousness as a form of consciousness before any content of consciousness. We might note that Husserl says that donation in general depends on the donation (*gebende*) of time (*Erfahrung und Urteil*, § 38, *La Passion*, pp. 95–98). This means just what is given or donated as an intuition (*Anschauung*) to phenomenological consciousness. Likewise Kant (A 19/B 33) says that objects are given (*gegeben*) in our intuition as sensibility. The givenness depends on transcendental reduction and not vice versa.[67] The question was focused on: what does the reduction leave outside?

The space of transcendental aesthetics could be defined as a contradiction between the impossibility of phenomenological living experience and the impossibility of intersubjective experience in a space of impossibility. Death remains a chiasmatic transcendental subjectless place, like writing. "My death" also indicates community and the place of meaning. This space of death would also open a place for ethics or impossible community such as Blanchot and Nancy have been describing.[68] A community would be in Husserlian terms unsymetric, inter-monadic, transcendental and intersubjective.

Community is also the place of writing or literature. These conditions for the possibility of an experience (not an object) turn out to be an experience of impossibility. Literature, then, is no existing object, a futural space of a "no-thing", the experience of nothing, an impossible experience of literature.

What then should we say finally about death and life? Death has been thought of as a border of negation which is always present as a negation. The affirmation of death seems to be forbidden: what is spoken is still foreclosed. As Husserl reflects, life in itself does not die (*Leben stirbt nicht*) as a universality and an inner unity of life (*Husserliana* XXIX, p. 344) and without life dying would have no meaning (*Sinn*, *ibid.*, p. 338).[69] Yet death would be then, perhaps, the origin and affirmation of life.

University of Helsinki
Finland

NOTES

[1] J. Derrida, *La Voix et le phénomène, Introduction au problème du signe dans la phénoménologie de Husserl* (Paris: PUF, 1967), p. 60. English translation by David Allison: *Speech and Phenomena and Other Essays on Husserl's Theory of Signs* (Evanston: Northwestern UP, 1973), p. 54. Hereafter *La Voix/Speech* in the text.

[2] I use these editions of books by Husserl: *Logische Untersuchungen* (The Hague: 1984), *Hua XIX 1*; *Ideen* (The Hague: 1971), *Hua III*; *Formale und Transzendentale Logik* (Halle: 1929), *Zur Phänomenologie des Inneren Zeitbewusstseins* (1893–1917), (The Hague: 1966), *Hua X*; *Cartesianische Meditationen und Pariser Vorträge* (The Hague: 1950), *Hua I Die Krisis der Europäischen Wissenschaften und die transzendentale Phänomenologie*, (The Hague: 1962), *Hua VI*; *Erfahrung und Urteil*, ed. Landgrebe (Hamburg: 1985), *Analysen zur passiven Synthesis (1918–1926)* (The Hague: 1966), *Hua XI*; *Zur Phänomenologie der Intersubjektivität. Dritter Teil: 1929–1935* (The Hague: 1973), *Hua XV*; *Die Krisis der Europäischen Wissenschaften und die transzendentale Phänomenologie. Ergänzungsband. Texte aus den Nachlass 1934–1937*, (Dordrecht: 1993), *Hua XXIX*.

[3] This corresponds to what Leibnitz says in *The Principles of Philosophy, or, the Monadology* §§ 4–6 and in *The Principles of Nature and Grace, Based on Reason* § 2: "Since the monads have no parts, they can neither be formed nor destroyed. They can

neither begin nor end naturally /---/". See also §§ 73–76 in *The Monadology* where Leibnitz writes about soul, death and birth. In G. W. Leibnitz, *Philosophical Essays*, edited and translated by Roger Ariew and Daniel Garber (Indianapolis and Cambridge: Hackett, 1989), pp. 207–225.

4 This comment was found by Natalie Depraz and Jean-Marc Mouillie in *Manuscript A VI 14a3* from 1930. See N. Depraz and J.-M. Mouillie, "Se donner 'la mort'" in *Alter* 1/1993: "Naître et Mourir", p. 126. According to them this means that I cannot have an experience of my own death; but only in the way of the other's death. I can have my death as an experience constituted by the others' death; but my death is not a givenness (*ibid.*, pp. 126–127). They conclude (p. 127) that: "[. . .] mon expérience de la mort est celle d'une non-expérience de ma propre mort. Autrui, comme mourant, *constitue* pour moi le sens de la mort comme *ma* mort [. . .]". I thank Dan Zahavi for pointing out this reading of Husserl.

5 Husserl says about transcendental intersubjectivity before starting to speak about death, in *Anthropologischen Welt*: "Zur Wesensform der transzendentalen Subjektivität gehört ihre transzendentale Intersubjekitivität in Selbstobjektivierung als offen-unendliche Menschheit und damit eine Objektivität der Geltung dieser Welt für alle transzendentalen Subjekte und für alle Menschen als Menschen" (*Hua XXIX*, p. 331).

6 Cf. Depraz and Moullie, p. 128.

7 See for example D. Zahavi, *Husserl und die transzendentale Intersubjektivität. Eine Antwort auf die sprachpragmatische Kritik* (Dordrecht: Kluwer 1996), especially p. 93, where Zahavi states that death, as dreamless sleeping I, is a limited concept (*Grenzbegriffe*), without any *Leiblichkeit*.

8 In J. Derrida, *L'Écriture et la différence* (Paris: Seuil, 1967). (English Translation: *Writing and Difference*, trans. Alan Bass, Chicago: University of Chicago Press, 1978.) In the question of the noema I thank here the discussion of Fregeanism that started from the school of Føllesdal. See also to Robert J. Solomon: "Husserl's Concept of Noema", in Frederick Elliston and Peter J. McCormick (eds.), *Husserl: Expositions and Appraisals* (Notre Dame: Notre Dame University Press, 1977).

9 See J. Derrida, "The Time of a Thesis: Punctuations", in Alan Montefiore, *Philosophy in France Today*. Trans. Catherine McLaughlin (Cambridge: Cambridge University Press, 1983).

10 This attempt can be found in J. Derrida, *Le Problème de la genèse dans la philosophie de Husserl* (Paris: PUF, 1990) which was Derrida's first "work" on Husserl, written in 1953–1954.

11 This is of course widely noticed. Look for example at Silvano Petrosino, *Jacques Derrida et la loi du possible*. Tranduit de l'italien par Jacques Rolland (Paris: La nuit surveillée, 1994), p. 29 or at Stephan Strasser "Von einer Husserl-Interpretation zu einer Husserl-Kritik. Nachdenkliches zu Jacques Derridas Denkweg", in *Phenomenologische Forschungen* 18 (1984), p. 156; and Peter Kemp, *Döden och maskinen. En introduktion till Jacques Derrida*. Översättning Ulf I. Eriksson (Stockholm: Symposion, 1990), pp. 129–130.

12 M. Heidegger, *Einführung in die Metaphysik* (Tübingen: Niemeyer, 1987). Cf. p. 49 for *fônê* and grammatics, pp. 59–60 for the difference between speaking and writing: English translation by Ralph Mannheim, *An Introduction to Metaphysics* (New Haven: Yale University Press, 1959).

13 J. Derrida, *De la grammatologie* (Paris: Minuit, 1967); English translation by Gayatri

Spivak; *Of Grammatology* (London: Johns Hopkins University Press, 1976), *Gramm.* in the text.

[14] For a commentary see for example Patricio Peñalver Gómez: "The Deconstruction of Sense" in *Analecta Husserliana*, Vol. XXXXVI (Dordrecht: Kluwer Academic Publishers, 1991), pp. 44–45.

[15] Usually cited from Maurice Blanchot, "Littérature et le droit du mort", in *La Part du Feu* (Paris: Gallimard, 1949), p. 313, *The Gaze of Orpheus and Other Literary Essays*, trans. Lydia Davis, Ed. P. Adams Sitney [New York: Station Hill, 1981], p. 43.

[16] The ego (I am) means here a transcendental apodeictical ego as an apodeictical self-evidence which is determined by presence and absence. As an adequate evidence death would not found ego and the sciences. (I thank Juha Himanka for clarifying this distinction and insisting that Husserl says so little says about the death because of the importance of apodeictical evidence.) On apodeictical evidence and ego see E. Ströker, "Husserls Evidenzprizip. Sinn und Grenzen einer metodischen Norm der Phänomenologie als Wissenschaft", *ZPF* 32, (1978), pp. 24–26 and on the partition of transcendental subjectivity see L. Landgrebe, *Der Weg der Phänomenologie. Das Problem einer ursprünglichen Erfahrung* (Mohn: Gütersloh, 1963), pp. 195–202.

[17] This "case" served first as an example for Lacan in 1955 and it is well known that Roland Barthes referred to the same text in the famous Johns Hopkins colloquium. On *The Signs and Languages of Man* in 1966. Later Derrida wrote about this in "Les Morts de Roland Barthes" (in *Psyché* [Paris: Galilée, 1987]), pp. 273–304, as an impossible enunciation (*énonciation*), referring to *Plaisir du Texte*. Death is there beyond the pleasure of the text. Derrida address the deaths of Roland Barthes. Death is meaningful in the plural, there is not just one death. Derrida confirms this interest in Mr. Valdermar's enunciation as as example in "Demeure. Fiction et témoignage" in *Passion de la littérature* (Paris: Galilée, 1996), p. 34.

[18] J. Derrida, *Spectres de Marx* (Paris: Galilée, 1994), p. 207.

[19] J. Derrida "Introduction" in Edmund Husserl. *L'Origine de la géometrie* (Paris: PUF, 1962), pp. 62–68. English translation by John P. Leavey, J. Derrida, *Edmund Husserl's* Origin of Geometry (Stony Brook: N. Hays, 1978), pp. 70–73.

[20] J. Derrida, "This Strange Institution Called Literature. An Interview with Jacques Derrida", in *Acts of Literautre*, Derek Attridge, trans. Geoffrey Bennington and Rachel Bowlby (London: Routledge, 1992), pp. 44–45.

[21] A phenomenological account of this is in Jugoslav Vlaisavljević: "Husserl's Legacy in Derrida", in Anna-Teresa Tymieniecka (ed.), *Husserl's Legacy in Phenomenological Philosophies, Analecta Husserliana*, Vol. XXXVI (Dordrecht: Kluwer, 1991), pp. 101–117.

[22] Derrida "This Strange Institution", *op. cit.*, p. 73.

[23] Topology of being refers also to Heidegger's "Zur Seinsfrage (1955)", in *Wegmarken* (Frankfurt am Main: Klostermann, 1978). Topology also covers "topography".

[24] J.-J. Rousseau, *Essai sur l'origine des langues où il est parlé de la mélodie et de l'imitation musicale. Édition critique*. Avertissement et notes par Charles Porset (Paris: Nizet, 1970), p. 68.

[25] D. Giovannangeli, *La Passion de l'origine. Recherches sur l'esthetique transcendantale et la phénoménologie* (Paris: Galilée, 1994), pp. 27–39. Hereafter *La Passion* in the text.

[26] D. Giovannangeli, *Écriture et répétition. Approche de Derrida* (Paris: U.G.E., 1979).

The third book of Giovannangeli is *La Fiction de l'être. Lectures de la philosophie moderne* (Brussels: De Boeck, 1990). The impact of the Giovannangelian interpretation of Derrida is recognized in Rudy Steinmetz, *Les Styles de Derrida* (Brussells: De Boeck, 1994).

[27] Further on Giovannangeli maintains that the Husserlian aesthetics would be objective and disinterested aesthetics in addition to Kant's *Third Critique*. Whatever Husserl's concrete thoughts about aesthetics may be, the notion of aesthetics would not be one of naïve aesthetic consciousness making judgements. Aesthetic judgement may even require a phenomenological reduction (*La Passion*, p. 36). The original prejudicative aesthetic sentiment is in relation to the exteriority, not to pure interiority (*La Passion*, p. 39).

[28] I. Kant, *Kritik der reinen Vernunft* (Hamburg: Meiner, 1990).

[29] J. Derrida, *Introduction*, p. 171; ET, p. 153; see also *Le problème de la genèse*, pp. 5–9.

[30] R. Gasché in his book *The Tain of the Mirror* (Cambridge, MA: Harvard University Press, 1986), pp. 156–157, points out that when Derrida in *De la grammatologie* proposes "a new transcendental aesthetics", he seems to be aiming at a "theory of originary constitution", but the notion of inscription displaces the concepts of phenomenology and linear genesis as constitution itself is deconstructed.

[31] Note that the preceding citation above from *De la grammatologie* should be inserted between these two citations from the same page; Derrida proposes and omits the term "new transcendental aesthetics".

[32] M. Theunissen, *Negative Theologie der Zeit* (Frankfurt am Main: Suhrkamp, 1991), pp. 342–443. For commentary on *Urimpression* see Rudolf Bernet, "Differenz und Anwesenheit. Derridas und Husserls Phänomenologie der Sprache, der Zeit, des Geschichte, der wissenschaftlichen Rationalität", *Phänomenologische Forschungen* 18, (1986), pp. 77–83.

[33] M. Heidegger, *Kant und das Problem der Metaphysik* (Frankfurt am Main: Klostermann, 1991), § 34; English translation by James S. Churchill, *Kant and the Problem of Metaphysics* (Bloomington: Indiana University Press, 1962).

[34] In J. Derrida, *Marges de la philosophie* (Paris: Minuit, 1972).

[35] F. Dastur, *La Mort. Essai sur la finitude* (Paris: Hatier, 1994), p. 38; English translation by John Llewelyn, *Death: An Essay on Finitude* (London: Athlone Press, 1996): "But may one not object that no phenomenology of death is possible because death is *par exellence* a non-phenomenon, and that is in vain to undertake the search for its essence because it is absolute non-being" (p. 40).

[36] S. Critchley. "Derrida: Private Ironist or Public Liberal", in Critchley *et al.*, *Deconstruction and Pragmatism*, ed. Chantal Mouffe (London: Routledge, 1996), p. 32. Critchley contrasts there quasi-phenomenology to the theoretical and the performative possibilities of style.

[37] "Se donner 'la mort'", *Alter* 1 (1993), p. 107.

[38] See for example Anne Montavont, "Passivité et non-donation"; J.-M. Moullie "Naissance, Mort et phénoménologie", in *Alter* 1 (1993).

[39] As an introduction of these aspects in Derrida's work see to Robert Smith, *Derrida and Autobiography* (Cambridge: Cambridge University Press, 1995), pp. 129–171.

[40] P. Ariès, *Western Attitudes toward Death from the Middle Ages to the Present*, trans. Patricia M. Ranum (Baltimore and London: Johns Hopkins, 1974). I have also looked

to *L'homme devant la mort* (Paris: Seuil 1977) and *Images de l'homme devant la mort* (Paris: Seuil, 1983), where the argument is developed further on.

[41] See for example Norbert Elias, *Über die Einsamkeit der Sterbenden in unserem Tagen* (Frankfurt am Main: Suhrkamp, 1990).

[42] J. Derrida, "Apories: Murir – s'attendre aux 'limites de la vérité'", in *Le Passage des frontières: Autour du travail de Jacques Derrida* (Paris: Galilée, 1993), p. 320; (English translation by Thomas Dutoit, *Aporias* [Stanford: Stanford University Press, 1993], p. 25).

[43] *Ibid.*, p. 317; ET, pp. 19–20.

[44] *Ibid.*, p. 337; ET, p. 79: "I hope that I have convinced you my purpose was not to justify a passage beyond knowledge, anthropo-thanatology, biology, or the metaphysics of death toward a more radical, originary, or fundamental thought, as if the limit were a known edge between, on the one hand, anthropology (be it even fundamental anthropology) and, on the other hand, ontology, an existential analysis, and more generally a more questioning thought of death in general. On the contrary, based on the example of Heidegger or of the virtual debate among Heidegger, Levinas and Freud, my discourse was aimed at suggesting that this fundamentalist dimension is untenable and that it cannot even claim to have any coherence or rigorous specificity".

[45] One might note that Husserl does not limit life to Human life. In *Anthropologischen Welt*, for example, it is a question of the possibility (of evidence's) self-givenness in animals and plants (*Hua*, XXIX, p. 229).

[46] I think that this is not just a Levinasian argumentation against Heidegger but also a consequence of thinking through Rosenzweig's *Der Stern der Erlösung* (Frankfurt am Main: Suhrkamp, 1988).

[47] M. Heidegger, *Sein und Zeit* (Tübingen: Niemeyer, 1987); English translation by John Macquarrie and Edward Robinson, *Being and Time* (London: Blackwell, 1962).

[48] The interpretation of these terms are in J.-L. Nancy, "La Décision de l'existence", in *Une Pensée infinie* (Paris: Galilée, 1990) and J. Greisch, *Ontologie et temporalité. Esquisse d'une interprétation intégrale de* Sein und Zeit (Paris: PUF, 1994), pp. 300–304.

[49] Look for the notion of the impossibility of a possibility in E. Levinas, *Le Temps et l'autre* (1947) (Paris: PUF, 1991, p. 92, n. 5. and J. Llewelyn, *Emmanuel Levinas. The Genealogies of Ethics* (London: Routledge, 1995), p. 10.

[50] E. Levinas *Dieu, La Mort et le Temps* (Paris: Grasset, 1993), pp. 58–59.

[51] *Ibid.*, pp. 34–42.

[52] Cf. E. Levinas, "Mourir pour . . .", in *Entre nous. Essais sur le penser-à-l'autre* (Paris: Grasset, 1991). Levinas says that the mineness (*Jemeinigket*) of *Sein und Zeit* (§ 9) is substituted to appropriation of Being as authenticity (*Eigentlichkeit*).

[53] E. Fink, *Tod und Metaphysik* (Stuttgart: Kohlhammer, 1969).

[54] See for example J. Lacan, *Le Séminaire, Livre VII, L'éthique de la psychanalyse* (Paris: Seuil, 1986), Chaps. XII–XVI.

[55] See for instance M. Blanchot, "La Voix narrative, le 'il', Le neutre", in *L'entretien infini* (Paris: Gallimard, 1969); English translation by P. Adams Sitney, *The Gaze of Orpheus* (Barrytown, N.Y.: Station Hill Press, 1981).

[56] J. Derrida, "Demeure", pp. 72–73; M. Blanchot, *L'instant de ma mort* (Montpellier: Fata Morgana, 1994), p. 20.

[57] M. Blanchot, *L'espace littéraire* (Paris: Gallimard, 1955).

58 M. Blanchot, "Le regard d'Orphée", in *L'espace littéraire*; ET, *The Gaze of Orpheus op. cit.*

59 It is not just a coincidence that the notion of Heideggerian thrownness in the world (*Geworfenheit*) and *Dasein's* openness (*Ent-schlossenheit*) as a projection (*Entwurf*) are a translation of the Kantian transcendental schematics. For uses of the word "schema" in this sense see J. Llewelyn, *Emmanuel Levinas*, p. 47.

60 In J. Sallis, *Echoes. After Heidegger* (Bloomington: Indiana University Press, 1990), pp. 97–117.

61 M. Ferraris, *Mimica. Lutto e autobiografia da Agostino a Heidegger* (Milan: Bompiani, 1992), pp. 113–114.

62 *Ibid.*, pp. 115–116.

63 I examine preliminarily a little of Heidegger's *Beiträge zur Philosophie* (*Vom Ereignis*) (Frankfurt am Main: Klostermann, 1989), because Ferraris also is tending to say that the *Beiträge* continue this way toward inwardness (Ferraris, *Mimica*, pp. 71ff.). He thinks on the context and development of the *Beiträge* where the same could as well said about the *Kantbuch* and *Sein und Zeit*. The *Beiträge* also open a different field for questions, not only where Heidegger writes using again the concepts of *Sein und Zeit* but when he is trying to elaborate new notions for thinking on death. Besides, Heidegger's approach does not change until he formulates the fourhold of immortals and mortals and the sky and the earth. In the *Beiträge* death still is *Dasein's* being towards death but that does not mean the denial of death as a denegation (*Verneinung*) nor death as no-thing (*Nichts*). Death is the highest and furthest witnessing of Being (*Der Tod das högste und äusserste Zeugnis des Seyns, Beiträge*, § 161). Death means here *Dasein's* relation to the originary truth of Being. Being towards death means temporalising and the truth of being as spatio-temporal (*Beiträge*, § 162). Death seems to belong here to the discourse of witnessing as Derrida says at the end of *Apories* (p. 338, ET, pp. 79–81). The interlinked themes like witnessing and attestation or the terms designating the second beginning (*Verhaltenheit, Scheu*) or habitation (*Aufenthalt*) also have meaning in terms of testimonial sacrality (in "Foi et Savoir", in G. Vattimo and J. Derrid, *La Religion* [Paris: Seuil, 1996], pp. 80–81). Being towards death is a *Dasein* without any foundation (*ab-grundiges Da-sein*) but this does not mean *Dasein* as groundlessness. *Dasein* is there between the Gods and men, *Dasein* is the place between (*Zwischen*) as a being that recognises the appropriation of Being as in a moment (*Augenblick*), *Beiträge*, § 163. Heidegger invents in the *Beiträge* a new description of *Dasein* as Being on the Way (*Weg-Sein*), where being there and here is being on a way (*Weg*). Death makes being there the last possibility of *Dasein's* being somewhere (*ausserste Mölichkeit des Da*). Cf. Also W. Richardson "Dasein and the Ground of Negativity; A Note on the Fourth Movement in the *Beiträge*-Symphony", *Heidegger Studies* (1993), pp. 37–52, about the distinction between *Ab-grund* and *Un-grund*.

64 The reference is to Heidegger's "Vom Wesen des Grundes (1929)", in *Wegmarken* (Frankfurt am Main: Klostermann, 1978) and to *Was Heisst Denken?* (Pfullingen: Neske, 1957) and also Derrida's "Che cos'è la poesia?", in M. Ferraris, *Postille a Derrida* (Truin: Rosenberg et Selier, 1990).

65 The background for Ferraris and Giovannangeli's discussion is Derrida's "Philo-polemology" in John Sallis (ed.), *Reading Heidegger. Commemorations* (Bloomington: Indiana, 1992). The French version appeared in *Politiques de l'amitié* (Paris: Galilée, 1994).

[66] D. T. Tran, *Phenomenology and Dialectical Materialism*, trans. Daniel J. Hermans and Donald W. Morano (Dordrecht: Kluwer Academic Publishers/Boston Studies in the Philosophy of Science, 1986), p. 121.

[67] J.-L. Marion, *Réduction et donation* (Paris: PUF, 1989) and J.-L. Marion, "Esquisse d'un concept phénoménologique du don", in M. M. Olivetti (ed.), *Filosofia della Revelazionie* (Milan: CEDAM, 1994). Marion is pointing to the originary givenness. For a critique of this look see Andrea Gonzi: "'Donation' e parola per una fenomenologica dell'anteriorita", in Olivette (ed.), *Filosofia della Revelazionie* and Peñalver Gómez, "The Deconstruction of Sense", *op. cit.* This discussion also supposes Derrida's writings, *Donner le temps* (Paris: Galilée, 1992) and *Donner là mort* (Paris: Métaillié, 1992).

[68] M. Blanchot, *La communauté inavouable* (Paris: Minuit, 1980) and J.-L. Nancy, *La communauté désoeuvrée* (Paris: Bourgois, 1986).

[69] See also Depraz and Moullie, p. 129.

JERZY GRZESZCZUK

A POSSIBLE REASON FOR THE 'FATAL VISION' OF THE FAMOUS AMERICAN SURGEON JEFFREY MACDONALD

S. N. Rumyantsev and J. Grzeszczuk have examined patients who, when in the presence of certain persons (inductors), develop various symptoms, among them nausea, psychic disturbances, hallucinations, loss of vision, lumps in the throat, numbness of hands and feet and fear of death.[5, 6] Patients are afraid of such encounters and avoid them. In cases where an individual is married to such an inductor, the spouses frequently quarrel or fight with each other without obvious reason. The marriage often ends in divorce, but sometimes in murder.

The phenomenon of the mysterious influence that some persons exert on others has been known since ancient times. The "Book of Tobias" (III–II century B.C., one of the deuterocanonical books of the Bible) describes how Raguel attempted to find a husband for his daughter, Sarah. The first candidate went to her bedroom and died the same night. Six other men lost their lives in the same way. Sarah was suspected of killing her husbands, but this was not true. The unlucky woman could not withstand these allegations and tried to hang herself. Similar cases continue to happen.

The cause of these symptoms and abnormal behavior may lie in the chemical interaction between humans. Body water (sweat, saliva, milk, urine, tears . . .) contain water-soluble antigens: A, B, H, Le^a, Le^b, I, Sd^a . . . Approximately 38% of the random United States population has in body water antigens A; 20% – antigens B; 78% – antigens H; 94% – antigens Le^a; 72% antigens Le^b.

Body water containing various antigens is secreted into the surrounding environment. In this way, an antigen cloud is formed around the secretor.[2, 5, 6, 9]

These antigens are not proteins but much smaller molecules (haptens), capable of leaving the organism. Liberated antigens are diluted outdoors, but their concentration might be high indoors or in the vicinity of secretors. When inhaled, these small antigens will eventually appear in the blood of sensitized recipients and precipitate a pathological reaction.

A possessor of anti-B antibodies will react to B antigens.

A-T. Tymieniecka (ed.), Analecta Husserliana LXVI, 349–356.

A possessor of anti-A antibodies will react to A antigens.

A possessor of anti-Le antibodies will react to Le antigens.

Approximately 55% of the population has anti-A antibodies (blood groups B and O); 86% – anti-B (blood groups A and O). Lewis antibodies (anti-Le^a and/or anti-Le^b) are quite common. They are present in as many as 3% of prenatal sera.

The introduction of foreign antigens into the blood through breathing can cause the same symptoms as a transfusion of mismatched blood, i.e., nausea, psychic disturbances, hallucinations, loss of sight, lumps in the throat, numbness of hands and feet, and even death. The more antibodies, the stronger the sensitivity to foreign antigens.

Antigens in body fluids have not attracted the attention of scientists so far.

Since 1981 I have studied a few hundred patients. The blood groups in both the inductor and the sensitized person (recipient) have been determined in 75 cases. The pathogenic interaction appeared most frequently in married couples (57 cases). It could also be seen among relatives (8 cases) and strangers (10 cases). The results are compiled in Tables 2 and 3.

I have investigated the cause of the sudden death (May 1988) of a

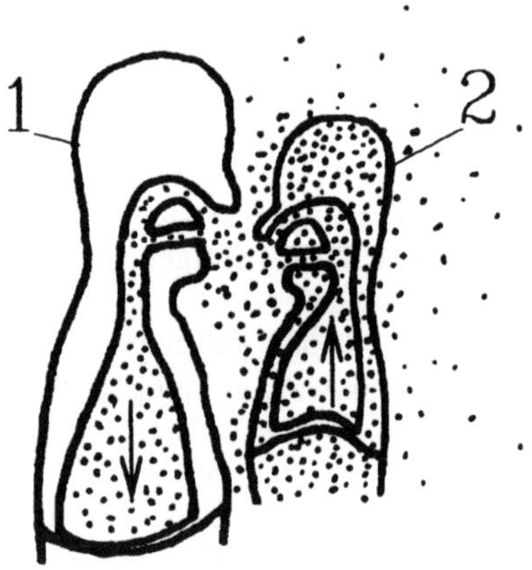

Figure 1. Antigens from secretor (inductor) *2* entering the recipient's *1* blood through the lungs.

married woman, Gabriela P. (33 years old), previously healthy, who after three immunizations, for 1) a complicated delivery – a caesarean operation (30), 2) a miscarriage (31) and 3) a miscarriage (32), complained of weakness when near her husband. For this reason the woman was afraid of sleeping or living together with him, left her home and filed an application for divorce. One conciliatory session and two regular curt sessions were held during which the woman insisted on divorce but could not explain her motives. Her husband loved her and refused to accept divorce.

In Poland there are lobbying groups exerting pressure on lawmakers to reduce the number of divorces. The court refused to dissolve this marriage, and they continued to live in separate apartments. One day the woman was visited by her husband and they had intercourse. Immediately afterwards she suddenly died, with no signs of violence in evidence on her body. The cause of death remained mysterious, but despite that, because their marriage was unsuccessful, the husband,

Table 1. L Secretors of A, B, H, I, Le or Sd^a antigens may exert a harmful influence on sensitized persons (recipients).

Antigens in body water of secretors (inductors) 	Antibodies of sensitized people (recipients) 
A	anti-A
B	anti-B
H	anti-H
Le^a	anti-Le^a
Le^b	anti-Le^b
I	anti-I
Sd^a	anti-Sd^a

Table 2. Comparison between the number of inductors and sensitized recipients relative to blood group.

Blood groups	Inductors	Sensitized recipients
0	ffpppppppp.p	ffffffffff.ffffffffff. ffpppppppp.ppppppp
A_1	ffffffffff.ffffffffpp. pppppppppp.	ffffffpppp.
A_2	fp	ffpppp
A	ppppp	ffp
B	ffffpppppp.ppp	fffffffffp.pppp
A_1B	fffppppp	fffp
A_2B		p
AB	ffpppp	
Total	30f + 45p = 75	44f + 31p = 75

f – one female; p – one male.

Table 3. Pathogenic influence of inductors on sensitized recipients relative to blood group.

Blood groups of inductors	Blood groups of sensitized recipients:
0	0: fp, A_1: fffpp, B: f, A_1B: fff
A_1	O: fffffffppp.pppppp, A_1: f, A_2: fppp, B: fffppppp, A_2B: p
A_2	0: f, A_1B: p
A	0: f, B: ffff
B	0: fffffpppp, A_1: ff, A: fp
A_1B	0: fffp, A_1: p, A_2: fp, B: f
A_2B	
AB	0: ffff, A_1: p, A: f
Total	44f + 31p = 75

f – one female; p – one male.

Mieczysław P., was charged with murder. He was accused of planning the crime and using a pillow to strangle his wife without leaving any injury. The husband pleaded innocent but was sentenced to fifteen years in jail.

This case of mysterious death attracted my attention, and I have collected evidence that after three immunizations the woman died because of strong sensitivity to antigens exuded by her husband. I have presented my results in a popular journal[2] and in a brochure.[3] The case has also been studied by Professor Tadeusz Marcinkowski, M.D., who accepted the theoretical possibility of sudden death due to serological incompatibility between the spouses.[10] Marcinkowski has published works[11, 12, 13, 14, 15, 16, 17] in which he describes the possibility of pathogenic interaction between humans mediated by their antigens.

In this short report I wish to mention that in my book[9] there are presented the basic facts of immunology and about seventy examples in which liberated antigens could have been the cause of abnormal behavior, social pathology, antigen diseases and sudden death. The book describes methods of blood examination and tests for antigens dissolved in the fluids of a person, as well as ways to interpret the results and avoid tragic consequences or problems in marital life.

I have attempted to explain the reason for the "fatal vision" of the famous American surgeon, Dr. Jeffrey MacDonald, who killed his wife and two daughters. This case was described in the famous book *Fatal Vision*[8] and treated on television in a program bearing the same title. The story is authentic. I have never read the book, but I saw a fragment of the film only.

Jeffrey MacDonald was watching television at home when his young daughter came to sit in his lap. After a while Jeffrey fell asleep and lost all contact with reality. He had hallucinations, felt strong pressure on his throat and dreamt that he was attacked by strangers who attempted to strangle him. In the face of mortal danger, he fought back and killed them. Soon he regained consciousness and was shocked to see his wife and both daughters dead. He called the police. A thorough investigation was carried out. It was demonstrated that there was no stranger in Jeffrey's home. No one was attempting to strangle him. That was his "fatal vision".

The case was studied by psychologists and psychiatrists, but could not be explained.

I believe that again we are dealing with interaction between humans through antigens. Jeffrey's daughter could have been a secretor of antigens to which he was strongly sensitized. When watching television his daughter came to sit in his lap. He inhaled a large quantity of her antigens. This caused psychic disturbances, hallucinations, loss of vision, numbness of hands and feet, a lump in the throat sensation and feelings of mortal danger. He imagined that someone was attempting to strangle him. He fought with the imaginary assailant, killing his two daughters and his wife, who had run to the rescue. He was sentenced for murder and is serving a sentence in an American jail.

CONCLUSION

Liberated antigens could be among the causes of the sudden death of previously healthy adults, the "fatal vision" of Jeffrey MacDonald, problems in marital life, divorces, abnormal behaviours. . . .

NOTES

1 Kagan, E. "Fundamentals of Immunology for Blood Bankers", in Denise Harmening (ed.), *Modern Blood Banking and Transfusion Practices* (Philadelphia: F.A. Davis Company, 1989), 47, ISBN 0-8036-4597-X.

2 Grzeszczuk, J. "Zabójcza miłość", *Wprost* 37 (Warsaw: 1992), pp. 59–60.

3 Grzeszczuk, J. *Allergy to a Husband* (in Polish) (Goleniów: Oficyna Wydawnicza BIOS, 1993).

4 Hirszfeld, L. *General Immunology* (in Polish) (Warsaw: Spółdzielnia Wydawnicza Czytelnik, 1948), p. 448.

5 Rumyantsev, S. N. and Grzeszczuk, J. "Nasilno mil ne budiesh", *Chimija i Zizn* 1 (Moscow: USSR Academy of Science, 1990), pp. 34–37.

6 Rumyantsev, S. N. and Grzeszczuk, J. "Pathogenic Effects of the Human Chemical Biofield", *Medical Hypotheses* 45, 1 (1995), pp. 94–98.

7 Ślopek, S. *Ilustrowany Słownik Immunologiczny* (Warsaw: PZWL, 1983), pp. 181, 193, 198, 578, 608.

8 McGinniss, J. *Fatal Vision* (New York: New American Library, 1985).

9 Grzeszczuk, J. *Allergy to People, Animals, Plants, Microbes and Objects Containing Antigens* (in Polish) (Goleniów: Oficyna Wydawnicza BIOS, 1994), ISBN 83-902270-0-2.

10 Marcinkowski, T. "Theoretical Possibility of the Arising of the Immune Complexes in the Cases of Serological Incompatibility Between Married Persons" (in Polish), *Postepy Medycyny Sadowej i Kryminologii* 2 (1996), pp. 487–489.

11 Marcinkowski, T. *Circulating Immune Complexes and Their Role in Some Pathologic Processes* (in Polish) (Szczecin: Wydawnictwo Promocyjne "Albatros", 1995), ISBN 83-85796-22-3.

[12] Marcinkowski, T. "Alzheimer and Pick Diseases from the Viewpoint of Prevention", *Medical Hypotheses* 46 (1996), pp. 180–182.
[13] Marcinkowski, T. *Multiple Sclerosis and Immunity (A New Presentation)* (in Polish) (Goleniów: Oficyna Wydawnicza BIOS, 1996), ISBN 902270-8-8.
[14] Marcinkowski, T. *Cluster Headache and Immunity* (in Polish) (Goleniów: Oficyna Wydawnicza BIOS, 1992).
[15] Marcinkowski, T. *From the Problems of Rheumatic Diseases – An Aspect of Prophylaxis* (in Polish) (Szczecin: Wydawnictwo Promocyjne ALBATROS, 1995).
[16] Marcinkowski, T. *Considerations on the Etiology and Pathogenesis of Schizophrenia* (in Polish) (Szczecin: Wydawnictwo Promocyjne ALBATROS, 1995).
[17] Marcinkowski, T. *Epilepsy – from the Aspect of Prophylaxis* (in Polish) (Szczecin: Wydawnictwo Promocyjne ALBATROS, 1995).

Adam Pawlak, Louis T. Houthakker and Mieczysław Migoń at the *Analecta* book exhibit and the NATA soft drink stand.

ALEXANDER KUZMIN

REFLEXION AND THE UNIVERSAL STRUCTURES OF CONSCIOUSNESS

In the first book of his *Ideas for a Pure Phenomenology and Phenomenological Philosophy*, which is the most important introduction to transcendental phenomenology, Husserl chooses the universal structures of pure consciousness as the object for research. He proposes to consider the bases for the realization of a pure phenomenology. This new science, according to Husserl, is based on the special method of phenomenological reduction. As a result of applying this method we acquire transcendental consciousness, which, in a sense, is "absolute being".[1] Thus "a phenomenological residue"[2] becomes the field of a new science.

Among the most general essential characteristics of the pure sphere of an experience as the basic element of the phenomenological consciousness, the first place is given to reflexion "through its universal methodological function".[3]

A transcendentally purified proper essence of an experience acquired by the use of phenomenological reduction will be manifested both actually and intentionally as an experience of pure Ego. An experience is caught by the Ego in different ways: it may be an immanent experiment, or some other immanent contemplation or presentation. And so, there is an ideal possibility that any experience will be caught and considered by the Ego. What the Ego directs reflexion towards is this process of catching an experience, and then for the Ego an experience becomes an object with respect to both its integrity and its components and possible intentionalities. As a matter of fact, reflexion is also experienced by us. So it can serve as a substratum for subsequent reflexions extending to the fundamental totality merging all possible experiences into one stream of consciousness.

Then, as a spontaneous action, reflexion distinguishes between the Ego and its experiences which are important during the transition from the natural aims of consciousness to the phenomenological project. True, phenomenology is more interested in the acts by which the relation of any experience to pure Ego is manifested than in the very fact of formation of an experience by an object of the Ego.

Therefore, making a distinction between the Ego and its experiences

A-T. Tymieniecka (ed.), Analecta Husserliana LXVI, 357–365.

is not yet a sufficient condition for speaking about the nature of phenomenological reflexion as such. The methodological universality of reflexion, owing to its phenomenological identity, is initiated by the designation, formation and manifestation of all the subsequent universal structures of consciousness. Being manifested, they start functioning as the basic, prerequisite knowledge for realizing the spontaneous possibilities of reflexion.

While considering reflexion as the question at issue, we recapitulate the principal, structural relation of consciousness to consecutively discover the temporal horizons of the realization of "immanent reflexion".[4] Actually experienced experiences being actually present appear to us as existing "now". This is instantly fixed by the reflective gaze of consciousness, owing to which possible relevant distinctions are established. Together with the experiences that "are now" appear the experiences that "have just been" – those which have not yet been considered by consciousness and, because of that, are non-reflectively quondam ("have been").

It should be noted that for natural consciousness experiences exist not only when we turn our attention to them but also because they have been experienced and thence existed even before the moment when we started to recognize them as experiences that "have just been". So immanent reflexion does not constitute temporal distinctions in order to consider experiences; it presupposes their presence in natural consciousness. Husserl offers the following explanation: "Reflexion . . . informs us of our earlier experiences which 'then' were present, were there, and yet had not been, immanently perceived".[5]

In exactly the same way we can anticipatorily reflect on our not yet existent experiences which we, however, recognize as belonging to pre-recollection. Here, the reflective gaze is fixed on a "future" experience of perception.

Discovery of the subsequent premises of the realization of reflexion is continued within the bounds of phenomenological and eidetic reduction. The results of the previous research become examples of "essential universality".[6] Then our task is to appropriate these results by means of "pure intuition" and to systematically study them. Here is where the eidetic investigation of the stream of experiences comes to the fore.

The eidetic or ideative background of the realization of reflexion is formed as direct contemplation takes part in creating the phenomenon of some act of experience such as that of the pleasure derived from fruitful, freely flowing theoretical meditations.[7] With that, we first of all, fix the conditions of the changes relevant to flowing meditations since this experience of pleasure is then considered and immanently perceived under the gaze of reflexion. True, that freeness of a flowing meditation is infringed upon and modified as meditation is now recognized, with its accompanying pleasure being caught and subject to phenomenological description.

The realization of the possibilities opening before us can result in our finding out the eidetic meaning of reflexion. In the first place, reflexion directed toward an experience of pleasure will make certain that this experience actually exists, but "not as one that has just begun".[8] The duration of an ongoing experience of pleasure is not yet observed in the focus of reflexion. At the present moment reflexion can focus only on past duration, on the mode an experience of pleasure is given; it can focus on earlier flows of meditation, and gazes fixedly on what went before; further, it can focus on the changes in an experience of pleasure; and, finally, it can focus on the defects of shifting one's gaze on a flowing phenomenon. The great number of modifications of the reflective gaze is also increased by the possibilities of focusing reflexion on its own capacity to objectify an experience of pleasure, thus "distinguishing between the pleasure being experienced but not considered and the considered pleasure".[9] With that, owing to shifts in viewpoint on an experience, the acts of apprehension and explication get modified.

The idea of an integral stream of experiences is prior to the possibilities of reflexion's manifestation and cannot be ascertained when the reflective gaze of consciousness is fixed on a distinct experience. The eidetic orientation of reflexion is manifested in the conduct of studies that are "scientific, aimed at systematic completeness, studies of essences".[10] The eidetic structure of consciousness is a sine qua non condition for emergence of reflexion and the possibilities for its realization. As regards the stream of experiences, its study and the ascertainment of the mode of its being are accomplished in acts of reflexion alone.

The last, but not least, matter of concern to us is the intentionality manifest in the objective aspect of experiences; it provides the prerequisites for finding out the mental meaning of reflexion as a phenomenological object. Phenomenological analysis of all the possibilities to be found in the stream of experiences has as its aim not just the detection of various moments of an experience but the detection of intentional moments. It also concerns experiences recognized as having been modified, the modifications of experiences arising out of varied reflexions, for example. Indeed, any recollection is modified in being "perceived", any expectation in "being perceived".[11] The difference between intentional and non-intentional experiences is not of particular concern to reflexion; therefore the intentional structure of consciousness precedes the possibilities of phenomenological reflexion.

Thus, the universal structures of consciousness are normatively set; but for them, consciousness might disintegrate into various mixtures of psychic and sensuous nature. The durability of the framework of consciousness is determined by its capacity to functionally adapt to changing situations, especially to temporal ones. Being constantly modified, like a live picture, consciousness is at the same time charged with an incredibly powerful stream of experiences. We have made an attempt to introduce reflexion into this stream, preserving distinctions between natural and phenomenological aims, leaving intact the structural organization of consciousness.

The meaning and purpose of the universal structure of consciousness are revealed by reflexion. From these brief sketches on reflexion we will now venture upon the study of reflexion as it is represented in Husserl's phenomenology.

Husserl's phenomenological philosophy distinguishes the following mental processes which can be denominated reflexion: (1) acts of analysis of the stream of experiences; (2) the methods of cognition of consciousness in general; (3) kinds of experiences.

Reflexion as a process of determining specifically distinct experiences becomes a leitmotif of the phenomenological analysis of the universal structures of consciousness. And here gnosiological strategy has to do with the shaping of the vast continuity in the specific peculiarities of experiences in accordance with the parameters of a discrete modus of reflexion. Reflexion as a method of cognition of absolute consciousness becomes objectified after having been studied in the phenomenological reduction. In the first instance all the modi of the immanent catching of essence and immanent experience proper will be treated as reflexion. Acts of analysis of the stream of experiences bears the name of reflexion so long as in essence they are immanent perceptions.

The statuses of reflexion as act, as method and as a kind of analysis of experiences seems to exhaust all the modifications of consciousness.

All possible modifications of consciousness result in the acquisition of an initial given, i.e., an invariable type of essence. But reflexion must prove its "absolute right" to consider such givenness.

It should be observed that from the phenomenological standpoint the adducing of arguments in favour of particular propositions under discussion is not capable of proof, for "the scale of truth is absolute lucidity".[12] So there is no point in pinning one's hopes on reflexion to guarantee the existence of experiences or the universal structures of consciousness. Pure intuition, initial in its significance, cannot be replaced by reflexion. Whatever arguments might be, only statements correctly expressing an initial givenness are meaningful and in accord with the truth. Pure intuition and immanent reflexion are intentionally different complementary means of acquiring phenomenological truth. They can be regarded as two questions: "What can we know?" and "What is it in fact?" If these questions were identical in their meaning, cognition would be senseless since its objective would be an achieved result. The meaning of phenomenological cognition is as it were in the narrowing or widening of the clearance between intuition and reflexion.

Side by side with the absolute right of reflexion to consider an initial phenomenological given, "the absolute right of immanent retention"[13] of continuing givens is recognized. Phenomenology aims at dealing with continuing givens whose nature is realized temporally: "still" living and having "just" formed. The phase of retention of the initiality of an experience is a constant essential type of existence of experience. Retention enables us to retain a continuing given in the living "now" of an experience. Just as reflexion in natural consciousness presupposes temporal distinctions, phenomenological consciousness guarantees retention as a continuing initial given. Since "every experience is a stream of formation in itself"[14] and its existence can be considered only by pure intuition, reflexion proves its absolute right to be "immanently perceiving reflexion"[15] in contrast with the relative right of immanent recollection. A recollection as a new product of memory cannot completely coincide with the being recollected. Here it is impossible to achieve exact correspondence, but it is the contents of memory as initially given yet not initially considered that reveal the true character of a recollection. Though relative, this right is a property of every recollection, and under certain circumstances this right can become predominant. Any lack of correspondence between a recollection and the being recollected is impossible for reflexion as immanent perception which represents the absolute with its qualities and intensities in apodictic authenticity.

Opening to reflexion, the apodictic authenticity of the absolute is based upon the assertion "that any 'reflexion' is by nature a modification of consciousness, the kind of modification any consciousness can essentially experience".[16] The absolute here is normatively given by the absence of the essential limits of the reflective modification of consciousness. Phenomenological studies of reflexion thus indicate the way to arrive at the apodictically authentic absolute. Reflexion as a kind of experience answers the question of how absolutely non-reflective experiences are possible. Reflexion as a method answers the question of how immanent experimentation is possible. Reflexion as an act answers the question of how immanent perception is possible. Thereby is the absolute constituted in the immanent sphere, tested by immanent means and contemplated immanently. In their turn, an experience, an experiment and contemplation when reduced to the immanent are given meaning and the right to exist. What proves to be absolutely immanent is "the field of free cogitations of one and the same pure Ego".[17]

The following schemes show the sequence of phenomenological description with the result to be sought.

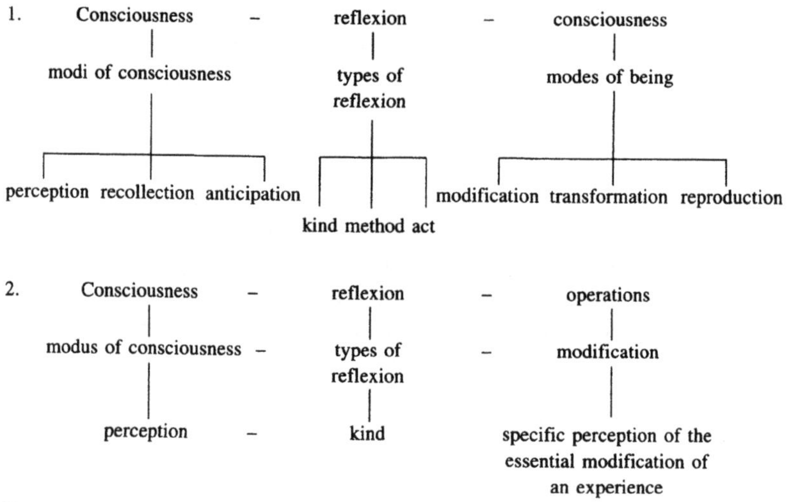

Etc.

3.

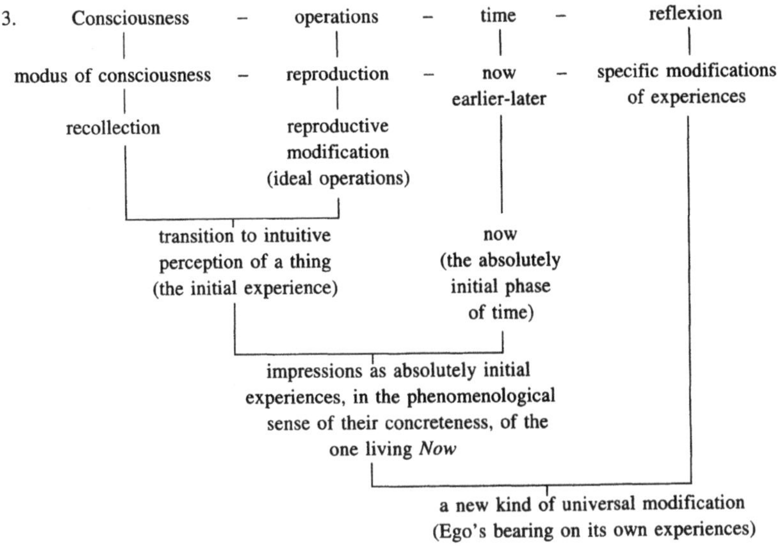

4.

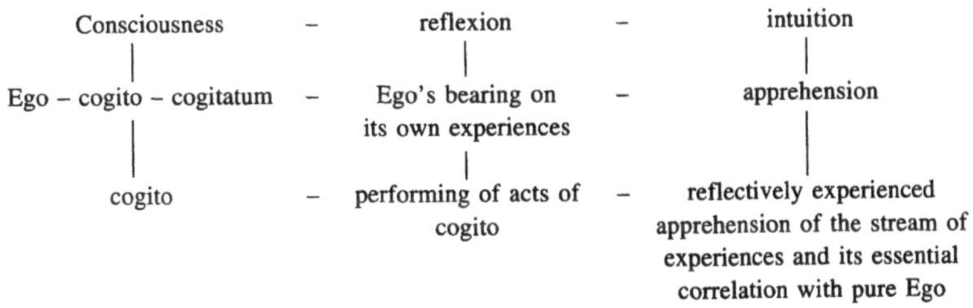

In Circle 1, "Consciousness – reflexion – consciousness", there arises the fundamental possibility of analysing and clarifying the diversity of the processes bound up with the stream of experiences. It is at the same time asserted that 1) any experience can be in this circle; 2) any reflexion can prompt such changes of consciousness; 3) any consciousness can enter into the modus of reflected consciousness. We speak of modifications in view of the fact that any reflexion involves certain transformations of a past experience or of experiences that are immediately present. A past experience can, in its turn, be specified to be a reflected experience, so that there is a modification of a higher rank. In this way we can pass to an absolutely non-reflected-upon experience and have to do only with its real and intentional givens.

What is of the greatest interest, though, is the passage of any experience along different lines and parameters to reflective modification. Every modus of consciousness, actually, can turn its reflective gaze upon itself and reveal a mode of being of consciousness characterized by inherent experiences. Besides, every experience has operational within it some kind of change. In order to define the various kinds of changes, which can in every experience be thought of ideally as operations, let us consider different types of reflexion. Specific reflexions are directly connected with modifications of some experience and in this sense could themselves be considered as experiences. Another type of reflexion is focused on essences. "All modi of the immanent apprehension of essence"[18] are of this type.

From the standpoint of reflexion on essences, every experience is the product of an invariable type of essence. This invariable essence is established in the ceaseless stream of retentions and protentions situating the living Now of an experience between "earlier" and "later". At the same time every experience has different forms of reproduction "which could be considered as ideal 'operational' transformations of initial experiences".[19] Thus, operations are a mode of being of consciousness in different forms of reproductive experiences. These are such modi as recollection, anticipation and reproductive fantasy.

All experiences paralleled in reflective modifications and reproductive operations presuppose, nevertheless, the total reality of essences. Therefore this parallelism does not affect intentional objects and the identical modes of givenness in experience. Operations are to a greater degree indicative of the ideal possibilities of changes and their iteration that are inherent in experiences.

Circle 2, "Consciousness – reflexion – operation", describes these ideal possibilities.

Circle 3 and Circle 4 allow us to answer the question asked above: How is an immanent experiment in the perception of an absolutely non-reflective experience of consciousness possible? Only after having reflectively experienced the acts of phenomenological consciousness, do we become convinced of the existence of the stream of experiences and of its binding relation to pure Ego. Only after having reflectively experienced different kinds of directedness in the stream of experiences, do we become convinced of the capacity of consciousness for different modifications including fixation on another Ego.

Thus, phenomenological reduction preserves the meaning of these experiments of consciousness and vindicates both their perceiving essential universality and experiences being contemplations of essences.

Reflexion actually confirms that immanent experiments arrive at pure and perfectly clear givens. That at any time consideration of an essence is attainable in an immanent experiment shows that the possibility of an objective given as such arises directly from the reflexion of consciousness and its subject.

Universytet Slazki

NOTES

[1] E. Husserl, *Ideen zu einer reinen Phänomenologie und phänomenologischen Philosophie*. Erstes Buch. *Allgemeine Einführung in die reine Phänomenologie* (The Hague, 1950) (Husseliana Bd III/1), p. 174.

[2] "Phänomenologisches Residuum". *Ibid.*, p. 59.

[3] "Um ihrer *universellen* methodologischen Funktion willen". *Ibid.*, p. 177.

[4] *Ibid.*, p. 178.

[5] "Reflexion . . . gibt uns Kunde von unseren früheren Erlebnissen, die 'damals' gegenwärtige, damals immanent wahrnehmbare, obschon nicht immanent wahrgenommene waren". *Ibid.*, p. 178.

[6] "Von Wesensallgemeinheiten". *Ibid.*, p. 179.

[7] See *ibid.*, pp. 179–180.

[8] *"Aber nicht als eben erst anfangende"*. *Ibid.*, p. 179.

[9] "Den Unterschied *erlebter* aber nicht erblickter, und *erblickter* Fredue". *Ibid.*, p. 180.

[10] "Einem Wissenschaftlichen, auf systematische Vollständigkeit abzielenden Wesensstudium". *Ibid.*, p. 180.

[11] See *ibid.*, p. 180.

[12] "Daß vollkommene Klarbeit das Maß aller Wahrheit ist". *Ibid.*, p. 185.

[13] "Das absolute Recht der immanenten Retention". *Ibid.*, p. 184.

[14] "Jedes Erlebnis ist in sich selbst ein Fluß des Werdens". *Ibid.*, p. 182.

[15] "Der immanent *wahrnehmenden* Reflexion". *Ibid.*, p. 184.

[16] "Daß *jederlei 'Reflexion'* den Charakter einer *Bewußtsciensmodifikation* hat, und zwar einer solchen, die prinzipiell *jedes Bewußtsein* erfahren kann". *Ibid.*, p. 181.

[17] "Ein Feld freien Vollzuges von Cogitationen des einen und selben reinen Ich". *Ibid.*,, p. 184.

[18] "Alle Modi immanenter Wesenserfassung". *Ibid.*, p. 181.

[19] "Die wie ideelle operative Umformungen des ursprünglichen Erlebnisses angesehen werden können". *Ibid.*, p. 182.

APPENDIX

Anna-Teresa Tymieniecka, President
in collaboration with
Gdansk Polytechnical University

PROGRAM

THE SECOND INTERNATIONAL CONGRESS OF PHILOSOPHY PHENOMENOLOGY OF LIFE AND THE SCIENCES OF LIFE

Gdansk Polytechnical University

June 5–9, 1997

Theme: ORIGIN OF LIFE INDIVIDUALISATION, EVOLUTION OF TYPES, NEW CRITIQUE OF REASON, HUMAN CREATIVITY

REGISTRATION

June 5: Registration from 10 am–4 pm, Congress Information Post in the Building of Gdansk Polytechnical University

SOLEMN OPENING OF THE CONGRESS

June 5: 11:30–1 pm: Main Hall of Polytechnical University, ul. Narutowicza 11, Gdansk

A-T. Tymieniecka (ed.), Analecta Husserliana LXVI, 367–376.

INVITED DIGNITARIES

Prof. Alexander Kołodziejczyk, President of the Polytechnical University
Mr. Henryk Wojciechowski, Governor of the Gdansk Region
Mr. Tomasz Posadzki, President of the City of Gdansk

WELCOMING RECEPTION

June 5: 1 pm: Welcoming Reception for the Participants of the Congress Offered by the *World Phenomenology Institute*

Opening Symposium: *THE TREE OF LIFE* IN FINE ARTS AND LITERATURE

Chair: Patricia Trutty-Coohill, USA

Presenters: Suzanna B. Simor, USA: THE TREE OF THE CREDO, SYMBOLISM OF THE TREE IN IMAGES OF THE CHRISTIAN CREED

Patricia Trutty-Coohill, USA: LEONARDO'S *ARBOR VITAE*: THE SALA DELLE ASSE

Nancy Goldfarb, USA: THE LIFE OF THE LYRE; POETS AND TREES

Tony Raczka, USA: THE TREE OF LIFE – A PARADIGM OF HUMAN GROWTH

Second Day, June 6, 1997 9 am–2 pm:

INAUGURAL LECTURE

Presided by Piotr Dominiak, Poland and Marlies Kronegger, USA
Anna-Teresa Tymieniecka, *World Phenomenology Institute*, USA
THE ORIGIN OF LIFE: PHILOSOPHY AND PHENOMENOLOGY OF LIFE AND THE NEW CRITIQUE OF REASON

Plenary Session I

The Dialogue Between Life Sciences and Philosophy

Marcello Sanchez Sorondo, Italy-Argentina: DIALOGUE WITH THE CLASSICS: HEIDEGGER, RICOEUR, TYMIENIECKA

Vijay Shankar Rai, India: ORIGINS OF LIFE, INDIVIDUATION AND EVOLUTIONISM

A.S. Antonites, South Africa: CREATIVE EMERGENCE AND COMPLEXITY THEORY

Horst Matthai, Mexico: ATOM AND INDIVIDUAL

Francesco Totaro, Italy: ON RECOMPOSING WORK, PRAXIS AND CONTEMPLATION

Bruno Caglieri, Italy: PSYCHIATRY AND PSYCHOLOGY

Mieczyslaw P. Migoń, Poland: LOGOS I ETHOS W MYSLI ANTROPOLOGICZNEJ ANNY-TERESY TYMIENIECKIEJ
(IN POLISH AND ENGLISH)

June 6, 2 pm–3 pm: lunch break
June 6, 3 pm–6 pm: afternoon sessions A,B,C

Session A, room 211: Presided by Francesco Totaro:

Mara Rubene, Latvia: FROM PHILOSOPHY OF CAVE TO PHILOSOPHY OF LABYRINTH – ON HUSSERL'S THEORY OF "LIVING NOW"

E. V. Altekar, India: THE DYADICS OF COMPLEMENTARITY: AN EPISTOMOLOGICAL REFLECTION

Gian Michele, Italy: THE TRIAL LIFE

Eldon Wait, South Africa: EXISTENTIAL PHENOMENOLOGY AND RECENT DEVELOPMENTS IN RESEARCH ON THE PERCEPTUAL ABILITIES OF CHILDREN

Ansis Zunde, Latvia: SIX PARA-PHILOSOPHICAL EXERCISES IN LATVIAN EURO(ONTO)POIESIS

Katarzyna Stark, Poland: EXISTENTIAL TIME AND ITS ROLE IN THE DEVELOPMENT OF PERSON IN NICOLAI BERDAYEV'S PHILOSOPHY

June 6th Session B, room 300, presided by Angela Ales-Bello:

Alberto Carrillo Canan, Berlin/Zacatecas, INDIVIDUATION AND TRUTH IN THE PHENOMENOLOGICAL TRADITION

Y.A. Pleckanov, Russia: EDUCATION AS CREATIVITY

Christine Berthold, Italy: THE INTERPRETATION OF THE CREATIVE PROCESS IN THE LITERARY BIOGRAPHIES OF STEFAN ZWEIG

Diane G. Scillia, USA: STIMULI TO INVENTION: NEW TECHNOLOGIES, NEW AUDIENCES, NEW IMAGES

Bogdan Ogrodnik, Silesian University, Poland: ABOUT METAPHYSICAL FOUNDATIONS OF LIFE

Rinalds Zembahs, Latvia: LIFE-SPACE AND LIFE-WORLD: MERLEAU-PONTY ON SITUATIONS

Session C (in Polish), room 264; presided by Prof. P. Dominiak

Grzegorz Graff, Karol Dziedziul, Politechnika Gdańska, Poland: MATEMATYCZNY HORYZONT PRZYSZLOSCI

Henryk Szabala, Uniw. Gdański, Poland: PROBLEM INDYWIDUALIZMU W FENOMENOLOGII

Włodzimierz Pawliszyn, Uniwersytet Gdański, Poland: CZY FENOMENOLOGIA JAKO NAUKA JEST MOŻLIWA? – STANOWISKO M. HEIDEGGERRA

Jaromir Brejdak, Poland: SAMOINTERPRETACJA CZASU JAKO ZASADA INDYWIDUALIZACJI FILOZOFII: DILTHEY, SCHELER, HEIDEGGER

Czeslaw Dmochowski, Poland: O LIKWIDOWANIU BLĘDNYCH WYOBRAŻEN

3rd Day, June 7

9 am–2 pm: Plenary Session II, room 211; presided by Tadeusz Buksinski and Jim Unah:

Creativity: The Specifically Human Sense of Life

Marlies Kronegger, USA: CHINESE GARDENS: THE RELATION OF MAN TO NATURE IN THE SEVENTEENTH CENTURY EUROPEAN CULTURE

Avelina Cecilia, Spain: HUMAN EXISTENCE AS A CREATIVE PROCESS (IN THE THOUGHT OF A-T. TYMIENIECKA)

Marc Van den Bossche, Belgium: MERLEAU-PONTY ON BODY, NATURE AND AESTHETICS

Piotr Mróz, Jagiellonian University, Poland: THE HUMAN ARTS AND THE NATURAL LAWS OF BIOS

J.J. Venter, South Africa: HUMAN DIGNITY: MAN IN THE UNIVERSE (ANIMAL-MAN-GOD)

Al. Husar, Romania: CREATIVITY IN BLAGA'S PHILOSOPHY

Elena Rogacheva, Russia: THE HUMAN CREATIVE CONDITION AND "FREEDOM TO LEARN" EDUCATION

June 7, 2 pm–3 pm: lunch break

June 7, 3 pm–6 pm: afternoon sessions A, B, C

Session A, room 211: Presided by Kim Rogers, USA

Maria Bielawka, Poland: TRANSCENDENTAL "TIMING" AS A MODE OF INDIVIDUALIZATION IN HUSSERL'S PHILOSOPHY

Marek Pyka, Politechnika Krakowska, Poland: REMARKS ON EMOTION AND SELFDETERMINATION IN MAX SCHELER AND ANTONI KEPINSKI

Cezary Mordka, UMCS, Poland: EX-CENTRICNESS AND COMMON SENSE. ABOUT THE STRUCTURE AND MANNERS OF BEING OF THE MAN

Leszek Pyra, Poland: MARTIN BUBER'S THEORY OF VALUES

Christer Bjurvill, Sweden: THE PHENOMENON OF FUTURE: BUILDING THE BRIDGE BETWEEN FUTUROLOGY AND PHENOMENOLOGY

Josef Sivak, Slovakia: DU MORTEL A L'IMPOSSIBLE ETERNEL

June 7th, 3 pm–6 pm

Session B, room 300: presided by Franciszek Blawat, Gdansk

Przemysław Gulda, Poland: HERMENEUTICS AND PHENOMENOLOGY IN POSTMODERN PHILOSOPHY

Gregg Anderson, USA: THE CENTRAL EXPERIMENT IN UNDERSTANDING LIFE

Anna Małecka, Poland: THE EMERGENCE OF A CARLYEAN HERO

Zofia Majewska, Poland: PHENOMENOLOGY AND THE CUBIST SPACE

Piotr Lenartowicz SJ, Jolanta Koszteyn, Poland: ABOUT CERTAIN SPECIFIC MEANINGS OF THE TERM "ADAPTATION" IN BIOLOGY

Session C, room 264: Presided by Marcello Sanchez Sorondo, Rome

Michelle Facos, USA: THE ECOSOPHICAL WORLD VIEW AND HOLISTIC HUMAN EXPERIENCE IN TURN OF THE CENTURY SWEDEN

Bianca d'Ippolito, ITALY: THE LIFE-WORLD AND THE TRANSCENDENTAL ILLUSIONS

Serg V. Komarov, Russia: PROBLEM OF "BODYNESS" CONSCIOUSNESS

Jari Kauppinen, Finland: DEATH IS A LIMIT OF PHENOMENOLOGY

Piero Trupia, Italy: PSYCHOLOGY: A DISCOURSE WITHOUT A SUBJECT?

4th Day, June 8th

June 8th, 9 am–2 pm: Plenary Session III, room 211: Presided by Christel Bjurvill, Sweden Piero Trupia, Italy

Origin, Evolution, Interplay

K.K. Misra, India: ORIGINS OF LIFE BEYOND REPLICATION

Spas Spassov, Bulgaria: CONTEMPORARY LIFE SCIENCES AND THE SCIENTIFIC WORLD VIEW

Frans Soontiens, The Netherlands: THE DUBIOUS ROLE OF CHANCE IN THE EVOLUTION OF LIFE

Carla Canullo, Italy: THE BODY AND THE SELF-INDIVIDUATION OF CONSCIOUS LIFE. PHYSIOLOGY AND PSYCHOLOGY IN MAINE DE BIRAN

Kim Rogers, USA: THE RECIPROCITY OF HUMAN ORGANISM AND CIRCUMSTANCE

Eva Syristova, Czech Republic: DIE SPRACHE DES TRAUMES UND DER TRAUMA DER SPRACHE

June 8th, 2 pm–3 pm: lunch break

June 8th, 3 pm–6 pm: afternoon sessions A, B, C

Session A, room 211: Presided by Aleksander Matejko, Politechnika Gdańska, and Rob Baum, Israel

Velga Velvere, Latvia: MULTIPLE SELVES IN SOREN KIERKEGAARD'S PSEUDONYMOUS WRITINGS

Bengt Tjellander, Sweden: THE CONSTRUCTION OF THE CONCEPT OF "THE OMNIVIDUAL"

Margarete Durst, Italy: FORMATION OF PERSONAL IDENTITY AND "NARCISSISM"

A. Kuzmin, Russia: REFLEXION UND ALLGEMEINEN STRUKTUREN DES BEWUSSTSEINS

Jerzy Grzeszczuk, Poland: A POSSIBLE REASON OF THE "FATAL VISION" OF THE FAMOUS AMERICAN SURGEON JEFFREY MACDONALD

V. Borodulin, Russia: THERAPEUTIC DIALOGUE IN PRACTISING PSYCHIATRY AND PROBLEMS OF CONSCIOUSNESS

Session B, room 300: Presided by Piotr Mroz, Krakow

Evgeny I. Arinin, Russia: PHENOMENON OF LIFE: PRINCIPLE OF UNDERSTANDING AND INTERPRETATION

Ignacy Fiut, Poland: POETRY AND ONTOLOGY

Liu Qinping, People's Republic of China: LIFE: THE TRUE, THE GOOD AND THE BEAUTIFUL

Jim Unah, Nigeria: POSTMODERNISM IS EXISTENTIAL PHENOMENOLOGY

Tony Raczka, USA: CREATIVE CONJUNCTION WITH A WILL-FED TREND

Rimma Kurenkova, Russia: HUMAN CREATIVE ACT AND SELF-INDIVIDUALISATION OF PERSONALITY IN THE EDUCATIONAL PROCESS

Ollie O. Oviedo, Mexico: A READER-RESPONSE APPROACH TO THE AESTHETIC EXPERIENCE OF OTHERNESS

Irene Angela Bianchi, Italy: FROM THE "EGO" TO THE WORLD OF A COMMUNITY ETHICS

Session C, room 264 (in Polish): Presided by A-T. Tymieniecka

Andrzej Lisak, Politechnika Gdańska, Poland: FENOMENOLOGIA NOEMATYCZNA, CIELESNOŚĆ I LOGOS PRZEDJEZYKOWY

Aleksandra Pawliszyn, Gdańsk University, Poland: HUSSERLOWSKIE A HEIDEGGERROWSKIE WIDZENIE CZASU

Mieczysław P. Migoń, Politechnika Gdańska, Poland: THE TIMING OF HUMAN CONDITION

Aleksander Matejko, Politechnika Gdańska, Poland: W OBRONIE WARTOSCI

5th Day, June 9th

9 am–2 pm: Plenary Session IV, room 211: Presided by Walter Lammi and Ella Buceniece:

Specifically Human Existence: The Aesthetic, the Moral, the Transcedent, the Continuity

Kolawole Ogundowole, Nigeria: CONCERNING CREATIVE UNFOLDMENT OF SPECIFICALLY HUMAN EXISTENCE

Angela Ales Bello, Italy: LIFE, PERSON, RESPONSIBILITY

Maija Kule, Latvia: PHENOMENOLOGICAL ASPECTS OF COMMUNITY: INTERSUBJECTIVITY, PARTICIPATION, BEING-AMONG-EVERYTHING-THAT-IS-ALIVE

Robert D. Sweeney, USA: CREATIVITY AND EVERYDAY LIFE

Maria Teresa de Noronha, Portugal: THE CREATIONISM OF LEONARDO COIMBRA AND THE "SAUDATE" AS MORAL "GOOD"

Daniela Verducci, Italy: "BILDUNG": GIVING FORM TO LIFE. PROCESS OF FUNCTIONALIZATION AND WORK IN THE PHILOSOPHY OF MAX SCHELER

June 9th, 2 pm–3 pm: lunch break
3 pm–6 pm: afternoon sessions A, B, C

Session A, room 211: Presided by Ignacy Fiut, Krakow

Walter Lammi, Egypt: HANS-GEORG GADAMER'S PLATONIC DESTRUCTION OF THE LATER HEDEGGER

Edyta Supińska-Polit, Poland: HEAVEN'S ANGELS GRINDING ORGANS: JOHN RUSKIN'S IDEA OF LIFE

Elga Freiberga, Latvia: SELF-IMAGINATION AS DRAMA OF AUTHOR

Rob Baum, Israel: ABSENCE, METAPHOR AND THE FLIGHT OF MEANING

Tadeusz Buksiński, Poland: HUMAN IDENTITY IN THE CONTEMPORARY WORLD

Session B, room 300: Presiding: Daniela Verducci, Macerata

Zaiga Ikere, Latvia: CHANGING ATTITUDES TOWARDS THE OTHERNESS

Joanna Ślosarska, Poland: L'EVENEMENT DANS L'ESPACE DE LA VIE COMME L'EFFECT DE L'INTERACTION DE L'INSTINCT, DU SENTIMENT, DE L'IMAGE ET DE LA TRANSCENDANCE SPIRITUELLE

Richards Kulis, Latvia: HISTORICITY OF LIFE AND HISTORICAL APRIORI

June 9th, 3 pm–6 pm

Session C, room 264: Presided by Adam Pawlak, Politechnika Gdańska

Ella Buceniece, Latvia: THE FRAGMENTS OF TOTALITY OF LIFE IN GEORG ZIMMEL'S PHILOSOPHY

Maria Zowisło, Poland: JUNG'S CONCEPT OF INDIVIDUATION AND THE PROBLEM OF ALIENATION

Eugeniusz Szumakowicz, Poland: DEATH AND ONTOLOGY
Gennady Brevde, Russia: RADIAL STRUCTURE OF THE CONSCIOUSNESS IN PLACE OF THE "DISCRETE VS. CONTINUITY"

Johann E. Hafner, USA: RECEPTION AND PERCEPTIBILITY OF LIVING BEINGS

June 9th, 6 pm: CLOSURE OF THE CONGRESS

Speakers: Marlies Kronegger, USA and Maria Bielawka, Poland

INDEX OF NAMES

–A–

Abelard 255
Adam 45–48, 50, 51
Alberti, Leon Battista 23, 24
Allan, J. 91
Ames, Roger 225
Andrea (St) 53
Anhart, Larry 98, 99, 103
Anjou, Robert of 52
Ariès, Philippe 336, 337
Aristotle 58, 91, 181, 293, 331
Asimov, Isaac 157
Azouvi, François 215

–B–

Bacon 303
Baertschi, Bernard 214
Beltrami, Luca 21, 26, 29
Bennett, J. 225
Bergson 9–10, 15, 187
Bernard, Claude 253
Berry, Wendel 35–37, 39, 43
Bianchi 253
Binswanger 254
Bjurwill, Christer 132, 137
Blanchot, Maurice 338, 342
Bloch, Ernst 157, 163
Boerhaave, Hermann 203, 207, 216
Bohm, David 77
Bohr, Niels 263, 264, 268–270, 281
Bonaventure (St) 46, 48–50
Bonnet, Charles 206
Bragdon, Claude 21
Bransford 226, 232, 233
Brentano 193, 256, 272
Brizio, Anna Maria 29
Broad, C. D. 273
Brooks 93
Brower 159
Bruzio 52
Buber 257
Buhler, Karl 272
Buytendijk, F. J. J. 85, 86

–C–

Cairns Smith, A. G. 68
Capra, Fritjof 285
Caritas 52
Castor 31
Catholica, Fides 53, 54
Chew, Geoffrey 285
Christ 46, 47, 50, 51, 53
Clytemnestra 31
Condillac 204, 212
Confutse 146
Convenevole 52
Coor, Gertrude 50
Copernicus 99
Corsander, Ann-Margret 147
Crescentio 25
Critchley, Simon 335

–D–

D'Andrea, Giovanni 52
D'Este, Beatrice 19, 25, 27
D'Ippolito, B. M. 254
Da Modena, Giovanni 48
Da Sesto, Cesare 29
Da Vinci, Leonardo 19–21, 23, 25, 27, 29–32, 142
Dalton 66
Dante 27, 48, 52
Darwin 3, 6, 9, 59, 61, 98
Dastur, Françoise 335, 337
De Biran, Maine 203–217
De Crescentius, Pietro 24
De Soissons, Yolande 49
De Tracy, Destutt 213, 214

De Voragine, Jacobus 45
De' Bartoli, Bartolomeo 52
De' Medici, Giovanni 23
De' Medici, Pietro 23
Depraz, Natalie 335
Derrida 4, 121, 122, 323, 325–336, 338, 340
Descartes Rene 99, 141, 165, 205, 270, 303, 304, 308
Di Bonaguida, Pacino 49
Di Giacomo da Bologna, Niccolò 52, 54
Di Giovanni, Paolo 48–50
Di Sassoferrato, Bartolo 52
Dilthey, Wilhelm 187, 188, 192–197, 199
Dobzhansky 93
Dorsey Grice 44
Dostoyevsky 97
Dunne 143

–E–

Eicken 93
Einstein, Albert 59–61, 64
Emboden 30
Eskola 142
Etlin, Richard 19
Eve 47, 48, 50, 51

–F–

Ferraris, Maurizio 339, 340
Feyerabend 85, 263
Fichte 298, 304
Fides 51, 52
Fink, Eugene 338
Fiumanò, Marisa 257
Foucault, Michel 140
Freud, 338
Fromm 257

–G–

Gadamer 256
Gaddi, Taddeo 49, 50
Galen 31
Galileo 92, 99
Gall (Doctor) 206, 208
Gamow 62
Gell-Mann, M. 86
Giacanelli 254
Gibson, J. J. 230, 231, 233
Gillman, Alfred 70, 71
Giotto 52
Giovannangeli, Daniel 330–332, 340
Gödel, Kurt 162
Goldstein, Kurt 226, 229
Goodwin, B. C. 7–9, 91, 93
Göran, Martin 147
Gould, J. 5
Granel, Gérard 330
Griesinger 256
Grzeszczuk, J.

–H–

Hall, David 225
Hartley, David 206
Hartmann, Nicolai 74
Harvey, William 142
Hawking, Stephen 57–60, 62–64
Hegel 292, 308, 309, 327, 335
Heidegger, Martin 4, 132, 133, 140, 165, 169, 170–172, 174, 177–184, 187, 188, 191, 195–199, 312, 313, 324, 333, 334, 337–340
Heisenberg, Werner 266, 267
Helen of Troy 31
Helena 53
Henry, Michel 330
Höijer 152
Hölderin, Friedrich 173
Hull, Clark 272
Hume, David 108, 301
Husserl, Edmund 83, 84, 89, 91, 92, 121–126, 138, 141, 143–147, 151–153, 159, 165–167, 169–171, 187, 193, 254, 256, 257, 272, 275, 276, 287, 310–312, 318, 323–335, 339, 341, 342, 357, 358, 360
Hutteman, A. 84

–I–

Imperial, Francisco 52
Ingarden, Roman 73, 74, 165, 187
Isabella (Queen) 50

–J–

Jacob 100
James, W. 301
Jasper 169, 255
Johansson, Ivar-Lo 148
Jupiter 30

–K–

Kant 58, 280, 281, 290, 298, 300, 304, 309, 331, 334, 339, 341
Karamazov, Ivan 101
Kasparov, Garry 57
Katz, David 272
Kaufman, Stuart 7, 90
Keats, John 35
Kemp, Martin 25, 28, 30, 31
Kierkegaard 187
Kimura 92
Kuhn 85

–L–

Lacan, Jaques 338
Laing 139, 140
Lakatos 85
Lange, M. 86
Leda 30
Leiber, T. 91
Leibniz 3, 15, 79, 282, 300, 304
Leonhard (St) 49
Levin 231
Levinas, Emmanuel 337–339
Lewin, R. 89, 91
Locke 300, 304
Lorenz, E. 162
Łorenzetti, Ambrogio 50
Louis (St) 49
Lovejoy, A. O. 273
Lukasiewicz, Jan 161

–M–

MacDonald, Jeffrey 354, 355
Mace 230
MacIntyre, Alasdair 157, 158
Magdalene 51
Marcel, Gabriel 165, 256, 257
Marcinkowski, Tadeusz 354
Marx, Karl 59
Mary 47, 48, 50, 51
Maslow, Abraham 133, 142
Matz, Richard 132
Mauru, Rabanus 47
May, Rollo 195
Mead, George Herbert 137
Meinong, Alexius 293
Melzer 152, 153
Menabuoi, Giusto 53
Merleau-Ponty, Maurice 121, 123–126, 131, 136, 165, 230, 254, 272, 325, 326, 330
Michelangelo 23
Michelozzo 23
Mill, J. 301
Miller, Jim Wayne 35, 37–39, 43
Molchanov 85
Monod, Jacques 99–101
Morellet, Abbot 210
Moullie, Jaen-Marc 335
Moustaka 144
Müller Walde, Paul 25
Muller, G. E. 272
Müller, Johannes 66

–N–

Nadel 278
Nancy 342
Natanson, Maurice 257
Newman, D. V. 89, 93
Nietzsche 97
Nordenfeldt, Lennart 145

–O–

Oliver, Mary 35, 38, 39, 43
Olsson, Fredrik 147
Ortega 225, 233
Osborne, Harold 21

–P–

Pacino 50
Packer, Toni 135
Panofski 255
Paul (apostle) 195
Pavlov, Ivan 273

Pedretti, Carlo 25, 27
Penrose, Roger 57, 59, 61–64
Pfänder, Alexander 165
Philip III 49
Piercy, Marge 35, 40, 41, 43
Pirzig 135, 138
Pisano, Nicola 51
Place, U. T. 276
Planck, Max 263, 267, 281
Plato, 73
Plotinus 73, 78, 81
Poe, Edgar Allan 328
Poincare 160
Pollux 31
Popper, Karl 85, 141, 264
Prirogine, Ilya 86, 101

–R–

Raguel 350
Régis, Rey (Doctor) 208, 215
Rench 93
Richter, Jean Paul 30
Ricoeur, Paul 276
Riedl 93
Rodbell, Martin 70, 71
Roscellino 255
Rousseau, Jean-Jacques 329, 330, 339
Rumyantsev, S. N. 349
Rusca, Ernesto 26
Russell, Bertrand 158, 159, 162

–S–

Sallis, John 339
Sarah 350
Sartre 165, 169, 170, 326
Saxl, Fritz 48
Scheerer 226
Scheler, Max 85, 165, 167–170, 187–192, 196, 197, 199, 229, 257, 287–294
Schoffeniels 93
Schutz, Alfred 255
Seneca 255
Sforza, Ludovico 19, 25, 27, 29, 30
Skinner, B. F. 272, 274
Socrates 163
Soontiens, F. 90, 93
Spaemann, Robert 190, 197
Spencer 292
Spinoza 304
Stahl 205
Stevens, Wallace 35, 41–43
Straus 237
Sullivan, Louis 19, 21–23, 25, 32
Suzuki, D. T. 144

–T–

Taine, Hippolyte 22–23
Tanzi 253
Tellenbach 256
Thao, Tran Duc 341
Theunissen, Michael 334
Tilliette, Xavier 215
Tolman, E. C. 272
Trevi, Mario 257
Tuomi, J. 90
Turing 8
Turner, Tina 136
Tymieniecka, Anna-Teresa 83–85, 93, 94, 165
Tyndareus 30

–U–

Uexkull 232, 233

–V–

Varela, F. 4
Verrocchio 23
Visconti of Milan, Luchino 52
Von Bertalanffy, Ludwig 270
Von Ehrenfels, Christian 293
Von Weiszäcker, Victor 257

–W–

Waddington, C. H. 91
Wahl, Jean 165
Waldrop, M. M. 88
Watson, J. B. 272–274
Weber, Max 278
Weinberg, Stephen 61
Wertheimer 272
Whitehead, Alfred North 76, 77
Wiley 93

Wilkner 152
Willis (Doctor) 206, 208
Wilson, Edward 14, 98
Wilson, R. 5
Wittgenstein 58
Wojtyla, K. 92
Wolff 300
Wordsworth, William 35
Wright, Frank Lloyd 19, 21
Wundt, W. 301

–Y–

Yeats, William Butler 35, 43

Analecta Husserliana

15. Tymieniecka, A-T. and Calvin O. Schrag (eds.), *Foundations of Morality, Human Rights, and the Human Sciences.* Phenomenology in a Foundational Dialogue with Human Sciences. 1983 ISBN 90-277-1453-3
16. Tymieniecka, A-T. (ed.), *Soul and Body in Husserlian Phenomenology.* Man and Nature. 1983 ISBN 90-277-1518-1
17. Tymieniecka, A-T. (ed.), *Phenomenology of Life in a Dialogue Between Chinese and Occidental Philosophy.* 1984 ISBN 90-277-1620-X
18. Tymieniecka, A-T. (ed.), *The Existential Coordinates of the Human Condition: Poetic – Epic – Tragic.* The Literary Genre. 1984 ISBN 90-277-1702-8
19. Tymieniecka, A-T. (ed.), *Poetics of the Elements in the Human Condition.* (Part 1:) The Sea. From Elemental Stirrings to Symbolic Inspiration, Language, and Life-Significance in Literary Interpretation and Theory. 1985
For Part 2 and 3 *see below* under Volumes 23 and 28. ISBN 90-277-1906-3
20. Tymieniecka, A-T. (ed.), *The Moral Sense in the Communal Significance of Life.* Investigations in Phenomenological Praxeology: Psychiatric Therapeutics, Medical Ethics and Social Praxis within the Life- and Communal World. 1986 ISBN 90-277-2085-1
21. Tymieniecka, A-T. (ed.), *The Phenomenology of Man and of the Human Condition.* Part II: The Meeting Point Between Occidental and Oriental Philosophies. 1986 ISBN 90-277-2185-8
22. Tymieniecka, A-T. (ed.), *Morality within the Life- and Social World.* Interdisciplinary Phenomenology of the Authentic Life in the "Moral Sense'. 1987
Sequel to Volumes 15 and 20. ISBN 90-277-2411-3
23. Tymieniecka, A-T. (ed.), *Poetics of the Elements in the Human Condition.* Part 2: The Airy Elements in Poetic Imagination. Breath, Breeze, Wind, Tempest, Thunder, Snow, Flame, Fire, Volcano . . . 1988 ISBN 90-277-2569-1
24. Tymieniecka, A-T., *Logos and Life.* Book I: Creative Experience and the Critique of Reason. 1988 ISBN Hb 90-277-2539-X; Pb 90-277-2540-3
25. Tymieniecka, A-T., *Logos and Life.* Book II: The Three Movements of the Soul. 1988 ISBN Hb 90-277-2556-X; Pb 90-277-2557-8
26. Kaelin, E. F. and Calvin O. Schrag (eds.), *American Phenomenology.* Origins and Developments. 1989 ISBN 90-277-2690-6
27. Tymieniecka, A-T. (ed.), *Man within his Life-World.* Contributions to Phenomenology by Scholars from East-Central Europe. 1989 ISBN 90-277-2767-8
28. Tymieniecka, A-T. (ed.), *The Elemental Passions of the Soul.* Poetics of the Elements in the Human Condition, Part 3. 1990 ISBN 0-7923-0180-3
29. Tymieniecka, A-T. (ed.), *Man's Self-Interpretation-in-Existence.* Phenomenology and Philosophy of Life. – Introducing the Spanish Perspective. 1990 ISBN 0-7923-0324-5
30. Rudnick, H. H. (ed.), *Ingardeniana II.* New Studies in the Philosophy of Roman Ingarden. With a New International Ingarden Bibliography. 1990 ISBN 0-7923-0627-9

Analecta Husserliana

31. Tymieniecka, A-T. (ed.), *The Moral Sense and Its Foundational Significance: Self, Person, Historicity, Community.* Phenomenological Praxeology and Psychiatry. 1990 ISBN 0-7923-0678-3
32. Kronegger, M. (ed.), *Phenomenology and Aesthetics.* Approaches to Comparative Literature and Other Arts. Homages to A-T. Tymieniecka. 1991 ISBN 0-7923-0738-0
33. Tymieniecka, A-T. (ed.), *Ingardeniana III.* Roman Ingarden's Aesthetics in a New Key and the Independent Approaches of Others: The Performing Arts, the Fine Arts, and Literature. 1991
Sequel to Volumes 4 and 30 ISBN 0-7923-1014-4
34. Tymieniecka, A-T. (ed.), *The Turning Points of the New Phenomenological Era.* Husserl Research – Drawing upon the Full Extent of His Development. 1991 ISBN 0-7923-1134-5
35. Tymieniecka, A-T. (ed.), *Husserlian Phenomenology in a New Key.* Intersubjectivity, Ethos, the Societal Sphere, Human Encounter, Pathos. 1991 ISBN 0-7923-1146-9
36. Tymieniecka, A-T. (ed.), *Husserl's Legacy in Phenomenological Philosophies.* New Approaches to Reason, Language, Hermeneutics, the Human Condition. 1991 ISBN 0-7923-1178-7
37. Tymieniecka, A-T. (ed.), *New Queries in Aesthetics and Metaphysics.* Time, Historicity, Art, Culture, Metaphysics, the Transnatural. 1991 ISBN 0-7923-1195-7
38. Tymieniecka, A-T. (ed.), *The Elemental Dialectic of Light and Darkness.* The Passions of the Soul in the Onto-Poiesis of Life. 1992 ISBN 0-7923-1601-0
39. Tymieniecka, A-T. (ed.), *Reason, Life, Culture, Part I.* Phenomenology in the Baltics. 1993 ISBN 0-7923-1902-8
40. Tymieniecka, A-T. (ed.), *Manifestations of Reason: Life, Historicity, Culture.* Reason, Life, Culture, Part II. Phenomenology in the Adriatic Countries. 1993 ISBN 0-7923-2215-0
41. Tymieniecka, A-T. (ed.), *Allegory Revisited.* Ideals of Mankind. 1994 ISBN 0-7923-2312-2
42. Kronegger, M. and Tymieniecka, A-T. (eds.), *Allegory Old and New.* In Literature, the Fine Arts, Music and Theatre, and Its Continuity in Culture. 1994 ISBN 0-7923-2348-3
43. Tymieniecka, A-T. (ed.): *From the Sacred to the Divine.* A New Phenomenological Approach. 1994 ISBN 0-7923-2690-3
44. Tymieniecka, A-T. (ed.): *The Elemental Passion for Place in the Ontopoiesis of Life.* Passions of the Soul in the *Imaginatio Creatrix.* 1995 ISBN 0-7923-2749-7
45. Zhai, Z.: *The Radical Choice and Moral Theory.* Through Communicative Argumentation to Phenomenological Subjectivity. 1994 ISBN 0-7923-2891-4
46. Tymieniecka, A-T. (ed.): *The Logic of the Living Present.* Experience, Ordering, Onto-Poiesis of Culture. 1995 ISBN 0-7923-2930-9

Analecta Husserliana

47. Tymieniecka, A-T. (ed.): *Heaven, Earth, and In-Between in the Harmony of Life.* Phenomenology in the Continuing Oriental/Occidental Dialogue. 1995 ISBN 0-7923-3373-X
48. Tymieniecka, A-T. (ed.): *Life. In the Glory of its Radiating Manifestations.* 25th Anniversary Publication. Book I. 1996 ISBN 0-7923-3825-1
49. Kronegger, M. and Tymieniecka, A-T. (eds.): *Life. The Human Quest for an Ideal.* 25th Anniversary Publication. Book II. 1996 ISBN 0-7923-3826-X
50. Tymieniecka, A-T. (ed.): *Life. Phenomenology of Life as the Starting Point of Philosophy.* 25th Anniversary Publication. Book III. 1997 ISBN 0-7923-4126-0
51. Tymieniecka, A-T. (ed.): *Passion for Place. Part II.* Between the Vital Spacing and the Creative Horizons of Fulfilment. 1997 ISBN 0-7923-4146-5
52. Tymieniecka, A-T. (ed.): *Phenomenology of Life and the Human Creative Condition.* Laying Down the Cornerstones of the Field. Book I. 1997 ISBN 0-7923-4445-6
53. Tymieniecka, A-T. (ed.): *The Reincarnating Mind, or the Ontopoietic Outburst in Creative Virtualities.* Harmonisations and Attunement in Cognition, the Fine Arts, Literature. Phenomenology of Life and the Human Creative Condition. Book II. 1997 ISBN 0-7923-4461-8
54. Tymieniecka, A-T. (ed.): *Ontopoietic Expansion in Human Self-Interpretation-in-Existence.* The I and the Other in their Creative Spacing of the Societal Circuits of Life. Phenomenology of Life and the Creative Condition. Book III. 1997 ISBN 0-7923-4462-6
55. Tymieniecka, A-T. (ed.): *Creative Virtualities in Human Self-Interpretation-in-Culture.* Phenomenology of Life and the Human Creative Condition. Book IV. 1997 ISBN 0-7923-4545-2
56. Tymieniecka, A-T. (ed.): *Enjoyment.* From Laughter to Delight in Philosophy, Literature, the Fine Arts and Aesthetics. 1998 ISBN 0-7923-4677-7
57. Kronegger M. and Tymieniecka, A-T. (eds.): *Life. Differentiation and Harmony... Vegetal, Animal, Human.* 1998 ISBN 0-7923-4887-7
58. Tymieniecka, A-T. and Matsuba, S. (eds.): *Immersing in the Concrete.* Maurice Merleau-Ponty in the Japanese Perspective. 1998 ISBN 0-7923-5093-6
59. Tymieniecka, A-T. (ed.): *Life - Scientific Philosophy/Phenomenology of Life and the Sciences of Life.* Ontopoiesis of Life and the Human Creative Condition. 1998 ISBN 0-7923-5141-X
60. Tymieniecka, A-T. (eds.): *Life - The Outburst of Life in the Human Sphere.* Scientific Philosophy / Phenomenology of Life and the Sciences of Life. Book II. 1998 ISBN 0-7923-5142-8
61. Tymieniecka, A-T. (ed.): *The Aesthetic Discourse of the Arts.* Breaking the Barriers. 2000 ISBN 0-7923-6006-0
62. Tymieniecka, A-T. (ed.): *Creative Mimesis of Emotion.* From Sorrow to Elation; Elegiac Virtuosity in Literature. 2000 ISBN 0-7923-6007-9

Analecta Husserliana

63. Kronegger, M. (ed).: *The Orchestration of The Arts – A Creative Symbiosis of Existential Powers.* The Vibrating Interplay of Sound, Color, Image, Gesture, Movement, Rhythm, Fragrance, Word, Touch. 2000 ISBN 0-7923-6008-7
64. Tymieniecka, A-T. and Z. Zalewski (eds.): *Life - The Human Being Between Life and Death.* A Dialogue Between Medicine and Philosophy, Recurrent Issues and New Approaches. 2000 ISBN 0-7923-5962-3
65. Kronegger, M. and Tymieniecka, A-T. (eds.): *The Aesthetics of Enchantment in the Fine Arts.* 2000 ISBN 0-7923-6183-0
66. Tymieniecka, A-T. (ed.): *The Origins of Life, Volume I: The Primogenital Matrix of Life and Its Context.* 2000 ISBN 0-7923-6246-2; Set ISBN 0-7923-6446-5
67. Tymieniecka, A-T. (ed.): *The Origins of Life, Volume II: The Origins of the Existential Sharing-in-Life.* 2000
ISBN 0-7923-6276-4; Set ISBN 0-7923-6446-5
68. Tymieniecka, A-T. (ed.): *PAIDEIA.* Philosophy / Phenomenology of Life Inspiring Education of our Times. 2000 ISBN 0-7923-6319-1
69. Tymieniecka, A-T. (ed.): *The Poetry of Life in Literature.* 2000
ISBN 0-7923-6408-2

Kluwer Academic Publishers – Dordrecht / Boston / London

GPSR Compliance
The European Union's (EU) General Product Safety Regulation (GPSR) is a set of rules that requires consumer products to be safe and our obligations to ensure this.

If you have any concerns about our products, you can contact us on

ProductSafety@springernature.com

In case Publisher is established outside the EU, the EU authorized representative is:

Springer Nature Customer Service Center GmbH
Europaplatz 3
69115 Heidelberg, Germany

www.ingramcontent.com/pod-product-compliance
Ingram Content Group UK Ltd.
Pitfield, Milton Keynes, MK11 3LW, UK
UKHW020828121125
464983UK00011B/70/J

* 9 7 8 0 7 9 2 3 6 2 4 6 3 *